HOLT

EARTH SCIENCE

HOLT
EARTH
SCIENCE

Robert H. Fronk
Head, Science Education Department
Florida Institute of Technology
Melbourne, Florida

Linda B. Knight
Earth Science Teacher
Paul Revere Middle School
Houston, Texas

SENIOR EDITORIAL ADVISOR
Curriculum and Multicultural Education

John E. Evans, Jr.
Science Education Specialist
Philadelphia, Pennsylvania

HOLT, RINEHART AND WINSTON
Harcourt Brace & Company

Austin • New York • Orlando • Chicago • Atlanta
San Francisco • Boston • Dallas • Toronto • London

ACKNOWLEDGMENTS

Content Advisors

Mapi M. Cuevas, Ph.D.
Professor, Department of Natural
   Sciences
Santa Fe Community College
Gainesville, Florida

Don de Sylva, Ph.D.
Rosenstiel School of Marine and
   Atmospheric Science
University of Miami
Miami, Florida

George Greenstein, Ph.D.
Professor of Astronomy
Amherst College
Amherst, Massachusetts

James P. McGuirk, Ph.D.
Meteorology Department
Texas A & M University
College Station, Texas

Lisa Rossbacher, Ph.D.
Vice President for Academic
   Affairs and Professor of Geology
Whittier College
Whittier, California

Brian J. Skinner, Ph.D.
Professor of Geology and
   Geophysics
Yale University
New Haven, Connecticut

Curriculum Advisors

Leon J. Zalewski, Ph.D.
Professor of Science Education
Dean, College of Education
Governors State University
University Park, Illinois

Robert Frank
Earth Science Teacher
Jefferson Junior High School
Caldwell, Idaho

Frank Garcia
Science Department Chairman
Strack Intermediate School
Klein, Texas

Loisteen Harrell
Earth Science Teacher
Lake Braddock Secondary School
Fairfax County Schools
Burke, Virginia

Nancy Hampton Johnson
Earth Science Teacher
McCulloch Middle School
Plano, Texas

Jim Pulley
Science Teacher
Oak Park High School
North Kansas City, Missouri

Dana Ste. Claire
Curator of Science and History
The Museum of Arts and Sciences
Daytona Beach, Florida

Doris Tucker
Earth Science Teacher
Rutherford County Schools
Spindale, North Carolina

Thomasena H. Woods, Ed. D.
Science Supervisor
Newport News City Schools
Newport News, Virginia

Reading/Literature Advisors

Philip E. Bishop, Ph.D.
Professor, Department of
   Humanities
Valencia Community College
Orlando, Florida

Edward C. Turner, Ph.D.
Associate Professor
College of Education
University of Florida
Gainesville, Florida

Cover Design: Didona Design Associates

Cover: Plosky Tolachik Kamchatka, Russia. Photo by Robert Cassell/The Stock Market

Printed in the United States of America

ISBN 0-03-032514-5

2 3 4 5 6 7   041 97 96 95 94

For permission to reprint copyrighted material, grateful acknowledgement is made to the following sources:

Abbeville Publishing Group: From *Early Maps* by Tony Campbell. Copyright © 1981 by Cross River Press, Ltd.

Addison-Wesley Publishing Company: Quote by cosmonaut Vladimir Kovalyonok from *The Home Planet* by Kevin Kelley. Copyright © 1991 by Kevin Kelley.

American Heritage Magazine, a division of Forbes Inc.: From "Land Below, Sea Above" by William Wertenbaker from *Mysteries of the Deep*, edited by Joseph H. Thorndike, Jr. Copyright © 1980 by Forbes Inc. From "Rivers in the Ocean" and "The Wealth of the Sea" from *Mysteries of the Deep*, edited by Joseph J. Thorndike, Jr. Copyright © 1980 by Forbes Inc.

Atheneum Publishers, an imprint of Macmillan Publishing Company: "Fossils" from *Something new begins* by Lilian Moore. Copyright © 1982 by Lilian Moore.

Basic Books, Inc., a subsidiary of HarperCollins Publishers, Inc.: From *Galileo and the Scientific Revolution* by Laura Fermi and Gilberto Bernardini. Copyright © 1961 by Basic Books, Inc.

Boys' Life Magazine: From "They've Found the Titanic!" by Jon C. Halter from *Boys' Life*, vol. LXXIX, no. 11, November 1989. Copyright © 1989 by the Boy Scouts of America. From "The Titanic Has Gone Down!" by Jon C. Halter from *Boys' Life*, vol. LXXIX, no. 10, October 1989. Copyright © 1989 by The Boy Scouts of America.

Bradbury Press, an Affiliate of Macmillan, Inc.: From *Living With Dinosaurs* by Patricia Lauber and illustrated by Douglas Henderson. Text copyright © 1991 by Patricia Lauber. Illustrations copyright © 1991 by Douglas Henderson. From *Volcano: The Eruption and Healing of Mount St. Helens* by Patricia Lauber. Copyright © 1986 by Patricia Lauber.

Daniel C. Brandenstein: Quote by astronaut Daniel C. Brandenstein, captain of the space shuttle *Endeavor*, May 1992.

Curtis Brown Ltd.: From "The Deepest Trench" from *Seven Miles Down* by Jacques Piccard and Robert S. Dietz. Copyright © 1961 by Jacques Piccard and Robert S. Dietz.

Eugene Cernan: Quote by Eugene Cernan, *Apollo 10* astronaut.

Children's Television Workshop: From "Space Cadets: Contact Visits a Camp for Kid Astronauts" by Russell Ginns from *3-2-1 Contact*, September 1990. Copyright © 1990 by Children's Television Workshop. All rights reserved.

The Christian Science Monitor: "Post Early for Space" by Peter J. Henniker-Heaton from *The Christian Science Monitor*, January 10, 1952. Copyright © 1952 by The Christian Science Publishing Society. All rights reserved.

Michael Collins: Quote by Michael Collins, *Apollo 11* astronaut.

The Cousteau Society: From *The Ocean World of Jacques Cousteau: Volume 1, Oasis in Space* by Jacques-Yves Cousteau. Copyright © 1973, 1975 by Jacques-Yves Cousteau.

Thomas Y. Crowell Company: From *The Maya World* by Elizabeth P. Benson. Copyright © 1967 by Elizabeth P. Benson.

High Flight Foundation: Quote by James Irwin, *Apollo 15* astronaut.

HarperCollins Publishers: From "Death of a Young Diver," from *The Living Sea* by Captain Jacques-Yves Cousteau with James Dugan. Copyright © 1963 by HarperCollins Publishers. From *One Day in the Prairie* by Jean Craighead George. Copyright © 1986 by Jean Craighead George.

William H. Jordan, Jr.: From "The Great Australian Sun Race" based on an article by William H. Jordan, Jr. Copyright © 1989 by William H. Jordan, Jr.

Alfred A. Knopf, Inc.: From *Hal Borland's Twelve Moons of the Year*, edited by Barbara Dodge Borland. Copyright © 1979 by Barbara Dodge Borland, as executor of the estate of Hal Borland.

Little, Brown and Company: From *Life on Earth: A Natural History* by David Attenborough. Copyright © 1979 by David Attenborough Productions, Inc.

Macmillian Publishing Company: From "Mysterious Footprints in Stone" from *The Macmillan Book of Dinosaurs and Other Prehistoric Creatures* by Mary Elting. Copyright © 1984 by Mary Elting.

The Scott Meredith Literary Agency: Quote by Carl Sagan from *Mars Beckons: The Mysteries, The Challenges, The Expectations of Our Next Great Adventure in Space* by John Noble Wilford. Copyright © 1990 by John Noble Wilford.

Metheun & Company: From *The Physics of Blown Sand and Desert Dunes* by R. A. Bagnold.

National Parks and Conservation Association: From speech by Paul C. Pritchard, President of the National Parks and Conservation Association.

National Wildlife Federation: From "Trash Trek" by Deborah Churchman from *Ranger Rick*, September 1990. Copyright © 1990 by National Wildlife Federation. "Volcano!" by Caroline Wakeman Evans from *Ranger Rick*, June 1988. Copyright © 1988 by the National Wildlife Federation.

Penguin Books Ltd., England: From *The Odyssey* by Homer, translated by E.V. Rieu, revised by D.C.H. Rieu. (Penguin Classics, 1946, Revised edition, 1991). Copyright © 1946 by E.V. Rieu; revised translation copyright © 1991 by D.C.H. Rieu.

Marian Reiner on behalf of Myra Cohn Livingston: "Asteroids," "Comets," "Jewels," "Messages," "Moon," "Satellites," "Secrets," and "Sun" from *Space Songs* by Myra Cohn Livingston (Holiday House). Text copyright © 1988 by Myra Cohn Livingston.

Time Life, Inc.: From *The Emergence of Man: The Lost World of the Aegean* by Maitland A. Edey and Editors of Time-Life Books. Copyright © 1975 by Time-Life Books Inc.

Toledo Blade: Quote by astronaut Aleksei Leonov during a walk in space.

University of Nebraska Press: From *Dust Bowl Diary* by Ann Marie Low. Copyright © 1984 by the University of Nebraska Press.

Warner/Chappell Music, Inc.: From lyrics from "Both Sides Now" by Joni Mitchell. Copyright © 1967, 1974 by Siquomb Publishing Corp. All rights reserved. Used by permission.

CONTENTS

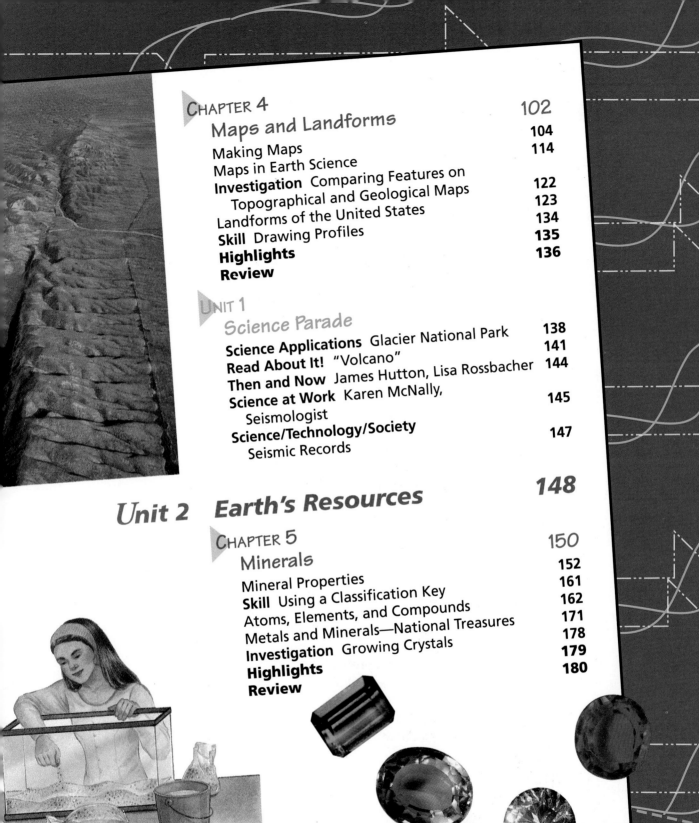

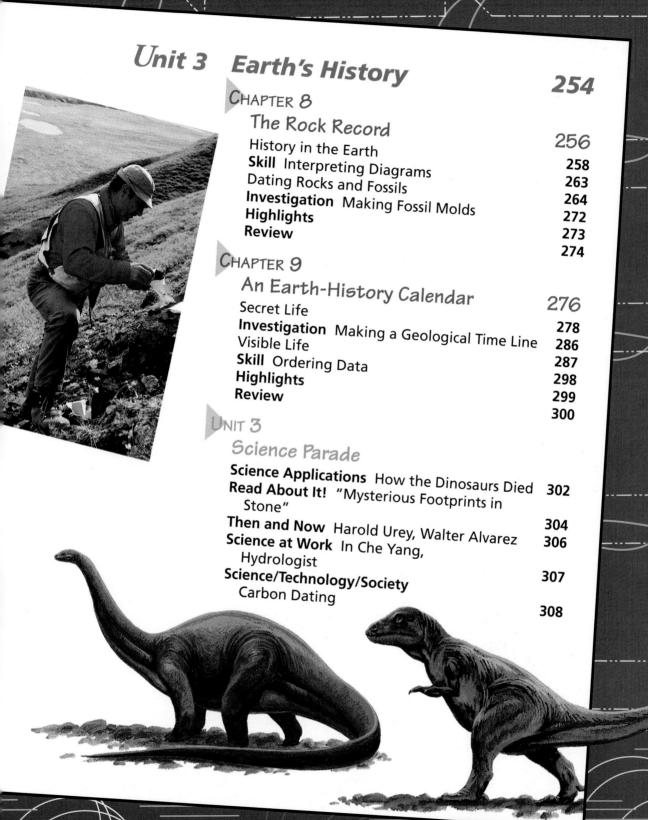

Unit 3　Earth's History　254

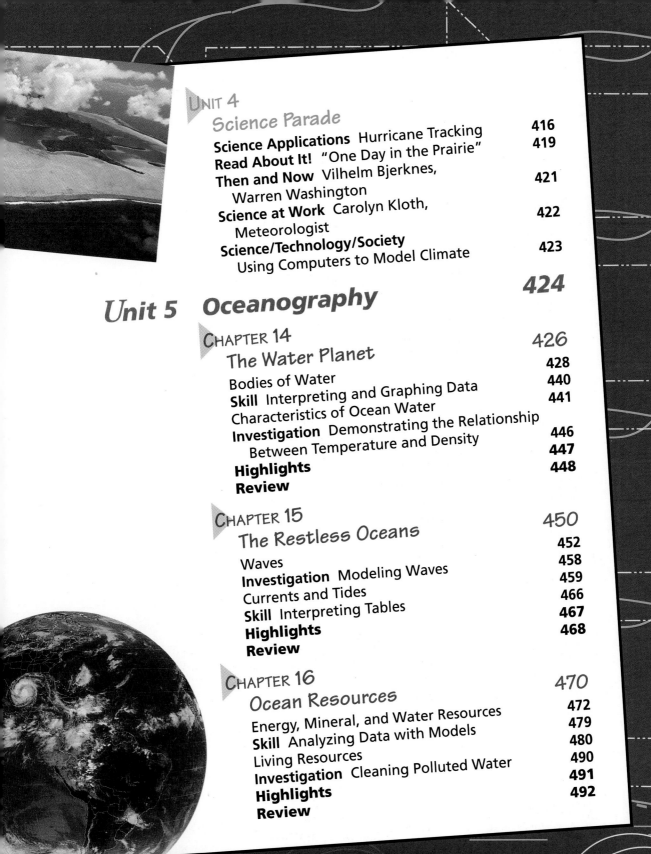

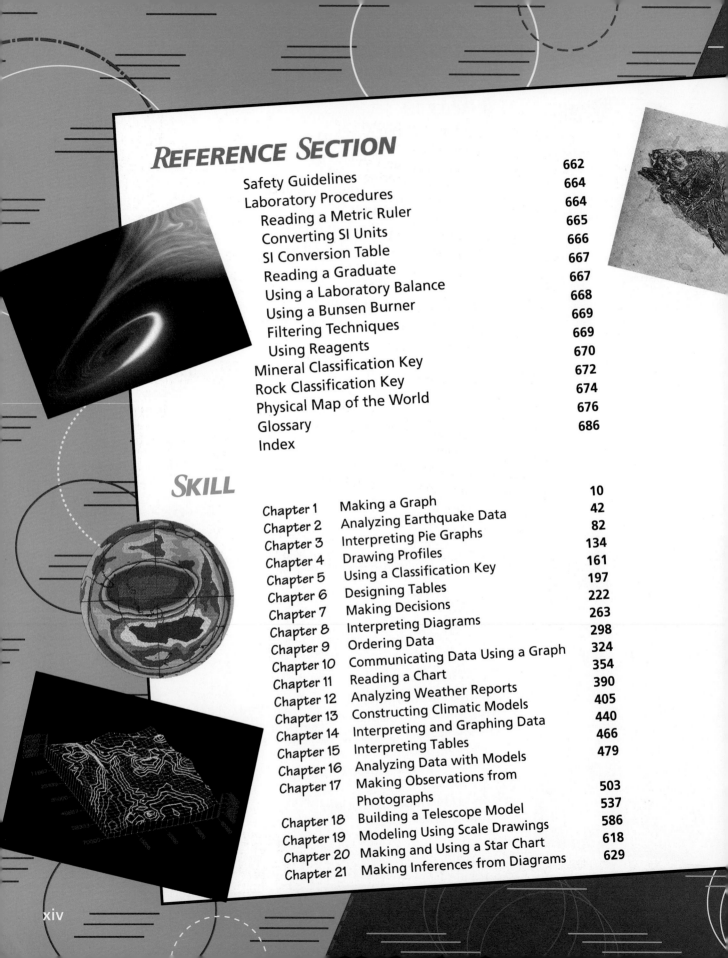

REFERENCE SECTION

SKILL

INVESTIGATION

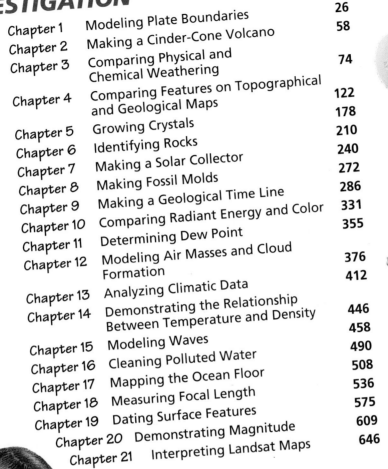

ACTIVITY

Discover By

About Holt Earth Science

Share the Wonder

Have you ever wondered why the seasons change? How weather affects the climate where you live? How the land around you has changed over time? How many stars there are in the sky? These are just a few of the questions that have been asked by earth scientists over the years. The answers to these and many of your own questions about the earth and space can be found in HOLT EARTH SCIENCE.

Explore the Nature of Science

Have you ever seen an ocean, a mountain, or a shooting star? Have you ever watched a thunderstorm? Have you ever examined a rock or a fossil? If you've ever done any of these things, you have experienced the study of earth science. The study of the earth and space is what earth science is all about. Science is knowledge gained by observing, experimenting, and thinking. Earth science is knowledge of the earth, its structure, its fresh water and oceans, its atmosphere and space. It is also knowledge of the forces that continue to change them over time. You probably have a lot of this knowledge already. HOLT EARTH SCIENCE will help you add to your knowledge through reading, discussion, and activities.

Work Like a Scientist

Much of science is based on observations of events in nature. From observations, you form questions about the things you see around you. By doing experiments, you will gather more information. During your study of earth science, you will be working and thinking like a scientist. You will develop and refine certain skills—such as your observation skills. In some cases, you will be asked to make predictions, and through activities you will have a chance to test the accuracy of your predictions. With new information, you will try to answer questions, to think about what you have observed, and to form conclusions. You will also learn how to communicate information to others, as scientists do.

Keep a Journal

One way to organize your ideas is to keep a journal and make frequent entries in it. Like a scientist, you will have a chance to write down your ideas and then, after doing activities and reading about scientific discoveries, go back and revise your journal entries. Remember, it's OK to change your mind after you have additional information. Scientists do this all the time!

Make Discoveries

Earth science is an ongoing process of discovery—asking and answering questions about the structure and forces that act upon the earth and space. If you choose a career in earth science, you may find answers to the questions of students and scientists who lived before you. Through curiosity and imagination, earth scientists are able to create a never-ending list of questions about the world in which we live and the universe beyond. You can also gain respect and appreciation for the earth's beauty and its vast resources.

Prepare yourself for discovery. Through your studies, you will learn many interesting and exciting things. Allow yourself to explore and to discover the ideas, the information, the challenges, and the beauty within the pages of this book. HOLT EARTH SCIENCE can help you understand how the world around you works and how it relates to the worlds beyond. You will begin to ask new questions and to find new answers. And you will discover that learning about the earth is fun.

EARTH'S STRUCTURE

Imagine you are standing here at the top of Triple Divide Peak in Glacier National Park. If you were to dump out a bucket of water, in which direction might the water flow?

At this location water falling on the ground can flow in three possible directions. Water falling on the northern slope makes its way to the Arctic Ocean. Water falling on the western slope flows to the Pacific Ocean. Water falling on the eastern slope eventually flows to the Gulf of Mexico and then into the Atlantic Ocean. This peak is just one of the many unique features to be discovered on this earth of ours.

Science PARADE

The Dynamic Earth

*T*here is nothing in the natural world quite like the fury and force of a volcano. It can spew superheated lava that engulfs everything in its path. Sometimes, like Mount St. Helens, its destructive power can transform an entire landscape in a single blast.

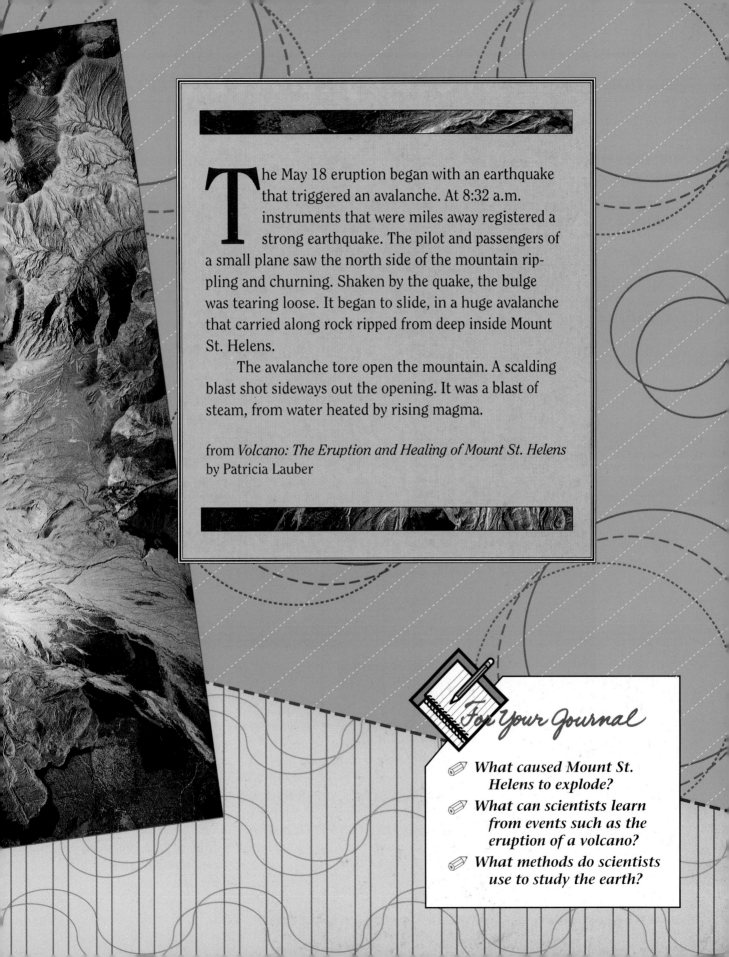

The May 18 eruption began with an earthquake that triggered an avalanche. At 8:32 a.m. instruments that were miles away registered a strong earthquake. The pilot and passengers of a small plane saw the north side of the mountain rippling and churning. Shaken by the quake, the bulge was tearing loose. It began to slide, in a huge avalanche that carried along rock ripped from deep inside Mount St. Helens.

The avalanche tore open the mountain. A scalding blast shot sideways out the opening. It was a blast of steam, from water heated by rising magma.

from *Volcano: The Eruption and Healing of Mount St. Helens* by Patricia Lauber

For Your Journal

- What caused Mount St. Helens to explode?
- What can scientists learn from events such as the eruption of a volcano?
- What methods do scientists use to study the earth?

A Geological Problem

Describe the work of geologists.

Compare and contrast the various methods of scientific research.

Explain how variables are used in scientific studies.

Few things are as dramatic as an erupting volcano. In 1980 Mount St. Helens erupted in Washington State. The eruption blew off the top of the mountain and completely destroyed everything for many kilometers around.

The Big Blast

Even before the "big blast," *geologists*—scientists who study the earth—raced to Mount St. Helens. The earth was "calling" them with its rumblings and swellings, revealing clues about its mysterious interior. For geologists, this was a great opportunity to gather data about the forces at work within the earth. Ms. Lauber continues with her description of "The Big Blast."

Before the eruption Mount St. Helens was like a giant pressure cooker. The rock inside it held superheated water. The water stayed liquid because it was under great pressure, sealed in the mountain. When the mountain was torn open, the pressure was suddenly relieved. The superheated water flashed to steam. Expanding violently, it shattered rock inside the mountain and exploded out the opening, traveling at speeds of up to 200 miles per hour.

The blast flattened whole forests of 180-foot-high firs. It snapped off or uprooted trees, scattering the trunks as if they were straws. At first, this damage was

puzzling. A wind of 200 miles an hour is not strong enough to level forests of giant trees. The explanation, geologists later discovered, was that the wind carried rocks ranging in size from grains of sand to blocks as big as cars. As the blast roared out of the volcano, it swept up and carried along the rocks it had shattered.

The result was what one geologist described as "a stone wind." It was a wind of steam and rocks traveling at high speed. The rocks gave the blast its great force. Before it, trees snapped and fell. Their stumps looked as if they had been sandblasted. The wind of stone rushed on. It stripped banches from trees and uprooted them, leveling 150 square miles of countryside. At the edge of this area other trees were left standing, but the heat of the blast scorched and killed them.

The stone wind was traveling so fast that it overtook and passed the avalanche. On its path was Spirit Lake, one of the most beautiful lakes in the Cascades. The blast stripped the trees from the slopes surrounding the lake and moved on.

Meanwhile the avalanche had hit a ridge and split. One part of it poured into Spirit Lake, adding a 180-foot layer of rock and dirt to the bottom of the lake. The slide of the avalanche into the lake forced the water out. The water sloshed up the slopes, then fell back into the lake. With it came thousands of trees felled by the blast.

The blast itself continued for 10 to 15 minutes, then stopped. Minutes later Mount St. Helens began to erupt upwards. A dark column of ash and ground-up rock rose miles into the sky. Winds blew the ash eastward. Lightning flashed in the ash cloud and started forest fires. In Yakima, Washington, some 80 miles away, the sky turned so dark that street lights went on at noon. Ash fell like snow that would not melt. This eruption continued for nine hours.

It had been obvious, even to early scientists, that volcanoes were release valves for the heat and pressure that builds up inside the earth. But why is there heat inside the earth? And what causes a mountain to suddenly destroy itself and everything around it in a violent eruption? You may have wondered about these things just as geologists have. If only the earth's interior could be examined and mapped directly, these questions could be answered. However, this isn't yet possible.

Much to their frustration, geologists are not able to do more than nick the outer layer of the earth's skin, or *crust*. So far, they have drilled about 13 km into the crust—the center of the earth is about 6400 km below the surface. This puts geologists at a distinct disadvantage compared with other scientists. Astronomers use telescopes to see the stars, meteorologists use satellites to see the clouds, and oceanographers use submersibles to see the ocean floor. Geologists can't see much more than the surface of the earth, yet this hasn't stopped them from investigating events like the eruption of Mount St. Helens.

Figure 1–1. Most scientists can see the events or phenomena they study.

 ASK YOURSELF

Why are geologists at a disadvantage when studying the interior of the earth?

Methods of Geological Research

Scientists often begin their investigations by making an educated guess, called a **hypothesis,** about what they think has occurred. Geologists on Mount St. Helens hypothesized that a "stone wind" had been responsible for much of the destruction.

Once a hypothesis is stated, the next step is to gather data that will either prove it or to disprove it. Scientists use several different research methods to do this, such as field study, laboratory experimentation, and model building.

Field Study

A field study, carried out in a natural setting, enables a scientist to gather information by *observation.* Field geologists might visit an area to observe the effects of an event, gather samples, draw maps, take photographs, and describe the study area in detail. Then they organize the information so that accurate comparisons of changes over time can be made. What observations do you think a field geologist might have made about the destruction on Mount St. Helens?

Laboratory Experimentation

Using information and samples from field studies, geologists extend their knowledge through *experimentation.* Scientists conduct laboratory experiments to isolate a small aspect of a larger phenomenon or event. By experimenting on a specific part of the whole in a controlled situation, scientists are more accurately able to explain relationships among variables that might be involved in the event. A **variable** is a part of an experiment that can be changed. In the following activity, you can do a simple experiment and decide what the variable is.

Figure 1–2. A geologist can make many observations about a field study area.

DISCOVER BY Doing

You will need sand, gravel, a cup, water, a bucket and an old soup can with holes in the bottom. Fill the can with gravel. Then, over the bucket, pour a cup of water into the can. How long does it take for the water to filter through the gravel? Replace the gravel with sand to see how long it takes for the same amount of water to filter through the sand. In this experiment, what is the variable? ✎

In the activity, you had just one variable. There are many variables in an event like the eruption of a volcano. However, geologists have no way of experimenting directly with a volcano. The challenge for the laboratory scientist is to design an experiment that tests the effects of each variable within the limits of a laboratory setting.

Figure 1–3. It is very challenging to design a valid experiment.

In order to see how one variable affects another, a valid experiment must test the effects of a single variable. Sometimes this is a difficult thing to do. For example, from observation and experimentation, geologists knew that superheated water could blow the top off a mountain. (Think about what happens to a lid on a pan full of boiling water.) But data showed that water and steam alone could not knock down a forest of large trees. **Data** is the information gathered by observation and experimentation. Geologists had to analyze additional data and construct a model to reach their goal of proving or disproving the "stone-wind" hypothesis.

Model Building   Scientists construct models to help them interpret events. Using data from laboratory experiments and observations gathered in the field, geologists constructed computer models to simulate the forces involved in the eruption of Mount St. Helens. Geologists often work with models to analyze the complex forces operating within the earth.

Models are only as good as the data on which they are based. Through the use of data from lasers, satellites, computers, field study observations, and remote-sensing equipment, geologists were able to construct an accurate model of the events in the eruption of Mount St. Helens.

Figure 1–4. Unlike this ship, some models do not look like the real thing. In science, it is more important that models act like the real thing.

The geologists studying Mount St. Helens concluded that their hypothesis of the "stone wind" was probably correct. **Conclusions** are statements based on all the information gathered; in this case from field studies, laboratory experiments, and models.

Hypotheses concerning isolated events, such as the "stone wind" of Mount St. Helens, fit like puzzle pieces into a larger picture of the structure of the earth. As more information is gathered and analyzed, hypotheses are strengthened, changed, or rejected. Geologists continue refining their hypotheses, eventually developing theories about the earth. A **theory** is a statement that explains why things happen the way they do. In the next section you will learn about several important theories in geology.

 ASK YOURSELF

Why do geologists use models?

SECTION 1 *REVIEW AND APPLICATION*

Reading Critically

1. Explain why geologists concluded that there was a "stone wind" on Mount St. Helens.

2. Why do variables have to be studied one at a time?

Thinking Critically

3. Compare and contrast the three types of scientific approaches used by geologists.

4. To map the wind patterns that affected the distribution of volcanic ash east of Mount St. Helens, which scientific approach would you use? Explain your answer.

SKILL Making a Graph

Scientists often record data in a table. However, this format may not always be the best way to present the data. Often a better interpretation can be obtained by placing data on a graph.

▶ **MATERIALS**
● graph paper ● pencil ● data from Table 1

▼ **PROCEDURE**

1. Give your graph a title, and write it on the top of your graph paper.
2. Draw two axes on the graph paper. The horizontal, or X, axis should be near the bottom of the paper. The vertical, or Y, axis should be near the left edge of the paper. Label both axes with the type of data you are going to plot.
3. Determine the scale for each axis. The scale should be such that the graph covers as much information as possible and is spaced evenly over the entire sheet of graph paper. Study the data carefully before deciding on the scale.
4. Plot the data from Table 1 on your graph. On one axis, plot the longitude of each island, starting with the island of Hawaii and moving to the west. On the other axis plot the age of each island from east to west, starting with Hawaii.

TABLE 1: THE HAWAIIAN ISLANDS AND SEAMOUNTS		
Island (or reef)	Approximate Age (in millions of years)	Longitude (in degrees and minutes)
Hawaii	0.0	155° 30′ W
Kanum	39.0	170° E
Kauai	4.1	158° 30′ W
Maui	0.6	156° 15′ W
Midway	18.0	177° 30′ W
Molokai	1.8	157° W
Necker	10.1	164° 30′ W
Nihoa	no data	162° W
Oahu	3.1	158° W
Pearl	20.1	176° W
Yuruaku	42.3	168° 30′ E

▶ **APPLICATION**
Use your graph to answer the following questions:
1. Which is the youngest island?
2. Which is the oldest island?
3. What is the relationship between the nearness of the islands to Hawaii and their relative ages?
4. The age of one of the islands does not seem to fit the pattern formed by the others. Explain.

✳ *Using What You Have Learned*
Using the graph and a map, predict the longitude of the next volcanic island to form in this chain. When do you predict that this might happen?

Developing a Theory of Plate Tectonics

Describe the theories of continental drift and sea floor spreading.

Explain how the theory of plate tectonics grew from earlier theories.

Predict how the positions of the continents may change in the future.

Did you know that there was once an ocean in what is now the midwestern part of the United States? And that part of what is now Alaska was once a tropical island? Did you know that ice used to cover much of what is now the Sahara desert, while there were once swamps in what has become Antarctica? These discoveries are like pieces in a giant puzzle—the earth puzzle.

Drifting Continents

Look at the jigsaw puzzle in the picture. It's hard to figure out what the whole picture looks like when the pieces are apart. The same is true for figuring out the earth puzzle. For many years, geologists did not see their discoveries as pieces of the same puzzle. They tried to explain each of their discoveries as a single event. They couldn't see how all the pieces of the earth puzzle fit together.

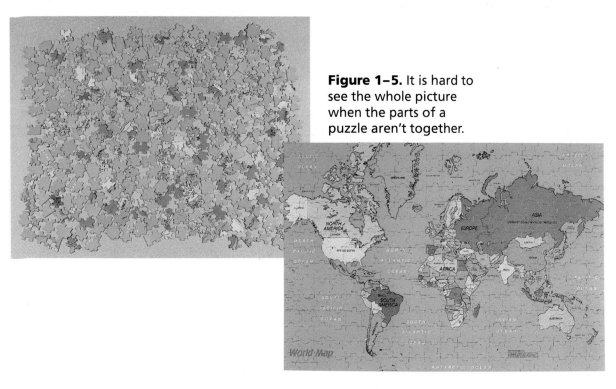

Figure 1–5. It is hard to see the whole picture when the parts of a puzzle aren't together.

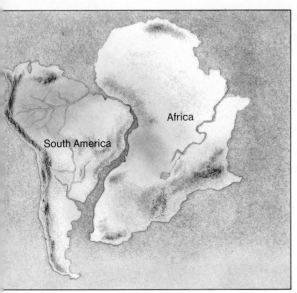

Figure 1–6. Africa and South America seem to be a perfect fit.

Have you ever noticed that on a world map the continents look a little like pieces of a jigsaw puzzle? For example, the continents of Africa and South America look as if they could fit together. Could these continents at one time actually have been joined together?

The idea that Africa and South America were once joined is part of a theory that says that, at one time, all the continents of the earth were connected. Then they drifted apart. This *theory of continental drift* was proposed in 1912 by German scientist Alfred Wegener. Wegener called the super-continent *Pangea,* which means "all earth." In the next activity, you can get some idea of what Pangea may have looked like.

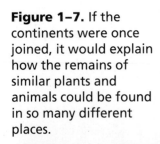 **VER BY** *Doing*

Your teacher will give you an outline map of the world. Cut around the outline of each of the continents, and try to fit them all together without the water that now separates them. Attach your version of Pangea to a blank sheet of paper, and compare it with those of your classmates. If all the continents were joined at one time, how many oceans would there have been? ✎

Figure 1–7. If the continents were once joined, it would explain how the remains of similar plants and animals could be found in so many different places.

The apparent fit of the continents wasn't the only evidence cited by Wegener as proof for his theory. He also noted that the remains of similar plants and animals were found in Africa and South America, as well as in India, Australia, and Madagascar. He wondered how these plants and animals could have been so similar if the continents and islands were always separated by wide oceans.

Finally, Wegener pointed to evidence showing that there were once glaciers in places like Australia, which are now too warm for permanent ice. He reasoned that such places must have been in colder locations at one time and that they had since moved to where they are now.

Wegener's theory of continental drift was not accepted by many of the scientists of his day. They found it hard to believe that massive continents could push their way through the solid rock of the ocean floor. They said it would be like ships trying to sail through a frozen sea. Some scientists said that the continents were once joined by a large land bridge that was now below the sea. There were still missing pieces to the puzzle.

 ASK YOURSELF

What is the theory of continental drift?

Sea Floor Spreading

If you look closely at the Atlantic Ocean basin, another puzzle piece can be found. In the center of the floor of the ocean—halfway between Europe and North America and between Africa and South America—is a chain of mountains called the *Mid-Atlantic Ridge.* These mountains, almost completely covered by water, are part of a larger mid-ocean ridge, the world's longest mountain range, circling the earth like the endless seam of a baseball.

Scientists have discovered three important things about the mid-ocean ridge. First, the layer of mud that covers most of the ocean floor is missing from the ridge. Second, the rocks get older as you move away from the ridge—the youngest rocks are in the center of the ridge. Third, the center of the ridge is much warmer than any other part of the ocean floor—hot material from deep inside the earth must be rising to the surface along the ridge. A new theory, the *theory of sea floor spreading,* was proposed to explain these discoveries.

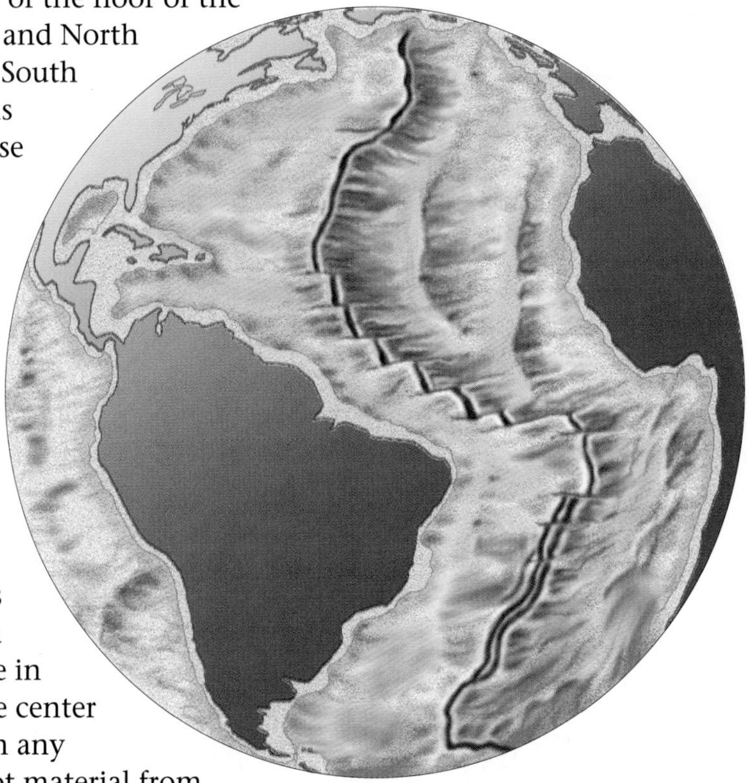

Figure 1–8. The mid-ocean ridge forms the longest mountain range on the earth.

The theory of sea floor spreading says that the center of the mid-ocean ridge—a deep, narrow valley—is a crack in the earth's crust. Hot, soft rock, called *molten rock,* constantly oozes from this crack. Cooled by the sea water, the rock becomes solid. Then it is pulled outward, away from the ridge, and replaced by new molten rock. Thus the sea floor spreads out along both sides of the mid-ocean ridge. The youngest rocks are always found near the crack in the mid-ocean ridge—the oldest rocks are found farthest from the ridge.

Figure 1–9. Molten rock rises in a crack in the earth's crust, because of sea floor spreading along the mid-ocean ridge.

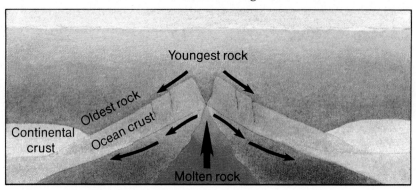

ASK YOURSELF

What is the theory of sea floor spreading?

Plate Tectonics

Sea floor spreading suggested a new theory to explain how continents move. What if the earth's crust is broken up into large pieces, or plates. Along the Mid-Atlantic Ridge, two of these plates must be moving away from each other.

Continents are usually parts of larger plates. They are carried along with moving plates, the way logs frozen into blocks of ice are carried along by a river. Thus, continents do not have to push their way through the solid crust to move—they move as sections of the crust move.

This new theory, called the *theory of plate tectonics,* combines the ideas of sea floor spreading and continental drift. *Tectonic* comes from the Greek word *tektōn,* which means "builder."

The theory of plate tectonics states that the earth's surface is divided into several major plates, also called *tectonic plates,* of varying sizes. Tectonic plates are often named for the main geographical feature they carry.

As you can see in the next diagram, the sizes and shapes of the continents are not the same as those of the plates that carry them. The rocks that make up the continents are lighter than

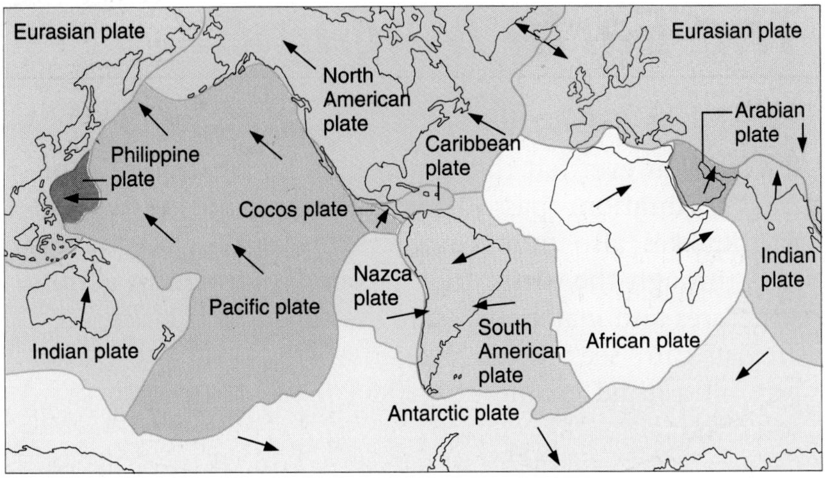

Figure 1–10. The earth's major plates.

those of the rest of the crust. Thus, the continents ride high on top of the moving plates. You can also see from the arrows that tectonic plates move in various directions. They also move at different speeds. In the next activity, you can see what might happen next.

DISCOVER BY *Problem Solving*

The North American plate is moving toward the northwest at about 2 cm per year, while the Pacific plate moves in the same direction, but at about 10 cm per year. What might eventually happen to southern California if these two plates continue to move as they do now? What will happen to the Atlantic Ocean if the North American plate, the South American plate, the Eurasian plate, and the African plate keep moving the way they are now? ✐

▶ ASK YOURSELF

How does the meaning of the word *tektōn* relate to plate tectonics?

SECTION 2   *REVIEW AND APPLICATION*

Reading Critically
1. What evidence is used to support the theory of continental drift?
2. How is the theory of plate tectonics built on earlier theories?

Thinking Critically
3. Explain how the theory of plate tectonics is built on a different set of hypotheses than the theory of continental drift.
4. Construct a model showing the possible position of the continents 50 million years in the future. Defend the design of your model with facts about plate movement.

The Earth Machine

Objectives

Identify *the layers of the earth.*

Explain *how convection currents move tectonic plates.*

Compare and contrast *the three types of tectonic plate boundaries.*

Have you ever felt the vibrations of a passing train or truck? Vibrations pass through the ground as waves. Earthquakes also produce waves. As earthquake waves pass through the earth, they provide information about structures and materials inside the earth. In this way, geologists can "see" inside the earth. This is similar to the way ultrasound enables a physician to see inside a person's body.

The Interior of the Earth

As waves pass through different materials, they travel at different speeds, and in some cases they change directions. To understand why waves are affected by the materials through which they pass, think about how hard it is to walk through water compared with air.

Some of the materials in the earth's interior slow earthquake waves and cause them to change direction. By carefully analyzing earthquake waves, geologists have learned that the interior of the earth is in layers. These layers are called the *crust,* the *mantle,* and the *core.* In the next activity, you can see how layers of different materials form.

Figure 1–11. Geologists can't just "look" inside the earth.

ACTIVITY

How do layers form?

MATERIALS

glass jar, water, soil, sand

PROCEDURE

1. Fill the jar nearly to the top with water.
2. Add soil and sand to the water and gently stir. What do you observe about the mixture?
3. Allow the mixture to sit undisturbed for about 5 minutes. Now what do you observe about the mixture?

APPLICATION

1. Which material ended up at the bottom of the jar? Why?
2. If the earth had once been a mixture of materials in a liquid state, which materials would have ended up in the center of the earth?

The Crust   The **crust** is a thin layer on the surface of the earth. Under the oceans, the crust is about 7 km thick. The continental crust is much thicker than the oceanic crust. It ranges from 25 km at the edges of the oceans to 70 km near the center of the continents.

The oceanic crust is composed of heavy, dark-colored rocks. Continents are made of various kinds of rocks that are lighter in weight and color than those of the ocean floor. The continental crust is also much older than the oceanic crust. Some continental rocks are 20 times older than the oldest rocks on the ocean floor.

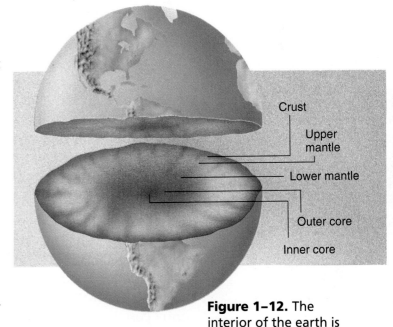

Figure 1–12. The interior of the earth is in layers.

There is a definite boundary to the earth's crust. Andrija Mohorovičić (ahn DREE jah moh hoh ROH vuh chihch), a Croatian geologist, discovered the boundary between the crust and the rest of the interior. Mohorovičić noted that earthquake waves did not pass directly through the earth. Instead, the waves were deflected at a point below the surface. This boundary, or *discontinuity*, is called the *Mohorovičić discontinuity* ("Moho," for short) in his honor. The Moho is about 7 km below the ocean floor and, on the average, about 35 km below the surface of the continents.

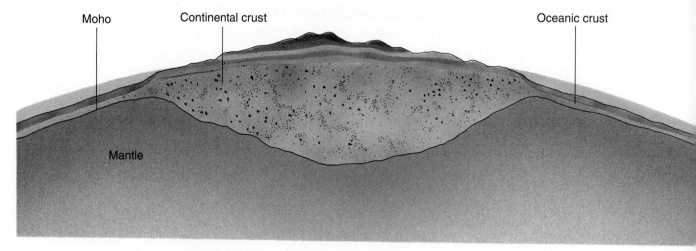

Figure 1–13. The continental crust is much lighter in weight and thicker than the oceanic crust.

The Mantle   The **mantle** is the layer immediately beneath the Moho. The mantle, which is about 2900 km thick, is divided into two parts. The upper mantle and the lower mantle are separated by an unnamed boundary. Below the boundary, the rocks of the mantle are hot and can deform (change shape) easily. The mantle is composed of rocks that are even heavier than those of the oceanic crust.

The Core   Below the mantle lies the **core,** the innermost layer of the earth. The core is separated from the mantle by a boundary called the *Gutenberg discontinuity.* This boundary is about 2900 km beneath the earth's surface.

Like the mantle, the core is also divided into two parts. The outer core is liquid and is probably made of iron. Scientists hypothesize that circulation within the outer core may be responsible for producing magnetic fields around the earth.

Inge Lehmann, a Danish scientist, discovered that the inner core is solid. It is probably a mixture of iron and nickel. The extreme pressure at this depth, due to the weight of overlying layers, probably keeps the inner core solid. Most materials expand when they melt, but because the pressure is so great, the materials of the inner core cannot expand or melt.

 ASK YOURSELF

What are the names of the layers of the earth's interior?

Currents Within the Mantle

The layers of the earth may also be grouped according to their consistency. The solid layer, called the *lithosphere* (LIHTH uh sfihr), is composed of the crust and the upper mantle. The lithosphere is broken up into tectonic plates.

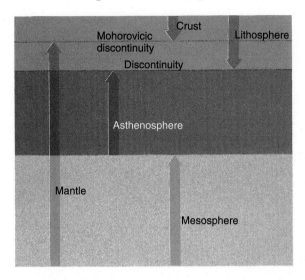

Figure 1–14. The plates of the lithosphere float on the partially molten rocks of the asthenosphere. The rocks of the mesosphere are mostly solid.

Below the lithosphere is the *asthenosphere* (as THEHN uh sfihr), a layer of the mantle in which temperature and pressure combine to cause a small amount of melting of the rocks. This makes the asthenosphere a bit softer than the rest of the mantle. Solid plates float on the asthenosphere.

Pressure on the asthenosphere comes from the overlying rocks of the lithosphere, but where does the heat come from to partially melt the asthenosphere's rocks? Some of the heat still lingers from the earth's formation. However, scientists hypothesize that most of the heat comes from the decay of radioactive material. Radioactive decay will be discussed later. In the next activity, you can make a model of the heating that occurs within the mantle.

DISCOVER BY *Doing*

CAUTION: Put on safety goggles, and keep them on during this activity. Half fill a large laboratory beaker or a glass baking dish with corn syrup. Add several drops of food coloring, but do not stir the mixture. Place the container on a hot plate or over a very low burner flame. As the syrup heats, a circulation pattern should begin to form. Observe the pattern by watching the movement of the food coloring. ✐

When rocks are heated, they become less dense and tend to rise, creating currents within the mantle. The currents carry warmer rocks up and cooler rocks down. These currents are very slow, moving rocks only a few centimeters per year. The currents create pockets of circulation called *convection cells*. **Convection** is the transfer of heat by this type of circulation.

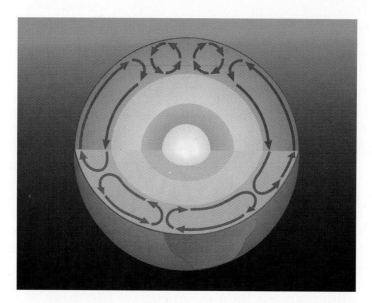

Figure 1–15. Convection currents within the mantle cause plates to move.

Although convection cells can be observed in a laboratory, the movement of rocks within the mantle is far more complicated than the laboratory model. Convection cells in the earth may have *eddies,* or currents within currents. The circulation within convection cells drives the tectonic plates moving on the asthenosphere. The following activity can show you how.

DISCOVER BY *Doing*

CAUTION: Be careful when using the hot plate. Using Figure 1–10 on page 15, sketch the shape of several adjoining plates on a sheet of paper. Cut out the plate outlines, separate them, and transfer the patterns to thin sheets of polystyrene. Cut out the polystyrene "plates," and float them in water in a glass baking dish. Gently warm the dish on a hot plate, and note the movement of the "plates" as convection cells begin to develop. Which material represents the lithosphere? Which material represents the asthenosphere? ✐

▼ **ASK YOURSELF**

How do convection cells move tectonic plates?

Plate Boundaries

Compare the map of the earth's plate boundaries, from page 15, with this map showing the location of many of the earth's earthquakes and volcanoes. As you can see, exciting geological events occur along the boundaries of plates. Although there are exceptions, most earthquakes and volcanoes occur near the edges of plates. Earthquakes are violent shakings of the earth's crust. Volcanoes are openings in the crust through which molten rock reaches the earth's surface. You will learn more about earthquakes and volcanoes in the next chapter.

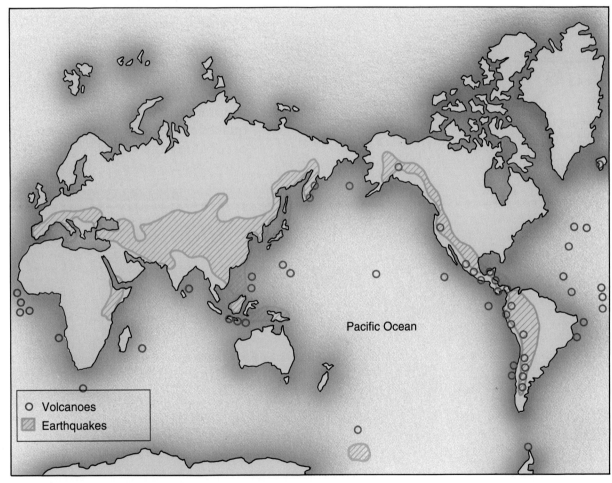

Pacific Ocean

○ Volcanoes
▨ Earthquakes

Figure 1–16. Why do so many earthquakes and volcanoes occur around the edge of the Pacific Ocean?

So many earthquakes and volcanoes occur around the edge of the Pacific plate that this area is called the *Ring of Fire.* Earthquakes, volcanoes, mountains, and ocean trenches are all associated with the boundaries between tectonic plates. Why is this so? If you think about a collection of plates, all moving independently of each other, it is easy to understand that some plates pull apart, some plates crash together, and others slide past each other.

Pulling Apart Imagine traveling to the middle of the Atlantic Ocean to investigate the Mid-Atlantic Ridge. Remember, this is a very young range of volcanic mountains. You might even see molten rock, or *lava,* spewing from the center of the ridge. Along this ridge, molten rock from below rises to the earth's surface through cracks, or *rifts,* in the ocean floor. These rifts are opened as plates pull apart. A boundary where tectonic plates are separating is a **divergent boundary.** The word *diverge* means "to move apart."

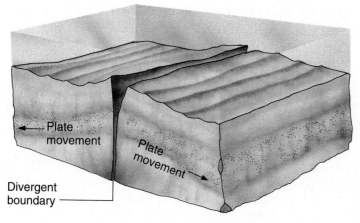

Figure 1–17. Plates pull apart along divergent boundaries.

Divergent boundaries are also found in a few places on land, such as the Great Rift Valley of Africa. This huge valley extends more than 4000 km, from the southern end of the Red Sea to Mozambique. The wide, flat valley is bordered by steep cliffs, more than 600 m high in some places. Frequent earthquakes and volcanic eruptions occur along this valley.

Figure 1–18. Africa may separate into two continents along the Great Rift Valley.

Africa is probably splitting apart along this rift. As the plates continue to separate, the valley floor drops. Many scientists hypothesize that in the next few million years, the valley will become a sea, with the cliffs forming the coastlines of two separate continents.

Crashing Together   If you were to construct a model of moving plates, in some places the plates would be pulling apart. But what would be happening on the other side of your model world? This represents another kind of boundary.

The place where tectonic plates come together is called a **convergent boundary.** The word *converge* means "to come together." Converging plates produce mountains, volcanoes, and ocean trenches.

Figure 1–19. Volcanoes are common along convergent boundaries.

Since some plates carry continents and others ocean floors, they respond differently when they converge. If an oceanic plate collides with a continental plate, the heavier oceanic plate is forced below the continental plate. An area where one tectonic plate is forced under another is called a *subduction zone.* As the oceanic plate is pushed under the continental plate and deep into the crust, the rocks heat up and melt. Some of this molten rock rises again, forming volcanoes along the continental edge of the boundary.

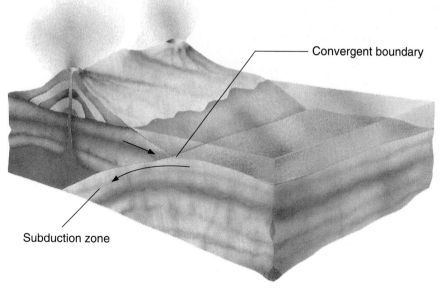

Convergent boundary

Subduction zone

Figure 1–20. A subduction zone forms where a continental plate and an oceanic plate converge.

During subduction, the oceanic plate pushes downward, and a trench forms between the two plates. Although trenches occur in many places, several are located around the edges of the Pacific plate. The greatest ocean depths are found in ocean trenches.

On the western edge of the South American plate is another subduction zone and trench. Here, the Pacific plate is being forced below South America, forming a deep trench called the Peru-Chile Trench. The Andes Mountains are also a result of these converging plates.

Figure 1–21. The rise from the bottom of the Peru-Chile Trench to the top of the Andes is greater than the elevation of Mount Everest, the world's highest mountain peak.

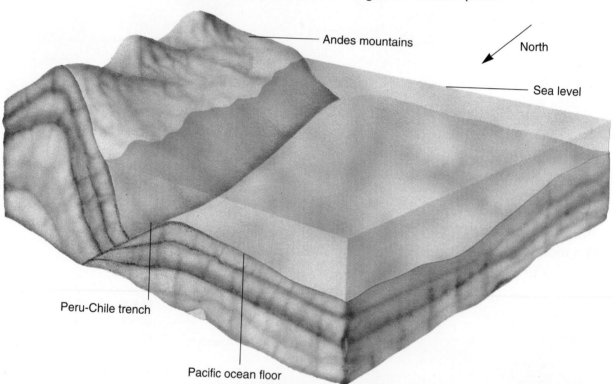

Several trenches in the Pacific and Atlantic oceans border chains of islands called *island arcs*. Many of the islands in these arcs are volcanic in nature. Island arcs and their associated trenches are the result of two oceanic plates converging. The Mariana Trench, located off the Mariana Islands, is the deepest trench at about 11 000 m. Look again at the map of plate boundaries on page 15. Where would you predict that this trench should be located? Confirm your prediction by looking at a world map.

When two continental plates converge, neither is forced under the other. Instead, both are pushed up, forming mountains. The rocks bend and break, often rising to great heights. The Himalaya Mountains, for example, formed when the Indian plate collided with the Eurasian plate. As these plates continue to move, mountains in this range rise by about 5 cm each year.

Figure 1–22. The Himalaya Mountains were formed by the collision of two continental plates.

Sliding Past   Sometimes plates slide past each other. Although this type of boundary, called a **transform boundary,** does not produce mountains or trenches, earthquakes are common. The famous San Andreas fault in California is located along the boundary where the North American plate passes the Pacific plate. A *fault* is a place where rock has moved on one or both sides of a crack in the crust. You will learn more about faults in the next chapter.

 ASK YOURSELF

Describe the three types of plate boundaries.

SECTION 3   REVIEW AND APPLICATION

Reading Critically
1. Diagram the layers and discontinuities of the earth's interior.
2. Explain the difference between the lithosphere and the asthenosphere.

Thinking Critically
3. How could the study of earthquakes be used to learn about the earth's interior?
4. Explain how convection cells cause the movement of tectonic plates.

INVESTIGATION

Modeling Plate Boundaries

▶ MATERIALS
- magazines (2) • desks or tables (2) • chalkboard erasers (2)

▼ PROCEDURE

1. You can make simple models of plate boundaries to study the effects of movements along the boundaries. To model diverging plates, place two desks together so that there is a narrow crack between them. Place two magazines into the crack. The magazines represent crust being formed as the plates pull apart. To simulate the movement of the plates, press down on the magazines and slowly pull the desks apart. What happens to the "plates"?

2. To simulate converging continental plates, place the desks about 5 cm apart, and lay the magazines side by side over the gap. Now push the desks together.

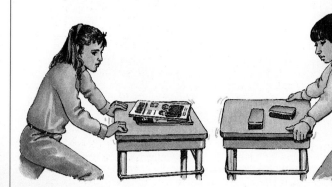

What happens as the "plates" converge?

3. To simulate an oceanic plate converging with a continental plate, tuck one of the magazines into the crack between the desks before you push them together. What happens as the "plates" converge?

4. To simulate a transform boundary, hold a chalkboard eraser in each hand and push your hands together as hard as you can. Now try to slide the erasers past each other in opposite directions. Do the erasers slide smoothly, or is the movement jerky? If the erasers represent the crust on opposite sides of a fault, what might the jerky motion produce?

▶ ANALYSES AND CONCLUSIONS
1. Which kind of boundary is most likely to produce mountains?
2. Which kind of boundary is most likely to produce ocean trenches?
3. Which kind of boundary is most likely to produce earthquakes?

▶ APPLICATION
1. Which procedure best represents the boundary between the North American plate and the Pacific plate?
2. What type of boundary is found along the west coast of South America?

3. Which of the three boundary types is found along the Mid-Atlantic Ridge?

✳ Discover More
Find out on which plates Melbourne, Australia; Hilo, Hawaii; Fairbanks, Alaska; San Juan, Puerto Rico; and Reykjavik, Iceland, are located. Find out in which directions the plates are moving. Determine which kinds of plate boundaries are nearby and find out what, if any, geographical features these boundaries have produced.

The Big Idea

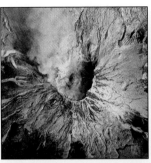

The earth is a dynamic structure composed of several distinct layers of solid and semisolid rocks. The lithosphere of the earth is made of rigid plates that move slowly in different directions, driven by convection currents. This allows new material to rise to the surface in some locations, while forcing other sections into the earth where they remelt. The formation of many of the earth's features can be explained by the interaction of these plates. Early supporters of the theory of plate tectonics relied predominantly on field studies and observational data. Later, there was an emphasis on model building and experimentation.

For Your Journal

Look again at the way you answered the questions at the beginning of this chapter. How have your ideas about volcanoes and what geologists can learn from them changed? Write about these changes in your journal.

Connecting Ideas

Review the evidence shown below, and then classify each piece as to whether it is from a field study, a model, or an experiment.

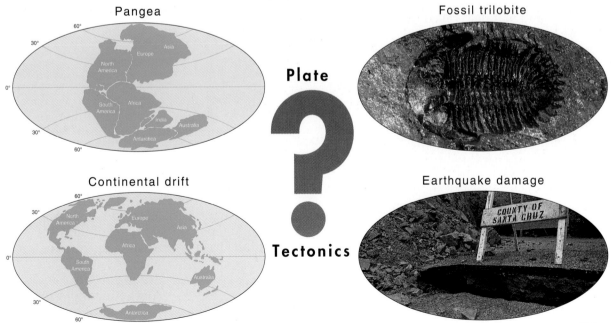

Pangea

Plate

Fossil trilobite

Continental drift

Tectonics

Earthquake damage

REVIEW

Understanding Vocabulary

1. For each set of terms, explain the similarities and differences in their meanings.
   a) crust (17), mantle (18), core (18)
   b) divergent boundary (22), convergent boundary (23), transform boundary (25)
   c) hypothesis (7), theory (9)

Understanding Concepts

MULTIPLE CHOICE

2. Scientists who study the earth are called
   a) astronomers.
   b) geologists.
   c) meteorologists.
   d) oceanographers.

3. The only portion of the earth thought to be entirely molten is the
   a) upper mantle.
   b) inner core.
   c) lower mantle.
   d) outer core.

4. The boundary between the crust and mantle is called the
   a) Mohorovičić discontinuity.
   b) Gutenberg discontinuity.
   c) convergent boundary.
   d) divergent boundary.

SHORT ANSWER

5. Explain why the continents float higher than the crust of the ocean basins.

6. What is the difference between a convergent boundary and a divergent boundary?

Interpreting Graphics

Use the diagram to answer the following questions.

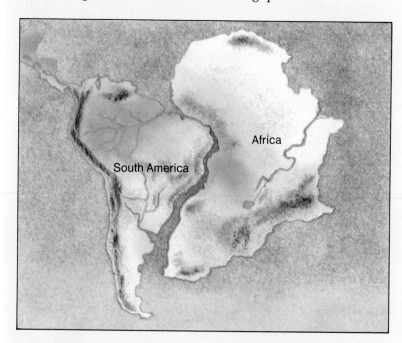

Africa

South America

7. Defend the theory that Africa and South America were once positioned as shown in the diagram.

8. How does an understanding of the Mid-Atlantic Ridge help scientists explain the drifting of Africa and South America to their present locations?

Reviewing Themes

9. *Changes Over Time*
Explain how convection cells affect the movement of continents over time.

10. *Systems and Structures*
In the system of study we call science, explain how a hypothesis can become a theory.

Thinking Critically

11. Explain how the theory of continental drift and the theory of sea floor spreading contributed to the theory of plate tectonics.

12. In studying the eruption of Mount St. Helens, describe what geologists learned from field study, from laboratory experimentation, and from model building.

13. Why do geologists hypothesize that the inner core of the earth is solid?

14. Some geologists have said that the section of California west of the San Andreas fault will someday end up near Alaska. Explain how this could happen.

Discovery Through Reading

Rona, Peter A. "Deep-Sea Geysers of the Atlantic." *National Geographic* 182 (October 1992): 105–109. Diving four kilometers to the Mid–Atlantic Ridge, a U.S.-Russian team investigates mineral-rich hot springs in a valley where the earth's plates separate.

Stager, Curt. "Africa's Great Rift Valley." *National Geographic* 177 (May 1990): 2–41. Limnologist Curt Stager reports on this geologically active area that cuts across East Africa, creating a landscape of extremes.

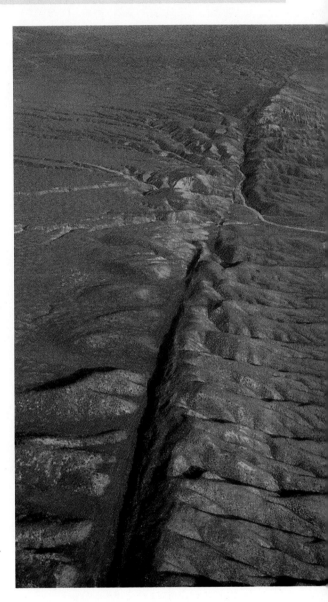

Building The Land

*E*arthquakes represent another force that can change the surface of the earth. An earthquake dramatically moves the earth's crust, sometimes with terrifying results. The following is a report on the New Madrid Earthquake, as recorded in the Philadelphia Gazette, February 12, 1812.

Philadelphia Gazette

Vol.II   No.42

February 12, 1812

Fifteen cents

Earthquake in West

[Monday, December 16th] - About 2 o'clock this morning we were awakened by a most tremendous noise, while the house danced about and seemed as if it would fall on our heads. I soon conjectured the cause of our troubles, and cried out it was an earthquake, and for the family to leave the house; which we found very difficult to do, owing to its rolling and jostling about. The shock was soon over, and no injury was sustained, except the loss of the chimney, and the exposure of my family to the cold of the night. At the time of this shock, the heavens were very clear and serene, not a breath of air stirring; but in five minutes it became very dark, and a vapour which seemed to impregnate the atmosphere, had a disagreeable smell, and produced a difficulty of respiration. I knew not how to account for this at the time, but when I saw, in the morning, the situation of my neighbors' houses, all of them more or less injured, I attributed it to the dust. . . . The darkness continued till daybreak; during this time we had EIGHT more shocks, none of them so violent as the first.

At half past 6 o'clock in the morning it cleared up, and believing the danger over I left home, to see what injury my neighbors had sustained. A few minutes after my departure there was another shock, extremely violent. I hurried home as fast as I could, but the agitation of the earth was so great that it was with much difficulty I kept my balance—*the motion of the earth was about twelve inches to and fro*. I cannot give you an accurate description of this moment; the earth seemed convulsed—the houses shook very much—chimnies [sic] falling in every direction. The loud, hoarse roaring which attended the earthquake, together with the cries, screams, and yells of the people, seems still ringing in my ears.

Fifteen minutes after seven o'clock, we had another shock. This one was the most severe one we have yet had—the darkness returned, and the noise was remarkably loud. The first motions of the earth were similar to the preceding shocks, but before they ceased we rebounded up and down, and it was with difficulty we kept our seats. At this instant I expected a dreadful catastrophe—the uproar among the people strengthened the colouring of the picture—the screams and yells were heard at a great distance. . . .

For Your Journal

- Why do earthquakes occur in certain places?
- What is the source of energy for earthquakes?
- What other processes change the surface of the earth?

Earthquakes

Objectives

Explain how the epicenter of an earthquake is located.

Compare and contrast the motions of P waves, S waves, and surface waves.

Correlate earthquakes with plate tectonics.

In the winter of 1811–1812, history was made as a string of earthquakes shook the area around New Madrid, Missouri. These earthquakes still rank among the strongest ever felt in North America. Shock waves from these earthquakes were felt as far away as Washington, D.C., and even Quebec City, Canada, nearly 1200 km away. These powerful earthquakes came from a major fault beneath the Mississippi River.

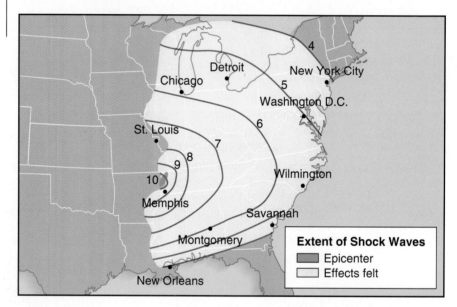

Figure 2–1. The New Madrid earthquakes were felt as far away as Canada. Larger numbers represent stronger effects.

Whole Lot of Shakin' Going On

As sections of the earth's crust slip past each other, friction keeps them from sliding smoothly. **Friction** is the force that opposes motion between two surfaces that are touching. Remember what is was like to skip along a sidewalk when you were younger? There was a lot of friction between your foot and the concrete. As your forward motion overcame the friction, your foot slipped, jerking you along. Similarly, friction within the earth's crust causes stress to build over time. The sudden release of this stress causes the jerking within the crust that we call an earthquake. In the Investigation in the last chapter, you simulated this jerky motion with chalkboard erasers. In the following activity, you can model earthquake effects with gelatin.

You can study earthquake motion by building a model with gelatin. Dissolve four packages of gelatin in 500 mL of hot water. Pour the dissolved gelatin into a flat-bottomed pan lined with plastic wrap. When the gelatin has cooled, remove it from the pan and cut it in half with a wet butter knife. Slowly slide one side of the gelatin past the other. Observe the friction build and the energy that is released as an "earthquake" shakes the gelatin. ✐

Just as ripples in a pond move away from the impact site of a thrown pebble, waves of energy radiate away from the source of an earthquake. The source of an earthquake is called the **focus.** Earthquake waves travel in all directions from the focus, which is often located well below the earth's surface. The energy from these waves shakes the surface of the earth from a point directly above the focus. This point on the surface is called the **epicenter.**

Figure 2–2. Waves of energy radiate from the focus of an earthquake, like ripples on a pond.

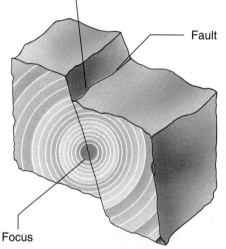

Epicenter

Fault

Focus

Today, an earthquake like the one in New Madrid would draw scientists and newspeople from all over the world. In the early nineteenth century, however, news traveled much more slowly, and the scientific understanding of such events was very limited. Verbal descriptions of an earthquake were the only measure available for comparing one event with another. Thus, accounts of motion and destruction were used to establish the size, or *intensity,* of an earthquake.

What information in the *Philadelphia Gazette* report of the New Madrid earthquake leads you to believe that this was indeed an earthquake of great intensity? The following activity can help you organize the data as a nineteenth-century geologist might have.

In your journal, record the descriptions of the New Madrid earthquake as related in the *Philadelphia Gazette*. Then design a word scale of descriptive terms and images that might be used to compare the intensities of different earthquakes. For instance, you could say that your house might "shake slightly," "shake a lot," "tilt," or even "fall apart." Each of these descriptions shows increasing intensity.

Today, the strength of an earthquake is measured with seismographs. A **seismograph** consists of a rotating drum wrapped with paper and a pen attached to a suspended weight. The pen presses gently against the paper-wrapped drum. The structure that supports the drum is connected to the solid rock of the earth's crust, so it will move only if the crust moves. Any vibration of the earth's crust causes the pen to produce a zigzag line on the paper. The recording made by a seismograph is called a *seismogram*. The first modern seismograph was built about 1870, but the first device that recorded a earthquake was invented in China in about A.D. 132 .

Figure 2–3. A seismograph records the trembling of the earth's crust during an earthquake.

Seismographs, set up in seismic stations all over the world, record even small tremblings of the earth's crust. By comparing information from at least three seismographs, it is possibe to locate an earthquake's epicenter by a process called *triangulation*. Look at the map. Why does it take at least three seismographs to pinpoint the location of an earthquake? In the next activity, you can make a simple seismograph.

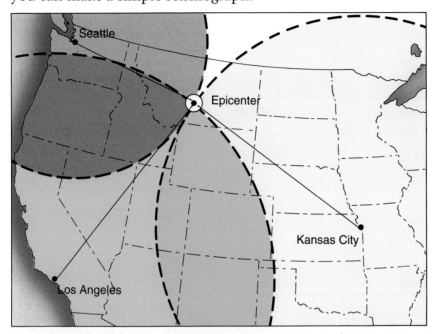

Figure 2–4. The location of an earth-quake can be calculated by triangulation.

ACTIVITY   *How does a seismograph work?*

MATERIALS
paper; tape; phonograph with a large, separate speaker; record; string; felt-tipped marker

PROCEDURE
1. Roll the sheet of paper into a cylinder, and tape the edges together.
2. Then place the paper cylinder and a record on the phonograph's turntable.
3. Carefully place the turntable on top of the speaker. Select the slowest speed for the turntable, and turn the bass all the way up.
4. Suspend a felt-tipped pen from a string, and gently touch it to the paper cylinder. Create a baseline for your graph by running the turntable without music for a few seconds. Then lower the needle onto the record to create a "seismogram."

APPLICATION
Compare your seismogram to the one in Figure 2-3. How are they similar? How are they different? How does a seismogram show large earthquake waves?

A measurement of an earthquake's intensity, called *magnitude*, can be determined by a seismograph. Small tremors produce small tracings on a seismogram, while more intense earthquakes produce much larger tracings. An earthquake's magnitude is reported using the **Richter scale.** This scale measures the relative amount of energy released by an earthquake. Table 2–1 shows the magnitudes of some especially intense earthquakes as rated on the Richter scale.

Table 2-1	Major Earthquakes	
Year	**Location**	**Magnitude**
1556	China	unknown
1812	New Madrid	8.0 (estimated)
1906	San Francisco	8.3
1906	Colombia	8.9
1923	Tokyo	8.3
1964	Alaska	9.1 (revised)
1976	China	8.0
1985	Mexico City	8.1

Originally it was estimated that each increase of one unit on the Richter scale represented a tenfold increase in earthquake intensity. However, further research has shown that the increase is not 10, but 32. For example, a magnitude 8 earthquake is 32 times more intense than a magnitude 7 earthquake, and 32×32, or about 1000, times more intense than a magnitude 6 earthquake.

 ASK YOURSELF

What is the focus of an earthquake? How is it different from the epicenter?

Earthquake Waves

There are two types of energy waves that originate from an earthquake focus. One is a *longitudinal wave.* A longitudinal wave travels by compressing the earth's crust in front of it and stretching the crust behind it. You can simulate longitudinal waves by compressing a section of a Slinky® and then releasing it. Energy will travel through the coil as a longitudinal wave.

Figure 2–5. Longitudinal, or P, waves stretch and compress the earth's crust.

Longitudinal waves are the fastest waves produced during an earthquake and, therefore, are the first waves to reach seismic recording stations. For this reason, longitudinal waves are also called *primary waves,* or **P waves.**

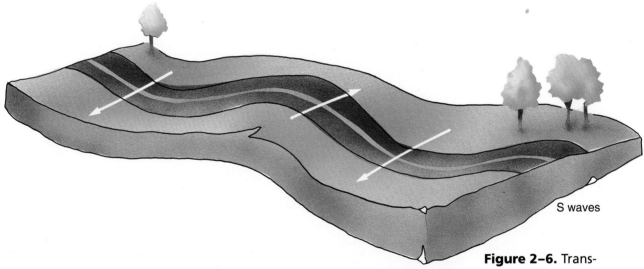

Figure 2–6. Transverse, or S, waves move the earth's crust from side to side.

The second type of wave originating from an earthquake's focus is a *transverse wave.* The motion of a transverse wave is similar to that of a rope shaken from side to side. Transverse waves move more slowly than P waves. Therefore, these slower waves are also called *secondary waves,* or **S waves.**

The interaction of P waves and S waves produces another kind of wave—**surface waves.** These cause the earth's surface to roll like the sea in a storm. Surface waves originate from an earthquake's epicenter, not its focus, and cause more destruction than either P waves or S waves. P waves and S waves shake buildings back and forth or side to side, but the rolling action of surface waves causes buildings to collapse.

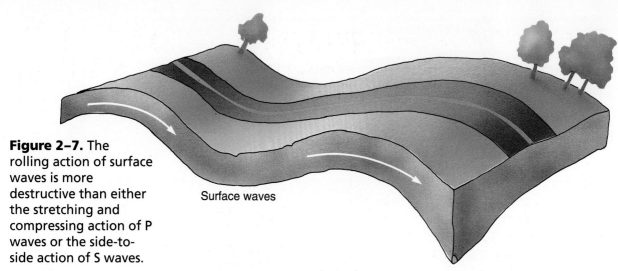

Figure 2–7. The rolling action of surface waves is more destructive than either the stretching and compressing action of P waves or the side-to-side action of S waves.

Surface waves

P waves are the first to be recorded on a seismogram, making a series of small, zigzag lines. S waves arrive later, appearing as larger, more ragged lines. Surface waves arrive last and make the largest lines. The difference in time of travel between P waves and S waves enables geologists to calculate the distance between the seismograph and the earthquake's focus.

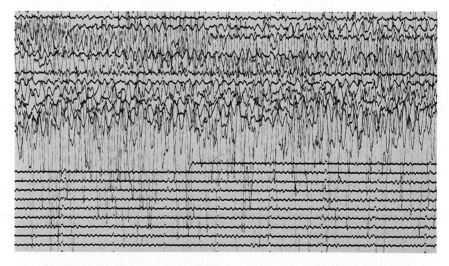

Figure 2–8. Notice the differences between the P waves (bottom) and the S waves (top) in this seismogram.

You might think of P waves and S waves as runners in a race. They both start from the same place (the focus) at the same time. At the halfway point, the faster runner (the P wave) might be 2 seconds ahead of the other runner (the S wave). If they each maintain their speed, the faster runner (the P wave) will be 4 seconds ahead at the finish line (the seismograph). You could use their rates of speed—a known quantity—and the difference in their arrival times to figure out how far they traveled.

 **ASK YOURSELF**

What type of earthquake wave does the most damage? Why?

The Causes of Earthquakes

According to an old Japanese legend, a giant catfish—the namazu—lives in the mud under the surface of the earth. Kashima, a brave warrior, keeps the namazu still by holding it down with a large rock. As long as Kashima keeps the namazu under the rock, the earth is quiet. But whenever Kashima relaxes his guard, the namazu thrashes around and the earth shakes and trembles.

There are many explanations of the causes of earthquakes. Some are legends, like the one you just read, while others are based on scientific data. You may recall from the last chapter that convection cells within the mantle are constantly moving the earth's tectonic plates. At transform boundaries, stress builds as the friction between plates stops their movements. Eventually, the plates slip, and the stress is released as great waves of energy—an earthquake. In the United States, many earthquakes occur along the San Andreas fault zone, where the Pacific plate slips past the North American plate. In some areas along the San Andreas fault, the plates move smoothly past each other, causing frequent, but minor, earthquakes. In other areas, friction causes the plates to stick to each other. Along these sections, frequent, minor earthquakes do not occur. However, these areas have a high probability of producing occasional major earthquakes. The cities of San Francisco and Los Angeles are located along such areas of the San Andreas fault.

Figure 2–9. The thrashing of the namazu is one legend about the cause of earthquakes.

Figure 2–10. The New Madrid fault zone is part of a rift in the rocks of the North American plate.

But what caused the New Madrid earthquake? How can plate movements cause earthquakes in the middle of a plate? Remember, the continents have not always been positioned or shaped as they are today. As the continents evolved into their present configurations, there have been times when other plates pushed hard against the North American plate, compressing the rocks of the continent. Later, when those plates pulled away, great tears, or rifts, split the rocks of North America. One such rift—the Reelfoot rift—contains the New Madrid fault zone, which is buried deep beneath the Mississippi River.

Some rifts continue to expand and stretch, eventually splitting apart, perhaps forming a new ocean. This occurred in the Middle East, where the Red Sea formed along a great rift. The Reelfoot rift, however, never split that far—it is a failed rift. Failed rifts are usually associated with low, sinking land. Thus the Reelfoot rift was a natural location for the formation of the Mississippi River.

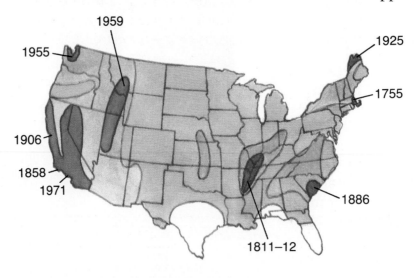

Figure 2–11. Dark areas mark places where large earthquakes *may* occur. Dates mark the locations where major earthquakes *have* struck.

Predicting earthquakes is much more difficult than locating an epicenter after one has occurred. There are several active earthquake zones in the United States, including the San Andreas fault zone and the Reelfoot rift. Geologists carefully monitor these areas, studying the patterns of small tremors and changes in the characteristics of the land in hopes of being able to predict a large-magnitude earthquake.

Since 1974 more than 4000 small earthquakes have been recorded in the region of the Reelfoot rift. In California hundreds of small earthquakes are recorded each year. Although the accurate prediction of earthquakes is not yet possible, people living in these areas must be prepared to take action in the event of a major earthquake.

Figure 2–12. Earthquakes can cause much property damage and loss of life in heavily populated areas.

Earthquakes usually occur without warning. If one occurs while you are outside, move into an open area to get away from buildings that might collapse. If you are inside, stand in a doorway or get under a desk or table that will protect you from falling objects. After the earthquake, stay away from fallen electrical wires and stay out of the way of emergency crews.

 ASK YOURSELF

How does the 1812 earthquake along the New Madrid fault relate to plate tectonics?

SECTION 1 *REVIEW AND APPLICATION*

Reading Critically

1. Compare and contrast P waves, S waves, and surface waves.
2. Explain why three different seismic readings are needed to find the epicenter of an earthquake.

Thinking Critically

3. Why are some areas along a fault likely to have stronger, but less frequent, earthquakes than other areas along the same fault?
4. Which type of earthquake wave probably did the most damage in the New Madrid earthquake? Explain your answer.

SKILL Analyzing Earthquake Data

▶ **MATERIALS**
● world map ● paper ● pencil

▼ **PROCEDURE**

TABLE 1: EARTHQUAKE LOCATION		
Number	Latitude	Longitude
1	60° N	152° W
2	45° N	125° W
3	35° N	35° E
4	30° N	115° W
5	30° N	60° E
6	20° N	75° W
7	50° N	158° E
8	40° N	145° E
9	15° N	115° E
10	15° N	105° W
11	10° S	105° E
12	5° S	150° E
13	0°	80° W
14	25° S	75° W
15	50° S	75° W

1. The table lists 15 locations where earthquakes have occurred. This is only a small sample of the earthquakes that occur each year.
2. Your teacher will give you a simple outline map of the world. Find the location of the first earthquake on the world map. Put a dot on the map, and label it 1. Do the same thing for each of the 15 earthquakes listed in the table. Describe any pattern of earthquake locations that seems to be developing.

▶ **APPLICATION**
1. Is there a relationship between the location of the earthquakes and the shapes of the continents?
2. How are the locations of the earthquakes related to the location of tectonic plate boundaries?

✳ *Using What You Have Learned*
From the information on your map, predict the likelihood of an earthquake occurring where you live. Which is easier to predict, where or when earthquakes will occur? Explain your answer.

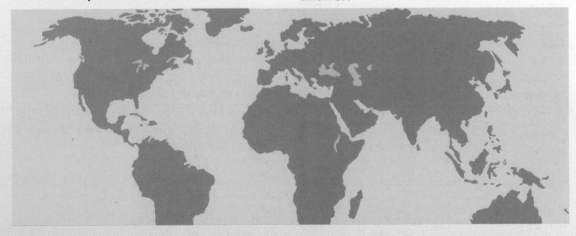

Folds, Faults, and Mountains

Mountains provide dramatic evidence of adjustments within the earth's crust. Rocks of the crust are constantly under stress. They are squeezed together, pulled apart, twisted, folded, fractured, and faulted. Much of the force needed for these changes comes from the movement of tectonic plates.

Do Not Bend, Fold, or Spindle

If you look at a map of the earth, you may get the idea that the earth's surface is fixed and cannot change. However, because of forces within the earth, the surface is slowly but constantly changing. Seashells, for example, can be found on the tops of some mountains. How did they get there? Could these mountains have once been under water and then risen to their present elevation?

Objectives

Describe the process that causes the crust to fold.

Classify faults by type.

Explain the relationships among folds, faults, and plate tectonics in the mountain-building process.

Figure 2–13. Looking at this mountain of rock, it is hard to imagine the force required to bend and break the earth's solid crust.

43

Figure 2–14. Under the right conditions, rocks can bend as if they were made of paper.

There is little doubt that the rocks in this photograph are bent. The question is, what conditions cause solid rock to bend like this? The next activity can give you an idea.

DISCOVER BY Doing

You can model the bending of rocks by pushing gently on one end of a sheet of paper while holding the other end in place. What happens to the paper? In the same way, slow, continuous force can bend rocks and build mountain ranges. ✎

Scientists hypothesize that some forces within the earth are released slowly and continuously, bending rocks like sheets of paper. When rocks bend this way, the process is called **folding.** Folds in rocks may be small wrinkles, or they may be large enough to form mountains. The Appalachian Mountains, for example, contain the remains of many great folds.

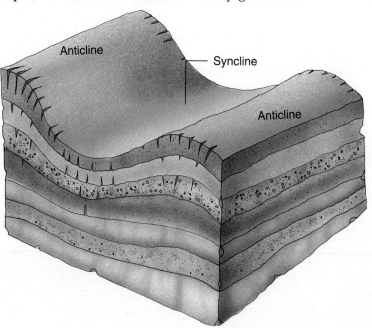

Figure 2–15. Anticlines often form parallel mountain ridges, with a syncline forming a valley between them.

During their long history, the Appalachians have been worn down. Today we see the remains of long, parallel ridges and valleys, which were formed as the rocks were folded. The ridges in folded mountains are called *anticlines,* while the lowered sections are called *synclines.* In most places, synclines form the valleys between the ridges.

 **ASK YOURSELF**

How do folded rocks form ridges and valleys?

Perfect to a Fault

Figure 2–16. Joints occur where rocks fracture without faulting.

Sometimes forces within the earth are strong enough to fracture rocks instead of folding them. If the rocks remain in place, the fracture is called a *joint*. Joints often form long columns in rocks. If there is additional force, a fault will form. **Faulting** occurs if there is significant movement along a fracture.

The rocks in the photograph below have clearly not been folded. Instead, they have been uplifted. Uplifting is one result of faulting. Faulting is caused by three different kinds of force—tension, compression, and shear. These three forces produce three different kinds of faults.

Figure 2–17. These rocks have been uplifted by faulting.

Normal Faults   There are three types of faults, but they all share some characteristics. For instance, all faults have two sides—the *footwall* and the *hanging wall*. One type of fault, called a *normal fault,* results from tension. Think of tension as a tug of war within the earth's crust. *Tension* stretches the crust, causing rocks to move and tilt. In a normal fault, the hanging wall drops in relation to the footwall, as the diagram shows.

Figure 2–18. These mountains are the result of normal faulting. The hanging wall has dropped, creating fault-block mountains from the footwall.

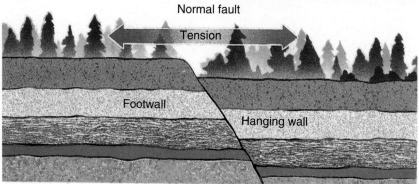

Normal fault

Tension

Footwall

Hanging wall

Large blocks of rock can be separated by normal faults. Then the blocks may be tilted, like a row of books that has fallen over. Mountains formed in this way are called *fault-block mountains.* The Basin and Range area of the southern Rockies has many examples of fault-block mountains. You can model the formation of fault-block mountains in the next activity.

DISCOVER BY Doing

Stand eight dominoes together on their short edges on a narrow strip of elastic. Slowly stretch the elastic, and observe what happens to the dominoes. How is the stacking of the fallen dominoes similar to the formation of fault-block mountains? 🖉

Reverse Faults   Some sections of the earth's crust may be pushed together, instead of being pulled apart. The force that squeezes the crust is *compression.* A fault formed by compression is called a *reverse fault.* In a reverse fault, the hanging wall moves up in relation to the footwall.

If the angle of the reverse fault is very low, the hanging wall may slide hundreds of meters over the footwall, forming over-thrust mountains. Many of the mountains in Glacier National Park are overthrust mountains. You can read more about this unique geological area in the Science Parade at the end of this unit.

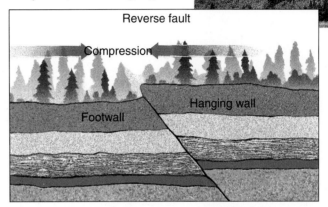

Figure 2–19. Overthrust mountains are the result of movement of a hanging wall over a footwall.

Lateral Faults   A third type of fault, called a *lateral fault,* is formed by shear. *Shear* forces run parallel to a fault. You can demonstrate shear by pressing the palms of your hands together, and then pushing one hand away from you while pulling the other hand toward you. If you tried the same thing with sandpaper in your hands, the friction would resist the sliding movement. But if enough force were applied, your hands might move suddenly. This type of jerky movement also occurs in normal and reverse faults, producing tremors and earthquakes.

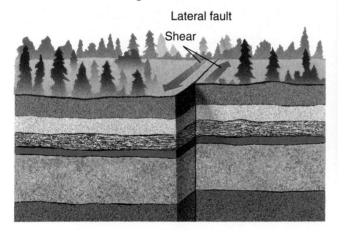

Figure 2–20. The San Andreas fault is a lateral fault.

In a lateral fault, the hanging wall moves parallel to the footwall, but in the opposite direction. The famous San Andreas fault in California is an example of a lateral fault.

 **ASK YOURSELF**

Name the different kinds of movements along faults.

Other Mountains and Plateaus

So far you have learned that mountains form from folds and faults in the earth's crust. Another type of mountain forms where a large body of molten rock rises close to the surface. The surface is lifted by the rock, resulting in a *dome mountain*. The Black Hills of South Dakota are dome mountains. Erupting volcanoes can also form mountains. You will learn more about volcanoes in the next section.

Also scattered over the earth are broad, flat areas that have elevations significantly above those of the surrounding area. These areas are called *plateaus* (plah TOHS).

Plateaus can be formed by the same forces that build mountains. The Colorado Plateau, for example, was formed as a large area of the crust was uplifted. The Colorado River slowly cut into the crust as it was lifting, forming the Grand Canyon. You will learn more about the cutting action of water in the next chapter.

Figure 2–21. Dome mountains are formed as molten rock lifts the surface.

Figure 2–22. The Colorado River cuts deep into the Colorado plateau in the area of the Grand Canyon.

Plateaus can also be formed in areas where lava pours out of the earth and covers a large part of the surrounding surface. The Columbia Plateau in Washington State is a lava plateau.

While large areas of the earth's crust are being uplifted, some small areas are actually sinking. Death Valley in California and Nevada is an example of such a place. The large block that makes up the floor of Death Valley is slowly tilting as one end sinks. Areas of the continents that are below sea level are indicators of sinking crust.

Figure 2–23. While some areas of the crust are uplifted, areas such as Death Valley are slowly sinking.

 **ASK YOURSELF**

How are plateaus formed?

SECTION 2 *REVIEW AND APPLICATION*

Reading Critically

1. Match the three forces that produce faults with the types of faults they produce.

2. How are dome mountains formed?

Thinking Critically

3. Explain how plate movement might result in the formation of a normal fault.

4. Why would folds be caused by compression rather than tension?

Volcanoes

Objectives

Explain why volcanoes usually form near plate boundaries.

Describe the relationship between hot spots and volcanic island chains.

Compare and contrast the different types of volcanoes.

On August 24, A.D. 79, Pompeii was a peaceful Italian city. Although Pompeii was located near Mount Vesuvius, a large volcano, the people were not worried. Mount Vesuvius had been quiet for hundreds of years. Suddenly, the mountain exploded with a force 10 times more powerful than the eruption that tore the side out of Mount St. Helens. Volcanic ash buried the city and its people, preserving forever this testimony to the earth's violent nature.

Magma and Lava

One of the most dramatic effects of plate tectonics is the formation of volcanoes. Most, but not all, volcanic activity occurs near the boundaries of tectonic plates. In the United States, for example, scientists have identified about 30 sites along the West Coast and in Alaska and Hawaii where a volcano could erupt at any time.

Figure 2–24. There are about 30 sites in the United States where active volcanoes occur.

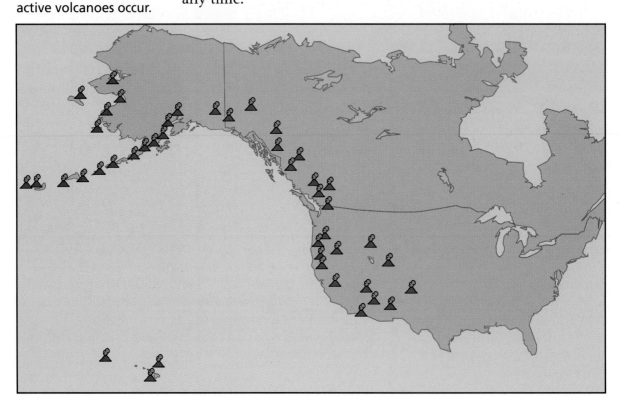

Forces around the edges of plates can cause pockets of magma to form. **Magma** is red-hot, molten rock within the earth. The temperature of magma is between 500°C and 1200°C. Magma is lighter than solid rock, so it tends to work its way toward the surface, like a hot-air balloon rising in the air. Sometimes rising magma cools and forms solid rock, called an *intrusion,* below the surface of the earth. Where magma reaches the surface before cooling, it forms a **volcano.**

Figure 2–25. A lava flow may cover a large area of the earth's surface.

Thin magma rises easily, forming *lava flows* that may stream great distances across the earth's surface. The Columbia Plateau, mentioned in the last section, resulted from just such a flow. Thick magma rises explosively to the surface. Lava is thrown into the air, building mountains. Some of this lava hardens before it hits the ground, forming *lava bombs*. At times, the force of a lava eruption is so violent that hunks of hardened lava, called *cinders,* are thrown hundreds of meters from the volcano. Particles less than 2 mm in diameter, called *ash,* can form great clouds that cover everything in wide areas as they settle back to the earth.

Figure 2–26. A violent eruption throws lava, cinders, and ash into the air. The ash often covers everything as it settles back to the ground.

ASK YOURSELF

What is the difference between a lava flow and a volcano?

The Ring of Fire and Other Hot Locations

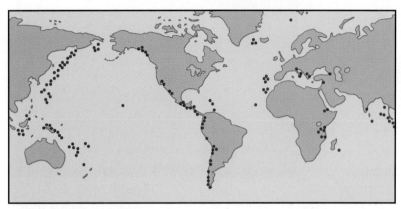

Figure 2–27. The dots show the location of many of the earth's active volcanoes.

Many of the earth's active volcanoes are found in a narrow belt that circles the Pacific Ocean. You may recall from the last chapter that this belt is often called the *Ring of Fire.* Other centers of volcanic activity are found along the mid-ocean ridge, around the Mediterranean Sea, in eastern Africa, Southeast Asia, and the Caribbean. With a little research, you can locate other geologically active areas of the earth.

DISCOVER BY *Researching*

Maintain a two-month earth watch by collecting newspaper and magazine articles on earthquakes, volcanic eruptions, and other noteworthy earth events. Keep track of where these events occur by plotting them on a world map. Note how their locations relate to plate boundaries. ✎

Converging Plates Volcanoes are found in the greatest numbers along two kinds of plate boundaries. First are zones of subduction associated with converging plates. Here, an oceanic plate is sinking into the asthenosphere, producing magma. Mount St. Helens, for example, was formed along the boundary where a piece of the Pacific plate is subducted beneath the North American plate.

Figure 2–28. Subduction is responsible for the formation of many volcanoes around the Pacific plate.

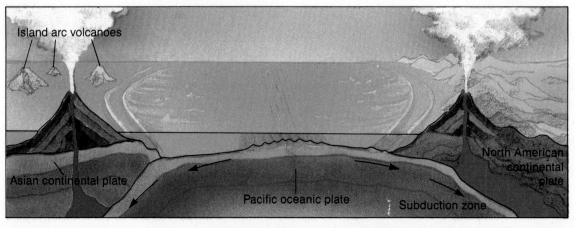

Island arc volcanoes

Asian continental plate

Pacific oceanic plate

Subduction zone

North American continental plate

Diverging Plates Volcanic activity is also found along boundaries where plates are pulling apart. Along the Mid-Atlantic Ridge, for example, constant eruptions take place beneath the sea. In some places, such as Iceland, the ridge rises above the ocean surface. The volcanoes found in southeast Africa are also caused by plates diverging.

Hot Spots Although most volcanoes are formed along plate boundaries, some develop in the middle of plates. Mid-plate volcanoes often occur in long, straight island chains, such as the Hawaiian Islands. You may recall that at one end of the island chain—Midway Island—the volcanoes are old and extinct. At the other end of the chain—the island of Hawaii— the volcanoes are young and active. Obviously, there must be a source of magma below these *hot spots*. But where does the heat come from? And why are the islands in a straight line?

Figure 2–29. Iceland was created by the separation of the North American and Eurasian plates.

Figure 2–30. The Hawaiian Islands are volcanic islands formed over a hot spot in the Pacific plate.

Figure 2–31. Yellowstone National Park is located over a hot spot under the North American plate.

The exact cause of mid-plate volcanoes is unknown; however, geologists hypothesize that these volcanoes develop from columns of magma rising over *hot spots* deep in the earth. A **hot spot** forms over an area between convection cells in the mantle. Hot spots melt the rocks of the mantle, creating a column of magma that is less dense than the surrounding rock and rises toward the surface. The magma often erupts through the lithosphere. A hot spot seems to remain stationary, but as a tectonic plate slowly moves across the rising column of magma, volcanoes are produced in a series. Since the volcanoes at the eastern end of the Hawaiian Islands are younger than those farther west, which way is the Pacific plate moving?

Hot spots don't always produce volcanoes. Sometimes the magma over a hot spot heats underground water, turning the water into steam. The steam is then released through cracks in the earth's surface, producing *geysers* (GY zuhrz). Yellowstone National Park, with the famous Old Faithful geyser, sits over a hot spot. Geologists predict that areas to the southeast of Yellowstone may soon begin showing signs of geyser activity, as well. Why would new geysers form southeast of Yellowstone National Park?

 ASK YOURSELF

Along what kinds of plate boundaries do volcanoes form?

Characteristics of Volcanoes

Figure 2–32. Most of the volcanoes of the Pacific islands are shield volcanoes.

Volcanoes are not all alike. Each seems almost to have a personality of its own. Some are quiet and gentle in nature. Others are violent. Volcanoes differ in their appearances and in the ways they erupt. Both of these characteristics are due primarily to differences in magma composition. In some volcanoes the magma may be heavy, dark, and thick, while in others it may be light colored and thin. Some magma contains steam and other gases that bubble up like a bottle of soda pop that has been shaken before opening.

Some volcanoes have rounded, gentle slopes and bases that cover a wide area. These volcanoes are called *shield volcanoes*. The lava of shield volcanoes is relatively fluid, and flows easily. Shield volcanoes are most common in the middle of an oceanic plate, such as the Hawaiian Islands, or where oceanic plates converge or diverge.

Thick, or viscous, magma traps gases inside the volcano until enough pressure builds up to push the magma out of the earth forcibly. If water is present, it quickly turns to steam as the lava erupts. The pressure of the steam blows the lava apart, which may shoot rocks, gases, lava bombs, and ash many kilometers into the air. In contrast to shield volcanoes, this kind of eruption builds a steep-sided volcano called a *cinder-cone volcano*.

Figure 2–33. Many of the volcanoes of Mexico and Central America are cinder-cone volcanoes.

Cinder-cone volcanoes consist mainly of ash and *tuff*. Tuff is formed from compressed ash, cinders, and lava bombs. Cinder-cone volcanoes are common where oceanic and continental plates converge. Many of the volcanoes in Mexico and Central America are cinder cones.

If the lava within a volcano changes, the nature of the eruption may change as well. Another type of volcano, formed from a series of alternating eruptions of different lavas, is called a *composite volcano*. A composite volcano consists of layers of ash, cinders, and tuff, alternating with lava flows. The slope of a composite volcano is steeper than that of a shield volcano but not as steep as that of a cinder-cone volcano. Mount Fuji in Japan is a composite volcano, although composite volcanoes are usually associated with continental plates.

Figure 2–34. Japan's Mount Fuji is a composite volcano.

An explosive eruption of an old composite volcano may blow away entire sections of the cone wall. Large sections of the wall then collapse into the hollow center, forming a *caldera*. A caldera is a wide basin, usually more than a kilometer across, in the center of a volcanic cone. Crater Lake in Oregon is in an old caldera, and Mount St. Helens in Washington formed a new caldera when it erupted in 1980.

Figure 2–35. Crater Lake in Oregon is formed within the caldera of a volcano.

 **ASK YOURSELF**

What are the characteristics of the three types of volcanoes?

Undersea Volcanoes

Volcanoes form not only on land but also under the oceans. In shallow water, volcanoes erupt violently, forming clouds of ash and steam. An underwater volcano is called a *seamount.* Seamounts look much like cinder-cone volcanoes on land. Another type of undersea volcano is called a *guyot* (gee OH). A guyot is an inactive undersea volcano with a flat top. How do you think the top of a guyot becomes flattened?

There are more than 10 000 seamounts and guyots on the Pacific Ocean floor. Most never reach the surface of the ocean. The Hawaiian Islands, for example, are just a small part of a much larger underwater volcanic-mountain chain. Eight of these volcanoes have become islands, but dozens more remain submerged. In the next activity, you can locate some of these and determine how many of them were formed.

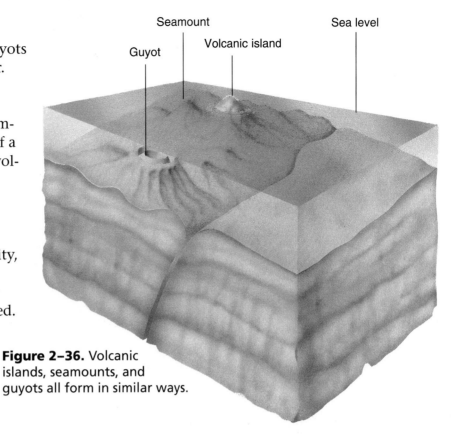

Figure 2–36. Volcanic islands, seamounts, and guyots all form in similar ways.

DISCOVER BY Doing

Using the world map on page 674, locate as many volcanic-mountain chains, guyots, and seamounts as you can. Where do most of these features occur? Which ones are associated with plate boundaries? Which ones are associated with hot spots? ✐

▶ ASK YOURSELF

How are volcanic islands, seamounts, and guyots the same? How are they different?

SECTION 3 REVIEW AND APPLICATION

Reading Critically
1. Explain how a chain of volcanoes can form in the middle of a plate.
2. Why is most volcanic activity located near plate boundaries?

Thinking Critically
3. What evidence supports the hypothesis that hot spots occur deep within the mantle?
4. Where would you expect any new volcanic islands in the middle of the Pacific plate to form? Explain your answer.

INVESTIGATION

Making a Cinder-Cone Volcano

▶ **MATERIALS**
- newspaper ● modeling clay ● scoop ● ammonium dichromate
- magnesium ribbon, 10 cm ● safety goggles ● laboratory apron
- matches

▼ **PROCEDURE**

CAUTION: This investigation should only be done in a well-ventilated area.

1. Spread newspaper over your work area. Working on the newspaper, shape the modeling clay into a 10-cm tall cinder cone. Make a crater at the top about 2 cm wide and 1 cm deep.

2. Using the scoop, fill the crater with ammonium dichromate. Insert the magnesium ribbon into the ammonium dichromate.
CAUTION: From this point on, everyone in the area should wear safety goggles and laboratory aprons.

3. Light the magnesium ribbon with a match. Stand well back as the magnesium ribbon acts as a fuse and ignites the ammonium dichromate. Observe the results.

▶ **ANALYSES AND CONCLUSIONS**

1. How does the composition of the erupted material affect the shape of the cinder cone?

2. Describe the distribution of the erupted materials after the simulation.

▶ **APPLICATION**

When would geologists use such a model for studying a volcano?

✳ *Discover More*

What changes could you make in the procedures that would enable you to simulate the effects of wind conditions on the ash fall?

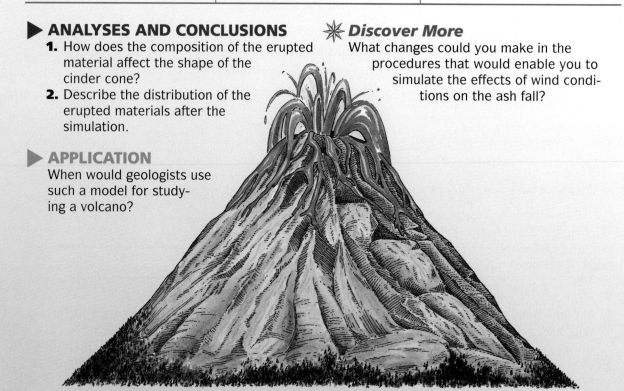

CHAPTER 2 | *H*IGHLIGHTS

The Big Idea

The surface of the earth is constantly changing. As tectonic plates move, certain locations experience dramatic events such as earthquakes and volcanic eruptions. Earthquakes are a result of energy being released along fault lines in the crust. Similarly, volcanoes result from a release of heat energy built up in the mantle and crust. Stresses within the earth have resulted in the uplifting of the crust into plateaus and mountainous regions.

For Your Journal

Look again at the way you answered the questions at the beginning of this chapter. How have your ideas changed? Be sure to record your new ideas in your journal.

Connecting Ideas

Using the tectonic plate map on page 15, locate the corresponding plates for some of the mountain ranges shown here. Predict the type of mountains you would find in each range, based on the movements of the various plates.

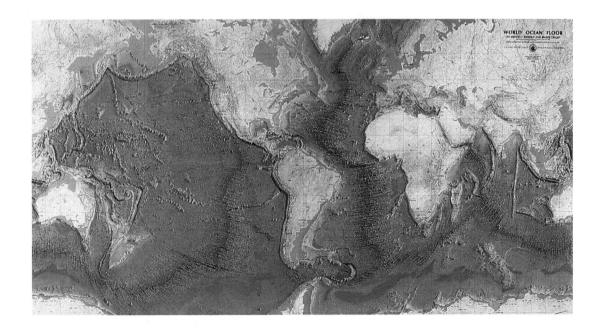

Understanding Vocabulary

1. For each set of terms, explain the similarities and differences in their meanings.
   a) faulting (45), folding (44)
   b) magma (51), lava flows (51)
   c) P waves (37), S waves (37), surface waves (37)

2. Explain how the terms in each set are related.
   a) volcano (51), hot spot (54)
   b) focus (33), epicenter (33)
   c) magnitude (36), Richter scale (36)
   d) tension (46), compression (47), shear (47)

Understanding Concepts

MULTIPLE CHOICE

3. An earthquake's epicenter can be located by using
   a) a map.   c) triangulation.
   b) a seismogram.   d) the Richter scale.

4. About how many times more intense is a magnitude eight earthquake than a magnitude six earthquake?
   a) 10 times   c) 100 times
   b) 32 times   d) 1000 times

5. The fastest waves produced during an earthquake are
   a) P waves.
   b) transverse waves.
   c) S waves.
   d) surface waves.

6. The type of earthquake waves that cause the most destruction are
   a) P waves.
   b) S waves.
   c) surface waves.
   d) longitudinal waves.

7. The San Andreas fault in California is an example of a
   a) normal fault.   c) lateral fault.
   b) reverse fault. d) thrust fault.

SHORT ANSWER

8. How does the focus of an earthquake differ from its epicenter?

9. In what three ways do volcanoes form?

10. Describe two ways that plateaus are formed.

Interpreting Graphics

Use these diagrams to answer the following questions.

11. Identify each type of fault shown here.

12. Name the forces that produce each of the faults shown here.

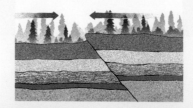

Reviewing Themes

13. *Energy*
Explain how the movement of tectonic plates is related to the release of energy in earthquakes and volcanic eruptions.

14. *Changes Over Time*
Explain why scientists predict that any new Hawaiian Islands will form to the southeast of the present chain.

Thinking Critically

15. Explain why Los Angeles and San Francisco are likely to have stronger, but less frequent, earthquakes than other locations in California.

16. What causes some volcanic eruptions to be explosive?

17. Explain how a chain of volcanoes can form in the middle of a plate.

18. Explain how geologists determine the distance between a seismograph and the epicenter of an earthquake.

19. Describe what the people in the picture should do after an earthquake.

Discovery Through Reading

Belt, Don. "Russia's Lake Baikal: The World's Great Lake." *National Geographic* 181 (June 1992): 2–39. Located in one of the most complicated and least understood fault zones on the earth, Lake Baikal is older and deeper than any other lake.

Grove, Noel. "Volcanoes: Crucibles of Creation." *National Geographic* 182 (December 1992): 5–41. This article describes a number of volcanoes and the effects of volcanic eruptions, which include land building, as well as destruction.

Eroding the Land

*B*ack in 1916, when a national park system was first envisioned, the parks were perceived primarily as aesthetic treasures. We still value these scenic vistas and their power to inspire and soothe, but today we recognize our national parks are more than wildlife zoos or collections of geological and historical treasures.

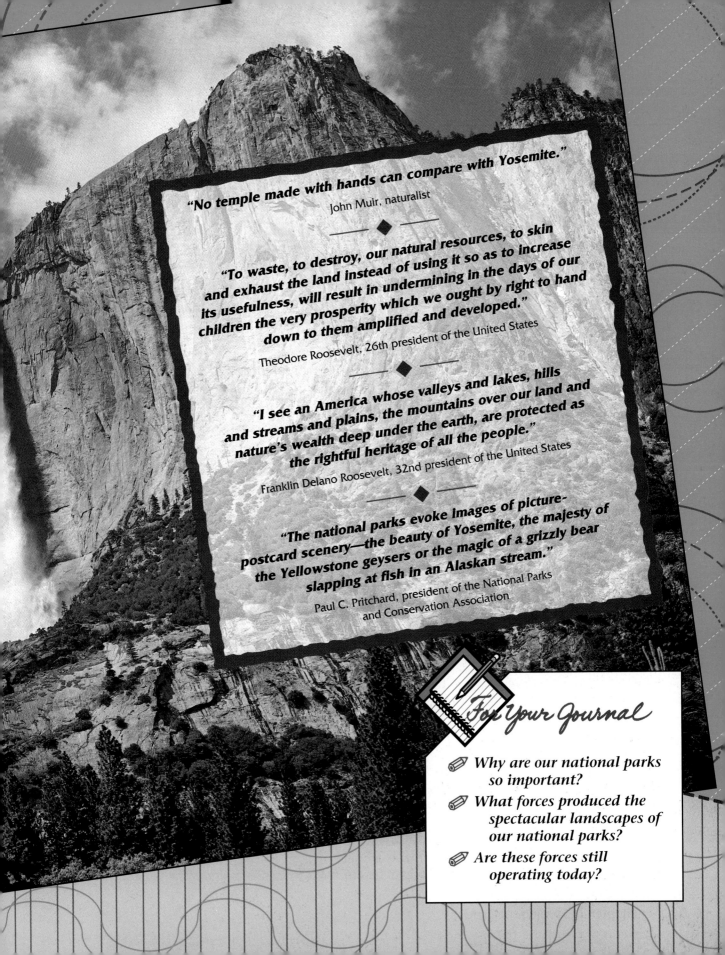

"No temple made with hands can compare with Yosemite."

John Muir, naturalist

"To waste, to destroy, our natural resources, to skin and exhaust the land instead of using it so as to increase its usefulness, will result in undermining in the days of our children the very prosperity which we ought by right to hand down to them amplified and developed."

Theodore Roosevelt, 26th president of the United States

"I see an America whose valleys and lakes, hills and streams and plains, the mountains over our land and nature's wealth deep under the earth, are protected as the rightful heritage of all the people."

Franklin Delano Roosevelt, 32nd president of the United States

"The national parks evoke images of picture-postcard scenery—the beauty of Yosemite, the majesty of the Yellowstone geysers or the magic of a grizzly bear slapping at fish in an Alaskan stream."

Paul C. Pritchard, president of the National Parks and Conservation Association

For Your Journal

- Why are our national parks so important?
- What forces produced the spectacular landscapes of our national parks?
- Are these forces still operating today?

Weathering and Mass Movement

Objectives

List the factors that affect the rate of weathering.

Compare and contrast the processes of physical weathering and chemical weathering.

Describe the forces involved in mass movement.

A system of National Parks and Monuments evolved out of John Muir's and Theodore Roosevelt's efforts in the early 1900s to preserve beautiful sections of the country. Concerned citizens continue to guard and preserve these national treasures. If you've visited one of America's national parks, you may have marveled at the beauty of falling water or sculptured landscapes. Did you ever wonder what processes shaped these landscapes or how long it took to create them?

Figure 3–1. Our national parks and monuments are set aside to preserve unique environments or works of nature.

Weather or Not

You may have noticed that, except for mountainous areas, very little rock is visible on the earth's surface. In fact, in places like the Great Plains you can travel for hundreds of kilometers and seldom see a rock. Most of the rock of the earth's crust is covered by soil. **Soil** is a combination of small rock fragments and organic material. Soil is formed when the underlying rock, or *bedrock,* of an area is broken into small fragments and mixed with organic matter from dead plants and animals.

Have you ever seen an un-painted house or barn that was weathered by rain, sun, and wind or read about an old sailor with a weather-beaten face? The phrase means that the house or the person has been changed by exposure to the weather. Where rocks are exposed to the environment, they also weather. The process that breaks rocks into smaller fragments, eventually producing soil, is called **weathering.**

Weathering processes are similar to the digestive processes in your body. As you chew, your teeth physically break food down into little chunks. The acid in your stomach then begins to digest the food chunks chemically, producing other materials. Weathering processes follow a similar sequence in breaking down rocks. Rocks are "chewed" by rain and wind into smaller chunks. This is *physical weathering.* Then acids in the air, rain, or soil chemically "digest" the rock chunks, producing other materials. This is *chemical weathering.* Both types of weathering help to change the earth's landscape.

Figure 3–2. This old barn has been weathered by sun, wind, and rain.

Although weathering is continuous, there are several factors that influence the rate at which it occurs. These factors include the composition of the rock, climate, topography, and the presence of vegetation.

The composition of rock affects the rate of weathering. Soft rocks are easily weathered, while hard rocks resist weathering. Certain features of rock formations, such as the presence of joints, faults, and layers, also affect the rate of weathering.

Climate also affects the rate of weathering. In places that receive little rainfall, such as the southwestern United States, the rate of weathering is slow. Water speeds up weathering, so the rates are usually higher in humid climates. Some areas have large temperature changes between seasons or even between day and night. Temperature changes that shift above and below the freezing point of water increase the rate of weathering. The next activity can help to show you why.

CAUTION: Do not use a glass jar or bottle. Fill a small hard-plastic jar or bottle all the way to the top with water. Secure the lid tightly, and place the container in the freezer overnight. What happens to the container? How does this relate to the process that produces potholes in street pavement during a cold winter? How does this process increase the rate of weathering? ✐

Some features of topography may increase the rate of weathering. For example, steep mountain slopes tend to have more exposed rock than do flat lands. Weathering occurs more rapidly if the rock is exposed to sun, wind, or rain. South-facing slopes receive more sunlight than north-facing ones. Mountains are also more likely to experience high winds and freezing temperatures. These conditions tend to speed up the weathering process.

Figure 3–3. What factors that could influence the rate of weathering do you observe in this photograph?

The presence of vegetation may also affect the rate of weathering. Plants often protect rocks from sun, wind, and rain. This slows the rate of weathering. However, dead vegetation may actually increase the rate of weathering because decaying organic matter produces acids that dissolve rock.

ASK YOURSELF

What factors increase the rate of weathering?

Physical and Chemical Weathering

Physical weathering breaks rocks mechanically—the composition of the rock stays the same, but the rock particles become smaller. Physical weathering is caused by the release of pressure, the wedging action of freezing water, plant roots, and abrasion. Chemical weathering changes the composition of rock. Chemical weathering is caused by the action of oxygen, water, and acids on rocks.

Physical Weathering There are four major types of physical weathering. One type begins shortly after a rock is exposed. Rock deep within the earth is under tremendous pressure from the weight of overlying materials. As this pressure is released by the removal of soil and rock above, the rock expands. Sudden expansion causes the surface of some rocks to crack or form joints. Cracks and joints serve as pathways for water and air to enter. This speeds up the weathering process. Expansion causes the surfaces of other rocks to flake off in sheets. This flaking is called *exfoliation*.

Another major type of physical weathering involves water and cold temperatures. Remember the bottle of water you put in the freezer overnight? Besides freezing the contents, you probably observed that the bottle was broken. As water freezes, it expands, exerting tremendous pressure. When water collects and freezes in rock cracks, it expands and enlarges the cracks. This process, called *frost action,* affects mostly layered and jointed rock.

Figure 3–4. Half Dome in Yosemite National Park is an example of a large dome of rock produced by exfoliation.

Ice

Figure 3–5. Frost action weathers layered or cracked rock.

A third type of physical weathering is caused by plant roots. If a seed germinates in a rock crack, pressure from the developing root enlarges the crack. This pressure is called *root pry*. You may have seen an example of root pry where a large tree growing close to a sidewalk or a driveway has lifted and cracked the pavement.

Figure 3–6. The pressure of growing plant roots can physically weather rock.

A fourth type of physical weathering occurs where rocks come in physical contact with other rocks. As rocks and rock fragments bounce down a hill, tumble in a stream, or blow in the wind, they bump against each other. This process, called *abrasion,* knocks off sharp edges, producing rounded fragments.

Figure 3–7. Rocks in a flowing stream tumble and bump into each other, rounding off sharp edges.

Chemical Weathering   While physical weathering breaks large rocks into smaller ones, chemical weathering changes the actual composition of rocks. You know what happens if you leave a bicycle out in the rain—it rusts. Rocks containing iron can also rust.

Figure 3–8. Rusting is an example of chemical weathering.

Figure 3–9. The red-clay soil of Georgia results from water combining with minerals in the soil.

Rusting is an example of a chemical weathering process called *oxidation*. But rain doesn't cause oxidation—air does. Water just speeds up the process.

Iron and other metals combine chemically with oxygen in the air to form oxides. Oxidation usually produces a color change. You know what iron oxide—rust—looks like. Aluminum oxide is black, and copper oxide is green.

Although water doesn't cause oxidation, it is the cause of another kind of chemical weathering. Water combines with certain materials in rocks in a process called *hydrolysis*. Both the rocks and the water break down and form new materials. The red-clay soils of some southern states are formed by the hydrolysis of materials in their parent rock.

Water is also involved in another kind of chemical weathering. Rain often contains dissolved chemicals from the air. Carbon dioxide occurs naturally in the atmosphere, and it increases with the burning of fossil fuels. Carbon dioxide and some air pollutants, such as sulfur dioxide, turn rainwater into a weak acid. The next activity can give you an idea of what an acid can do to rock.

DISCOVER BY *Doing*

CAUTION: Avoid getting any of the liquid on your skin or clothes. Put a few drops of a weak acid, such as lemon juice or vinegar, on a piece of chalk. What happens to the chalk? ✐

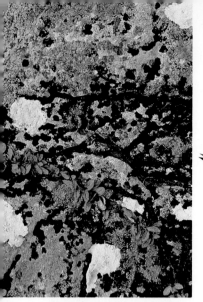

Figure 3–10. Lichens produce an acid that can weather bare rock.

Decaying organic matter also produces acids that chemically weather rock. Even bare rock can be weathered by the acids produced by the roots of lichens—organisms that are part alga and part fungus. The action of lichens is one of the most important factors in the formation of soil from bare rock.

▼ **ASK YOURSELF**

What is the difference between physical and chemical weathering?

Mass Movement and Landforms

If you live in an area with steep hills and snow, you may have rolled snowballs down a slope and watched them gain speed and pick up more snow. Rocks are affected by mass movement in much the same way. **Mass movement** is the downhill movement of rocks and rock fragments due to the force of gravity. There are four types of mass movement: creep, flow, landslide, and subsidence.

Creep occurs when rock materials saturated with water move slowly downhill. The movement is often so slow that changes are hard to see, but the long-term results can be observed. An example of creep is shown in the photograph.

Figure 3–11. What evidence of creep do you see in the photograph?

When water-saturated rock material moves at a noticeable rate, the process is called *flow*. There are several types of flows: debris flows, earth flows, and mudflows. All of them involve the movement of water-saturated rock fragments, clay, or soil down slopes. The distinctions between them are the amount of water mixed with the materials. Earth flows and mudflows may contain more than 80 percent water. Mudflows, also known as mudslides, are common along the California coast because of the clay soils found there.

Figure 3–12. The house in the picture (right) was caught in a mudslide, while the rock debris in the other picture (below) is the result of a landslide.

A *landslide* occurs when dry soil or rock moves down a steep slope. The movement is usually rapid and unexpected because unlike mud flows, which are usually caused by heavy rain, most landslides have no direct cause. The force of gravity simply overcomes the friction among rocks, and the material moves downhill. The fallen material collects at the base of the slope and forms a mass called *talus*.

Figure 3–13. This sinkhole in Winter Park, Florida, was caused by the dissolving of limestone beneath the earth's surface.

When rock and other material sinks to a lower level without sliding down a hillside, the process is called *subsidence* (suhb SY duhns). Subsidence can occur when certain rocks beneath the surface, such as limestone, are dissolved by water. Then the overlying rock collapses into the hollow, forming a sinkhole. Subsidence can also occur as newly deposited materials compact, or where wells pump out water or oil stored in rock. Many areas along the Gulf Coast are experiencing subsidence due to the compaction of materials.

Some other interesting landscapes, or *landforms,* are created by weathering processes. For example, the mesas and buttes of the American West are created by physical weathering. Mesas are large, flat-topped areas with nearly vertical sides, while a butte is a lone, flat-topped hill that rises abruptly above the surrounding landscape. In both mesas and buttes, the top rock, or cap rock, is more resistant to weathering than the rock around it. The less-resistant rock weathers away, leaving these landforms.

Figure 3–14. Mesas (left) and buttes (right) result from the physical weathering of soft rocks capped by a resistant layer.

The cities of Mesa, Arizona, and Butte, Montana, are named after prominent landforms nearby. What places in your area are named after a particular landform?

Sometimes resistant layers in a rock formation alternate with layers that are less resistant. More easily weathered rock will form a gentle slope, while resistant rock will weather into vertical cliffs. This difference in weathering of rock layers in the same formation is called *differential weathering.* Differential weathering has created many of the features of the Grand Canyon.

Figure 3–15. What examples of differential weathering can you see here?

 ASK YOURSELF

How are landslides and mudflows similar? How are they different?

SECTION 1 *REVIEW AND APPLICATION*

Reading Critically

1. What causes mass movement?

2. What factors affect the rate of weathering?

Thinking Critically

3. Compare and contrast physical and chemical weathering. Give examples of each.

4. What is the main agent of weathering in the formation of a mesa or a butte? Explain your answer.

INVESTIGATION

Comparing Physical and Chemical Weathering

▶ MATERIALS

- limestone ● balance ● glass jars with lids (2) ● soda water ● tap water
- gravel

▼ PROCEDURE

1. Select two small pieces of limestone, and determine the mass of each piece. Record the data.
2. Half-fill one jar with soda water and the other jar with tap water. Drop a piece of limestone into each jar. Describe what you see.

3. Let the jars sit overnight. Remove the limestone from each jar, and determine the mass of each piece again. Record the new data, and compare it with the original data.
4. Return the limestone to the jar of tap water, and add a few small pieces of gravel.

Put the lid on the jar, and shake it 50 times. Take care not to shake the jar so hard that it breaks.
5. Remove the limestone from the jar, and determine its mass again. Record the data, and compare it with the original mass and the mass after sitting in tap water overnight.

▶ ANALYSES AND CONCLUSIONS

1. Did the mass of either piece of limestone change? If so, under what conditions?
2. What types of weathering are demonstrated in this activity?

▶ APPLICATION

Which setup was similar to the process of physical weathering? Which was similar to chemical weathering?

✳ Discover More

Using the same materials, find out what would happen if physical and chemical weathering took place at the same time.

Soils

The formation of soil is one of the most important results of weathering. Life as we know it would not be possible without soil for growing grains, fruits, vegetables, and pasture grasses. Care must be taken to preserve and protect the soil, because the process of soil formation is very slow.

Soil Formation

It takes from 100 to 1000 years for 1 cm of soil to form. During the formation of soil, many factors are involved. The factors include parent material, climate, topography, organisms in the soil, and time.

Figure 3–16. The great variety of foods available to Americans depends on good soils.

Parent material is the original rock from which a soil is formed. Soils formed from shale, for instance, are different from soils formed from limestone.

The climatic factors that influence soil formation are temperature change and precipitation. Soils form more quickly in moist climates because the rate of weathering is greater. Soils also form more quickly where there are significant temperature changes. Remember the effects of freezing temperatures on rock.

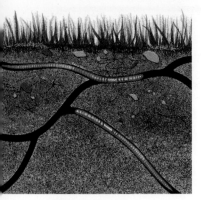

Figure 3–17.
Earthworm tunnels

Also related to climate is topography, or the shape of the land. You will not find much soil on a steep slope because mass movement removes newly formed soil. Flat land or a gradual slope allows soil to accumulate.

Plant and animal life in the soil also influences soil formation. Tunnels made by ants and earthworms, for instance, create spaces for air and water. These spaces cause additional weathering to occur, forming more soil. And dead plants and animals decay, adding to the humus of the soil. **Humus** (HYOO muhs) is the organic part of the soil.

Another factor in soil development is time. The longer the time available for soil formation, the more soil there will be. This will be true except in areas where environmental factors move the soil somewhere else.

 ASK YOURSELF

What are five factors involved in soil formation?

Soil Composition

Soil consists of inorganic (nonliving) and organic (once living) material. However, at least 80 percent of all soil is inorganic—weathered rock. In the next activity, you can make your own soil.

ACTIVITY
What is the composition of good soil?

MATERIALS
paper cups (3), peat moss, sand, bean or corn seeds, balance

PROCEDURE
1. Punch several small holes in the bottom of each of the cups.
2. Fill the first cup with peat moss, the second cup with sand, and the third cup with a mixture of both.
3. Plant 2 or 3 seeds in each cup, add water daily, and allow them to sprout.
4. In a few weeks, carefully remove each sprout from its cup, wash off any loose material, and determine its mass.

APPLICATION
1. In which cup did the plants grow best?
2. In which cup was the material most like rich, garden soil?
3. Which of the materials represented the inorganic part of the soil? Which material represented the organic part of the soil?

If you were to dig into the ground, you could see a profile of the soil. A soil profile shows that soil develops in horizontal layers, called *horizons*. Each horizon is labeled with a capital letter, starting with A at the top. Examine the soil profile shown here. Then try the next activity.

Figure 3–18. How do the horizons differ from each other?

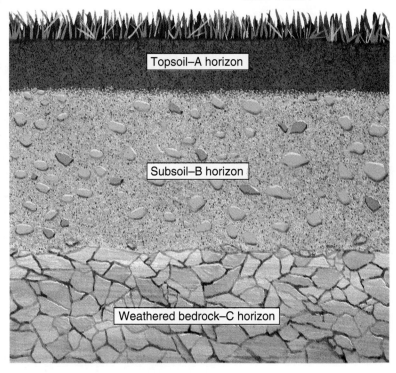

Topsoil–A horizon

Subsoil–B horizon

Weathered bedrock–C horizon

ᴅɪꜱᴄᴏᴠᴇʀ ʙʏ *Doing*

Dig a small sample of soil from your school grounds or your home. Then examine it under a dissecting microscope. List and describe all the different components of the soil that you see. 🖉

The A horizon contains humus and small amounts of clay and sand. The roots of many plants are restricted to this horizon. The next layer, the B horizon, contains coarse clay, sand, and a small amount of humus. The B horizon also contains dissolved materials that were washed down from the A horizon by water. The movement of materials from an upper horizon to a lower horizon is called **leaching.**

Below the B horizon is a thick layer of partially weathered rock—the C horizon. Some chemical weathering occurs in the C horizon, although the process is very slow. Most plant roots do not reach this layer, and exposure to air and water is also limited here. Beneath the C horizon is bedrock.

Although many soils have all the horizons just described, these layers are not always easy to identify. Soils also vary from region to region. Remember, climate is one of the factors that controls the rate of chemical weathering. Areas that are cold throughout most of the year have poorly developed soils because a frozen surface does not allow much weathering to occur. These areas also lack protective plant growth, so wind often blows the soil away.

Humid, tropical areas develop thick soils because of the high temperature and abundant rain they receive. In such areas, however, many elements are leached from the top layer of soil by water. Thus, the soil may not be rich enough to grow crops.

In temperate climates, there is usually less rainfall and, therefore, less leaching of soil nutrients than in the tropics. Many temperate areas develop thick, rich soils. Due to their richness, temperate soils are among the world's most fertile. Much of the world's food comes from soils of temperate areas, like those of the midwestern United States and Canada.

Figure 3–19. Which soil looks like it would grow the best crops? Explain your answer.

 ASK YOURSELF

Why are tropical soils usually poor in nutrients?

Land Use

Dry land covers less than 30 percent of the earth's surface. Of that, only about 13 percent can be farmed. The rest is either too rocky, has poor soil, or the climate is not good for growing crops or pasture grasses. In the United States, the same land is used nearly every year for crops. The nutrients used by plants are replaced by chemical fertilizers. So efficient are American farming practices that less than one person in ten is involved in farming, yet a surplus of food is produced.

Figure 3–20.
Technology is important to food production in America. However, many developing countries must rely on older farming methods.

In many parts of the world, farming is less efficient. In some developing countries, nearly everyone is involved in producing food. But the soil is often poor, and farmers cannot afford the chemicals needed to make it produce again. When crops can no longer be grown, people abandon the land. New land must be found each year to produce food for a growing population. This often means clearing tropical rain forests. But tropical soils are poor in nutrients, and in a few years the crops fail. Then the land is abandoned once again, and more forests are cleared.

Countries with advanced farming methods have problems of a different kind. Improper use of chemicals can poison the soil, making it unfit for growing crops.

Figure 3–21. Some chemicals threaten the survival of species such as the bald eagle.

Pesticides, chemicals used to kill crop-destroying insect pests, and *herbicides,* chemicals that kill unwanted plants, can accumulate in the soil. They can become a hazard to other animals—or humans. Some chemicals, such as DDT, have been banned from general use. DDT once threatened some species of birds, including brown pelicans and bald eagles, with extinction.

Figure 3–22. Marginal soils can support pasture land or forests.

Another use of land and soil is for the grazing of animals. Land not fertile enough to grow crops or too hilly for farming may be suitable for pasture. Some pasture land is planted with grasses for feeding cattle, sheep, or goats. In other pastures, the animals feed on the native plants. Forests also occupy large areas of marginal land, providing lumber for building materials, fuel for heating and cooking, and pulpwood for paper products.

With each new day, the demand for land increases. Worldwide population growth means more food must be grown, more trees must be cut for paper and building materials, and more land must be provided for people to live on. All land must be evaluated for its best possible use.

One way to manage land is to improve marginal soils, making them more productive. Plants called *legumes,* which include peas, beans, and clover, improve the fertility of some soils, because they produce nutrients that other plants can use.

The shape, or contour, of the land can be used to reduce the loss of soil. Soil easily washes off steep land unless crops are planted in rows that follow the land's natural contour. Many farmers also use minimum-till farming to reduce soil loss. The stubble of the previous crop is left in the field over the winter to protect the soil. New crops are planted in narrow strips between the stubble to reduce the exposure of the soil to drying winds.

Figure 3–23. Contour plowing keeps soil from washing away.

ASK YOURSELF

Why is it important to manage land use carefully?

SECTION 2 REVIEW AND APPLICATION

Reading Critically
1. What are the differences among soil horizons?
2. What is leaching?

Thinking Critically
3. If temperate climates produce good soils, where else in the world might you expect to find soils as good as those in North America?
4. Why are both inorganic and organic materials necessary for the development of good soils?

SKILL *Interpreting Pie Graphs*

▶ MATERIALS
- paper ● pencil

▼ PROCEDURE

1. Examine the five pie graphs. Each graph shows the uses of soil on a different continent.

2. Using the key, determine how much land on each continent is or could be used for growing crops, for grazing, for forests, and for other purposes.

▶ APPLICATION

1. Which continent has the greatest percentage of soil in forested land?
2. Which continent has the greatest percentage of soil in grazing land?
3. Which continent has the greatest percentage of soil in cropland?
4. Which continent has the greatest percentage of soil used for other purposes?

✳ *Using What You Have Learned*
Do some library research to find out the uses of land in your state; then make a pie graph of land use in your region. Label the percentage for each type of land use.

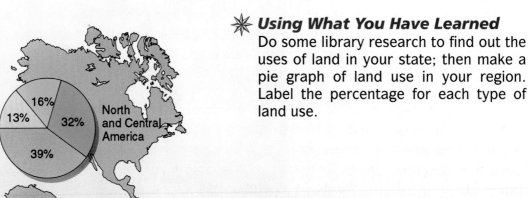

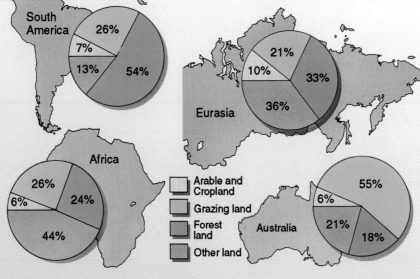

Erosion and Deposition

Flowing water, blowing wind, advancing ice, and crashing waves have one thing in common—the ability to move large quantities of rock and soil and deposit them somewhere else. In doing so, these agents of erosion slowly reshape the land. Many of the spectacular landforms of our national parks were shaped in this way. The Grand Canyon, for example, was created by the flow of the Colorado River.

Flowing Water

As water rushes downhill, it carries with it soil and small rocks, which are deposited at the bottom. The process that moves soil and rock and reshapes the land is called **erosion.** Water is not the only agent of erosion, but it is the most important.

Water has been called "the great leveler" because of its ability to erode the highlands and fill in the lowlands. Even a small stream can carry a tremendous quantity of material. The amount of material carried by a stream or a river is called a *load.* The load a stream carries depends on its velocity and the volume of water flowing into the stream. A large stream or a fast-moving stream carries a larger load than a small stream or a slow-moving stream.

Figure 3–24. Although small, this stream carries a large load.

Surface Erosion

Every piece of rock eroded by a stream or a river helps to reshape the land. As more and more material is eroded from the source of a stream, the stream grows in length. At the other end—the mouth—depositing of eroded material, or *deposition,* occurs. The mouth is the low point, or *base level,* of a stream. A stream cannot erode below the elevation of its base level. Sea level is the ultimate base level of all rivers and streams, and most deposition occurs in the sea.

Figure 3–25. Why can't a stream erode below the elevation of its mouth?

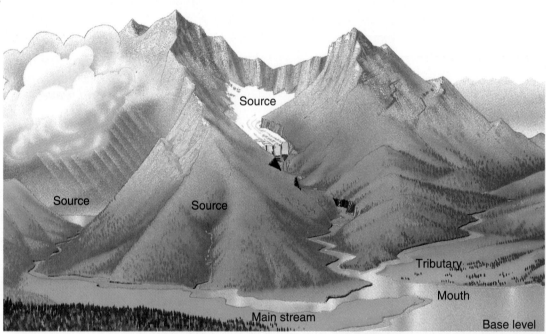

As rivers carve valleys and form plains where mountains once stood, they take on certain characteristics. A V-shaped valley with steep walls forms where there is a large difference in elevation between a river's source and its mouth. Such a river is called a *young river,* not because of its age, but because of its rapid erosion.

Figure 3–26. A young river has steep walls and a V-shaped valley.

Where a river encounters a layer of rock that resists erosion, rapids and waterfalls often result. The layer of resistant rock becomes a temporary base level for the water upstream. As the river works to remove the resistant rock, the waterfall slowly moves upstream.

Figure 3–27. Niagara Falls slowly moves upstream as the river erodes rock under the resistant rock.

A *mature river* works back and forth across its valley, widening it and developing a flood plain. *Flood plains* are flat, low lying areas over which a river spreads in times of high volume. When a stream overflows its regular banks, it deposits much of its load on its flood plain, enriching the soil. Flooding is a natural part of river flow. Before the development of the Aswan Dam, Egypt's Nile River flooded every spring, providing nutrients for summer crops.

Figure 3–28. A mature river has a wide, gently sloping valley.

Old rivers wander across their floodplains, forming broad curves called *meanders*. Meanders often form such broad loops that one bend cuts into another. During floods, most of the water takes the shorter route, leaving the abandoned meanders to form *oxbows*—temporary crescent-shaped lakes. Some rivers, such as the Mississippi, may be young near their source, mature in the middle, and old near their mouth.

Water Deposits As a river slows, it is no longer able to carry a large load. The load is then deposited. Sometimes a river deposits its load on its bed, forming a sandbar. Before channels were dredged to aid transportation, there were many sandbars along the lower Mississippi River.

Figure 3–29. Meanders form because old rivers are already near their base level, so erosion occurs mainly along their banks.

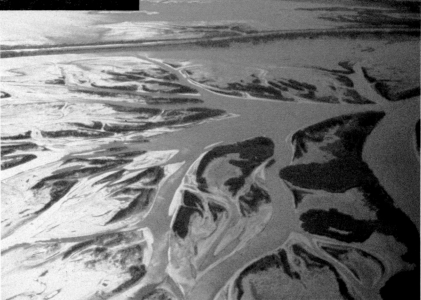

Figure 3–30. A delta is formed by deposition.

A river's load can be deposited at its mouth, forming a *delta*. Even small streams can form deltas, but the delta at the mouth of the Mississippi River is especially well developed.

Underground Erosion   As is the case with surface water, ground water is able, in some cases, to erode and deposit rock material. If the rock is limestone, ground water actually dissolves the rock, leaving large passageways that are caves when they reach the earth's surface. Water can move easily though these passageways. With continued erosion, the ceilings of some caves collapse, forming sinkholes. The landscape around Mammoth Cave National Park in Kentucky has many of these sinkholes.

Like the sandbars and deltas deposited by surface waters, underground water also forms deposits. Sometimes these deposits are spectacular. As water containing dissolved limestone drips from the ceilings of caves, the water evaporates, forming stone "icicles" called *stalactites*.

Figure 3–31. Spectacular deposits are common in limestone caverns, such as Carlsbad Caverns in New Mexico.

 ASK YOURSELF

Why do rivers have a different appearance depending on their stage of maturity?

Blowing Wind

Desert landscapes have long been known for their beauty and starkness. Landforms carved by the wind are memorable features of Zion National Park and Bryce Canyon National Park, both in Utah.

Figure 3–32. Wind erosion can create beautiful landforms such as these.

If you have ever been to a beach on a windy day and felt the sting of sand grains hitting your legs, you have some understanding of how blowing wind erodes rock. The effect of wind-blown sand on rock formations is similar to the process of sandblasting a building. As sand grains strike a rock surface, they chip pieces away, reshaping the surface and sometimes creating spectacular landforms.

When conditions are right, the wind can also erode large amounts of soil. Wet soils tend to resist wind erosion. Therefore, as the amount of water in the soil decreases, wind erosion increases. Part of the erosion problem in dry areas is due to the fact that lack of soil moisture means few plants can grow. Loss of this protective ground cover exposes soil and rock to additional erosion by the wind, forming *deflation basins*.

Figure 3–33. A dry, bare field can lose a lot of soil to wind erosion.

Wind erosion can be temporary. The "dust bowl" in the western United States in the 1930s resulted from several years of unusually dry weather. When the land dried out, the plants died and the exposed soil was blown away. However, as the area began receiving normal amounts of rain again, the protective plant cover returned and the wind erosion stopped. In the next activity, you can learn more about the drought of the 1930s.

DISCOVER BY *Researching*

Use your school or community library to learn about the problems associated with the dust bowl of the 1930s. You might want to view the classic movie, *The Grapes of Wrath,* or read John Steinbeck's book of the same title for additional background. Be sure to record your findings in your journal. ✎

As with running water, blowing wind also leaves deposits. Sand dunes are probably the best known form of wind deposit. R. A. Bagnold, a scientist who did extensive research in Northern Africa, describes a sand dune like this:

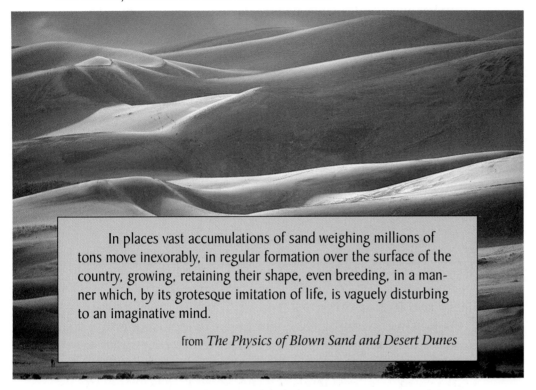

In places vast accumulations of sand weighing millions of tons move inexorably, in regular formation over the surface of the country, growing, retaining their shape, even breeding, in a manner which, by its grotesque imitation of life, is vaguely disturbing to an imaginative mind.

from *The Physics of Blown Sand and Desert Dunes*

Sand dunes are characterized by gently sloping sides that face into the wind, and steep sides that face away from the wind. Dunes move, or migrate, as sand is rolled up the gentle slope and deposited over the crest onto the other side.

 ASK YOURSELF

How is wind erosion similar to sandblasting?

Advancing Ice

Imagine all of Canada and much of the northern United States covered with ice more than 3 km thick. Imagine a summer day with temperatures less than 0°C and half a meter of new snow. Today most of the earth's ice is located in the polar regions. The rest is found on high mountains. But for much of the earth's recent history, glaciers covered large parts of North America and Europe. A **glacier** is a sheet of ice covering a large continental area or a mountain valley.

There are only two large *continental glaciers* remaining from the Ice Age: one in Greenland and the other in Antarctica. However, many mountain valleys around the world still have smaller glaciers called *alpine glaciers.* Glacier National Park in Montana has several large alpine glaciers.

Figure 3–34. Glaciers may be hundreds of meters thick, and cover areas as large as Antarctica.

A continental glacier may be as much as 4 km thick and so heavy that it causes the rock beneath it to sink into the mantle. If the ice sheet covering Greenland were to melt suddenly, the island would rise by nearly 1 km. The South Pole is under approximately 3 km of ice. But there is no glacier at the North Pole. Glaciers form on land, and the North Pole is in the Arctic Ocean. However, there are pieces of glaciers, called *icebergs,* in the polar seas. An iceberg is a great mass of ice broken off a glacier. Many icebergs form in Glacier Bay National Park in Alaska.

Figure 3–35. The formation of an iceberg from a glacier is shown in this series of photographs.

Glacier Formation   Why do glaciers form in some cold regions but not in others? Many areas of the world receive snow during the winter. However, in most areas the snow melts during the summer. Glaciers form only in areas where summer temperatures are too low for all the snow to melt.

As the snow accumulates, pressure increases on the bottom of the snow pile and the snow compacts. Any snow that melts quickly refreezes, forming ice. Fresh snow is about 90 percent air. Once half of the air spaces in the snow pile have been filled with ice, the pile is called *firn*. Firn takes about a year to form. You can make some firn in the next activity.

Discover by Doing

Make a snowball or a ball of crushed ice, and place it in a freezer until it is solidly frozen. Then take the snowball from the freezer, and apply heat and pressure to it by squeezing it tightly between your hands. When the snowball begins to melt, place it back in the freezer. Repeat this procedure several times until some of the air space in the snowball is filled with ice. How is this similar to the way firn is produced in a glacier? ✎

Eventually firn changes into solid ice. The transition from snow to firn to ice may take three to five years in places where summer temperatures cause some of the snow to melt. In colder parts of the world, this transition may take as long as 100 years. Why would this change take longer in places where it is colder?

Figure 3–36. These photographs show the transition from snow (left) to firn (center) to ice (right).

Even though glaciers are solid ice, they can still move. As pressure within a glacier increases due to its weight, some of the ice softens and begins to flow like hot plastic. Individual ice grains, responding to the pull of gravity, slide past one another, allowing the ice to move slowly downhill.

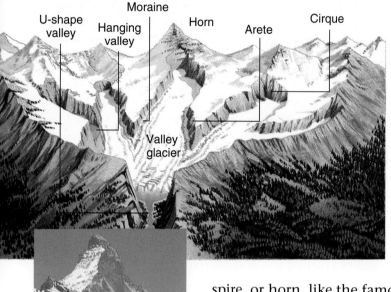

Labels on diagram: U-shape valley, Hanging valley, Moraine, Horn, Arete, Cirque, Valley glacier

Eroding and Depositing

Because of their mass, glaciers are especially effective at eroding rock. Some material simply falls from mountain slopes onto the surface of glaciers. But glaciers also pluck rocks from valley walls and bulldoze material in front of them as they move.

As an alpine glacier moves, it carves out a wide, U-shaped valley. Bowl-shaped depressions are often formed at the source of a glacier. An eroded mountain peak at the source of several glaciers forms a dramatic spire, or horn, like the famous Matterhorn in the Swiss Alps. Various alpine landforms are shown in the diagram.

When a glacier melts, it leaves piles of rocks behind in mounds and layers called *moraines*. Landscapes that have experienced erosion by glaciers are usually covered by heaps of boulders, sand, and fine silt mixed together. The shape of the piles of debris varies from ridges and hills to thick, lumpy blankets covering the land. The diagram shows some of the more common deposits left by glaciers. Many of these glacial features can be seen in Wisconson's Kettle Moraine Park.

Figure 3–37. Alpine glaciers create many notable landforms, including the famous Matterhorn in Switzerland.

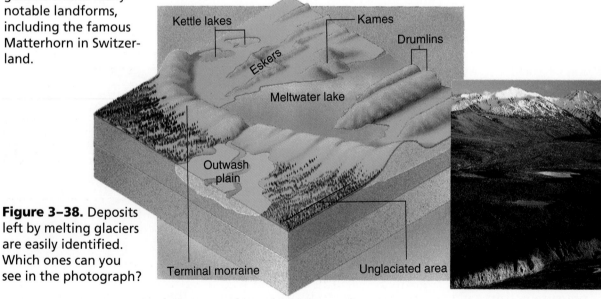

Labels on diagram: Kettle lakes, Kames, Drumlins, Eskers, Meltwater lake, Outwash plain, Terminal morraine, Unglaciated area

Figure 3–38. Deposits left by melting glaciers are easily identified. Which ones can you see in the photograph?

ASK YOURSELF

How do glaciers form?

Crashing Waves

Many of the features seen along coastlines are due to erosion and deposition by running water, wind, and glaciers. These are called *primary coasts*. Coastlines formed by erosion and deposition by waves are called *secondary coasts*.

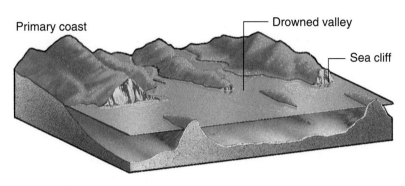

Even the most resistant rock can be eroded by pounding waves. If rocks along a coast are uniformly resistant to erosion, a straight coastline forms. If the rocks are not uniformly resistant, the waves form an irregular coastline. Headlands of hard rock jut into the sea, and softer rock erodes into sandy coves and inlets.

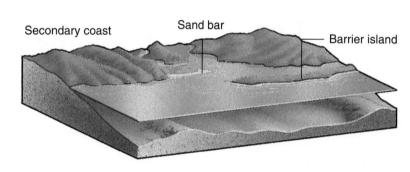

Sandy beaches are the most obvious wave deposits. In addition, sand washed from a beach and deposited offshore forms a sandbar or barrier island. There are many barrier islands along the Atlantic and Gulf coasts, from New York to Texas. Why do you suppose there are no barrier islands along the Pacific coast?

 ASK YOURSELF

How is a secondary coast formed?

SECTION 3 *REVIEW AND APPLICATION*

Reading Critically
1. List the main deposits of surface water, wind, glaciers, and waves.
2. Why is water called the "great leveler"?

Thinking Critically
3. Explain why temperature is a controlling factor in the amount of time it takes for a glacier to form.
4. Explain how slope, volume, and velocity of a river are related to its ability to erode and deposit materials.

Running Water

Describe *the water cycle.*

List *the factors that affect runoff.*

Compare and contrast *porosity and permeability.*

Most of the earth's water is stored in the oceans. Small amounts are found in glaciers, in rivers and lakes, underground, and in the air. Water constantly cycles from the oceans to the air, to the land, and back to the oceans again. The study of the water cycle—especially the effects of running water—is called *hydrology.*

Surface Water

The moment a raindrop falls to the earth, it begins its return to the sea. Water soaks into the ground or moves across the surface, always seeking a lower level. The movement of water across the surface of the earth is called **runoff.** The shape of the land and the amount of rainfall affect runoff. The speed at which water flows across the land is determined by topography. The steeper the slope, the faster the runoff. The volume of runoff is influenced by the amount of rain and by how much water soaks into the ground.

Figure 3–40. Sometimes rain falls so fast that its runoff causes flooding.

Runoff increases during a heavy rain because there is not enough time for much water to soak into the ground. Runoff also increases if rain falls on ground that is already saturated with water or if the water cannot get into the ground. Large paved areas, such as those found in cities, provide little or no opportunity for water to get into the ground. Therefore, all the rain that falls must run off. In the next activity, you can calculate the amount of runoff from a small area and estimate the runoff from a large area.

Calculating

Determine the area of a metal baking pan. Measure the volume of water that collects in the pan during a rain. Your measurement should be in L/m^2. Next, find the area of your school's outdoor basketball court or parking lot, and estimate the volume of runoff from a typical rainfall on that surface.

Runoff decreases in areas with extensive plant growth. Plants slow down water's movement, so more water soaks into the ground. Some types of soil can also reduce the amount of runoff. If soil is loosely packed or if it has a large humus content, water can more easily soak into the ground.

► ASK YOURSELF

What factors affect runoff?

Watersheds

As water runs off, it carries soil and rocks with it. If enough material washes away, a small channel, or gully, forms. Gullies usually form where the surface is not protected by a cover of plants, such as in a plowed field. With continued runoff, a gully develops into a stream. A stream carries water throughout most of the year, while a gully carries water only when it rains. Streams empty into rivers. The volume of water in most rivers is large because many streams, or *tributaries,* empty into them.

A river with its many tributaries creates a system for draining water from the land. The area drained by a river and its tributaries is called a **watershed.** Some watersheds are small and include only a single stream and the gullies that feed into it. There are also watersheds that drain whole sections of continents. The Amazon River watershed, for example, drains more than 6 million km^2 of South America and includes more than 15 000 tributaries.

Figure 3–41. Runoff may eventually form a channel, or gully.

An imaginary line, or *divide,* separates one watershed from another. For example, the Continental Divide, which runs the length of the Rocky Mountains, separates the watersheds that empty into the Pacific Ocean from the watershed that empties into the Gulf of Mexico. There are several large watersheds in the United States. The largest three are shown in the diagram.

Figure 3–42. This map shows several major watersheds in the United States.

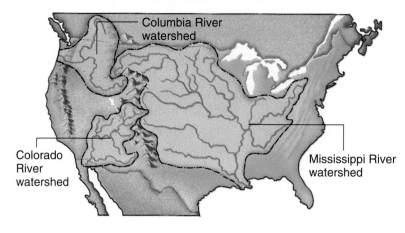

Columbia River watershed

Colorado River watershed

Mississippi River watershed

Figure 3–43. In swampy places, the water table is at the surface.

 **ASK YOURSELF**

What separates individual watersheds?

Ground Water

Remember, not all the rain that falls runs off into streams and rivers. Some water soaks into the ground, where it enters tiny air spaces in the soil and rocks. As water moves through these air spaces, some of it sticks to soil particles. Much of the water, however, continues moving through the ground. Eventually it reaches an area where all the spaces are filled with water. This is the top of the *water table.*

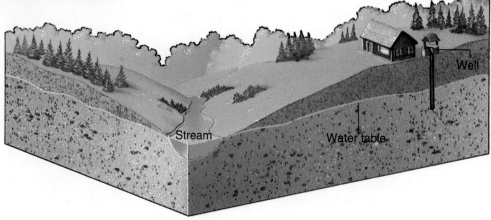

Well

Stream   Water table

The depth of the water table varies in different locations. Have you ever noticed that a hole dug on a beach quickly fills with water? In areas near the sea, the water table is at sea level, which may be just below the surface of the sand. In high desert regions, the water table may be hundreds of meters below the earth's surface. Wherever the earth's surface is below the water table, there is usually standing water, such as a lake, a pond, or a swamp.

In most places, soil holds only a small fraction of the water that soaks into the ground. Much more water is held in pores within the underlying bedrock. It may seem strange to you that a seemingly solid object, such as a rock, can be porous. But the structure of some rocks is similar to that of a sponge. If you try pouring a small amount of water into a sponge, you will find that most of the water will be trapped in the pores of the sponge. The amount of water that a rock can hold depends on its porosity. **Porosity** is the ratio of the volume of air space in the rock to the total volume of the rock. In the next activity, you can measure porosity.

DISCOVER BY *Doing*

To measure the porosity of a sponge, soak a rectangular sponge in a bowl of water and then wring as much water as possible out of the sponge into a graduate. Use the following formula to calculate the porosity:

$$porosity = \frac{volume\ of\ pore}{total\ volume} \times 100$$

A rock that is porous can hold a lot of water, but it will not necessarily allow the passage of water through it. If you look at a cork, you can see that it is very porous. However, if you put a cork in a bottle and turn the bottle upside down, none of the contents of the bottle will flow through the cork.

Figure 3–44. These diagrams show the construction of two rocks. Which rock is more porous?

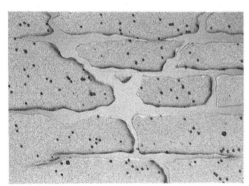

Like cork, some rocks are porous but water cannot move through them because they are not permeable. **Permeability** is the ability of rock to allow water to pass through. The permeability of a rock depends on the extent to which the pores are connected to each other.

A permeable rock layer forms an *aquifer.* Water flows slowly through an aquifer, moving downhill under the influence of gravity. Some aquifers transport large amounts of water over great distances. For instance, much of the midwestern United States lies on an aquifer that starts in the Rocky Mountains.

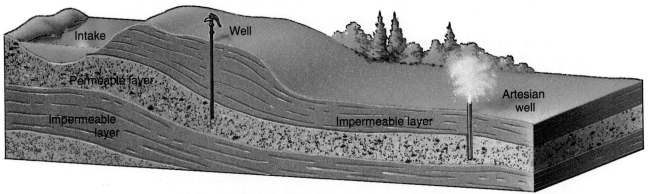

Figure 3–45. Aquifers help to transport water through the water cycle.

Water can be obtained from aquifers by drilling wells into them. Sometimes pressure in an aquifer is so great that water emerges without being pumped. This type of well is called an *artesian well.* Where a small aquifer comes to the surface naturally, it forms a spring.

 **ASK YOURSELF**

Why are some rocks porous?

SECTION 4 *REVIEW AND APPLICATION*

Reading Critically

1. What are porosity and permeability?

2. Explain how there might be several divides in the watershed of a large river.

Thinking Critically

3. Why is the building of a large shopping mall likely to increase flooding in an area?

4. Why wouldn't a swamp form in an area that is above the water table?

The Big Idea

Almost as soon as a rock is exposed, physical and chemical processes begin to weather it. Probably the most important result of weathering is the formation of soil.

In addition to weathering, the land is shaped by the action of water, wind, ice, and waves. Although each of these agents erodes and deposits, leaving characteristic landforms, water is the "great leveler."

For Your Journal

Look again at the way you answered the questions at the beginning of this chapter. How have your ideas changed? Be sure to record your new ideas in your journal.

Connecting Ideas

Copy the concept maps into your journal, and complete them with ideas from this chapter.

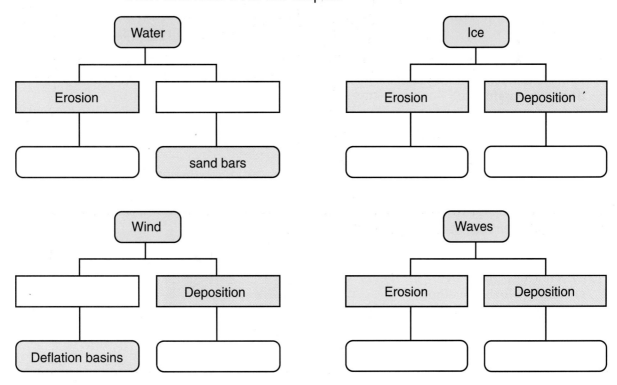

Water
— Erosion
— [ ]
— [ ]
— sand bars

Ice
— Erosion
— Deposition
— [ ]
— [ ]

Wind
— [ ]
— Deposition
— Deflation basins
— [ ]

Waves
— Erosion
— Deposition
— [ ]
— [ ]

Understanding Vocabulary

1. Explain how the terms in each set are related.
   a) weathering (65), mass movement (70)
   b) runoff (94), watershed (95)
   c) creep (70), flow (71)
   d) porosity (97), permeability (98)
   e) glacier (89), icebergs (90)
   f) horizons (77), leaching (77)

Understanding Concepts

MULTIPLE CHOICE

2. Which of the following does not contribute appreciably to soil formation?
   a) cycles of freezing and thawing
   b) physical weathering of a rock
   c) chemical weathering of a rock
   d) erosion

3. Which of these locations would probably experience the greatest amount of chemical weathering?
   a) Sahara Desert
   b) Amazon rain forest
   c) Bryce Canyon National Park
   d) Kettle Moraine Park

4. Which soil horizon is most important to plant growth?
   a) A horizon
   b) B horizon
   c) C horizon
   d) bedrock

5. What are often used to improve the fertility of soil?
   a) pesticides
   b) herbicides
   c) contour plowing methods
   d) legumes

SHORT ANSWER

6. Discuss the effects of physical weathering in the southwestern United States.

7. Compare and contrast the different horizons of a typical soil profile.

8. How can farmers conserve soil?

9. Briefly discuss how a stream forms.

10. What factors can increase the load of a river?

11. Compare and contrast river valleys with glacial valleys.

Interpreting Graphics

Discuss how the process of weathering and erosion occur in each of these environments.

12.

13.

Reviewing Themes

14. *Changes Over Time*
   Explain how a river matures.

15. *Environmental Interactions*
   Discuss the role of vegetation in weathering.

Thinking Critically

16. What kinds of human activities might cause subsidence? What can be done to prevent it?

17. How can you tell the direction from which glaciers have moved over an area?

18. Sand dune erosion is a problem on some beaches due to the use of motor vehicles on the beach. In most places, such activities are banned. What can be done to prevent beach erosion in places that allow vehicles to travel on the sand?

Discovery Through Reading

Bailey, R. *Glaciers*. Time-Life Books, 1992. The photographs in *Glaciers* are beautiful examples of how masses of ice form and change the land over which they move.

Brighton, Carol. "Beach Saver." *Popular Science* 240 (June 1992): 27. This article describes a beach watering system designed to reduce beach erosion.

CHAPTER 4

MAPS AND LANDFORMS

*P*ygmies living in the Arctic? California near the North Pole? Even map-making geniuses could make such outrageous errors, as Gerardus Mercator did on an influential map of the Arctic in 1595. Mercator, author of the famed "Mercator projection," showed the magnetic North Pole twice and pictured two islands that didn't exist. Early cartographers could not rely on satellite photographs, as map-makers do today. Instead, Mercator had to base his map on a mostly fictitious traveler's journal, as an historian explains:

Supposedly describing the experiences of an English visitor to Greenland, it painted a nightmarish picture of four large islands around the Pole, separated from one another by violent currents. These currents, it was believed, carried water back into the center of the earth where it was then recycled. This was not a prospect to encourage Arctic exploration. Yet by showing open sea to the south of the polar islands, Mercator's map offered explorers alternative routes to the Pacific, via the north of Russia or Canada.

from *Early Maps*
by Tony Campbell

Jan Vermeer, *The Astronomer*, 1668, Louvre, Paris, France.

For Your Journal

✏️ How do maps model the earth?

✏️ How does a map show elevations and landforms?

✏️ How do landforms relate to processes that shape the earth?

Making Maps

Objectives

Describe the development of early maps.

Compare and contrast latitude and longitude.

Explain how maps model the earth, and **describe** the shortcomings they have.

You have probably heard the expression "You can't get there from here." People often use that expression when "getting there" requires a complicated or roundabout path. It's a good thing we have maps. But maps haven't always been around, and some early maps wouldn't have helped much anyway. Map making has come a long way toward getting us "there."

Figure 4–1. Perhaps a map would be helpful.

"What do you mean "You can't get there from here!?""

Early Maps

Babylon was an early civilization in the Middle East, the site of present-day Iraq. The Babylonians believed that the earth was a flat disk, completely surrounded by an ocean they called Bitter Waters. Outside the disk were thought to be seven islands linking the world to an outer Heavenly Ocean, where the gods lived.

The ancient map on the next page shows the area around the Mediterranean Sea. You can probably identify some of the features of Africa and Arabia. The mud tablet shown is actually a map, too—the oldest known map. Dated to about 2500 B.C., it is thought to show the outline of a valley estate or village.

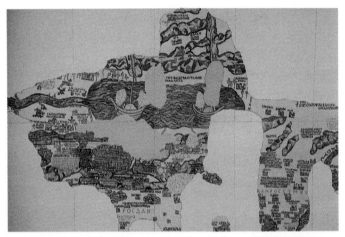

Early civilizations employed a variety of map making techniques, using materials that were handy. You can try the same thing in the following activity.

Figure 4–2. The photograph (left) shows one of the oldest maps of the western Mediterranean. The old clay map (right), was found in Iraq.

DISCOVER BY *Doing*

Imagine you are a Babylonian landowner and you wish to map the boundaries of your estate. Let your school be your ancient estate. Using whatever materials you wish—modeling clay, paper and crayons, or whatever, make a map of your property. Identify important buildings and the property lines. Label nearby buildings and features, such as streams, hills, and roads. Compare your map with those of your classmates. Do you think any of your maps will last 4500 years? ✐

Map making was a very creative activity. The Inuit of the Arctic, for instance, cut shapes of coastal islands out of dark-colored animal skins. To make a map, they sewed the shapes onto a light-colored skin that represented the ocean. The Egyptians engraved maps on gold, silver, and copper plates to iden-

Figure 4–3. Micronesians made maps from shells and bits of coral to help them locate the many nearby islands.

tify the locations of gold mines and other valuable properties. The people of Micronesia, in the Southern Pacific Ocean, made maps similar to the one shown here, to show the location of neighboring islands. *Micronesia* means "small islands." Why would such a map be useful to a group of islanders in the South Pacific?

Art or Science?

In early times, map making was an art rather than a science. It was based more on imagination, guesswork, and travelers' tales than on precise measurements of actual land and water features. Exploration was a risky undertaking, and early explorers reported many dangers, both real and imagined. The fanciful map shown here was drawn by a Swedish map maker in 1539. He depicted the fjords of northern Norway as being infested with monsters that made quick work of ships.

Early map makers constructed the best models they could, based on the small amount of information available. Each time adventurers returned from their trips, they brought with them new data. Map makers used this new information to improve the maps they had made.

But early maps also led to the development of **cartography,** the science of map making. In fact, as early as the second century A.D., map making had started to become more precise in some areas of the world. Ptolemy, an Egyptian astronomer and geographer, began to change map making from an art to a science. Many of his contributions still influence cartography today.

Figure 4–4. The map by Olaus Magnus shows many imagined threats to seagoing explorers. Real dangers, such as the whirlpool, however, could be just as fatal.

Sail Round to the North!

One of Ptolemy's innovations had to do with direction. Maps have not always been drawn with today's directional bias. Early map makers often put east—the direction of the rising sun—at the top of their maps. Others set the direction of the prevailing winds at the top. Ptolemy was the first to put north at the top of a map. Because of his work in astronomy, Ptolemy's maps were based on a round Earth. He also improved the understanding of the actual size of the earth and of the distances between landmasses.

Ptolemy's maps were so good that they were still being used well into the fifteenth century. At that time Johann Gutenberg, the German inventor of the printing press, produced copies of Ptolemy's maps, making them readily available. About the same time, interest in geography increased dramatically. Adventurers were busy exploring unknown parts of the world. With this new age of exploration, even more was learned about the sizes and positions of the continents.

Figure 4–5. Ptolemy combined his knowledge of astronomy and geography to make accurate maps.

 **ASK YOURSELF**

Why were early maps made from a variety of materials?

Latitude and Longitude

Imagine that you are an early explorer about to set sail for some distant port. You have a map of the port, made by the last explorer to visit that place, but the ocean between you and the port is uncharted. You aren't sure which direction to sail, or even how far to go.

As a result of this kind of difficulty, map makers developed two sets of imaginary lines to help locate places accurately on a map. One set of these imaginary lines, running east and west around the earth, are the lines of **latitude.** Lines of latitude are parallel—they never meet or cross—so they are also refered to as *parallels.*

The imaginary lines that run from the North Pole to the South Pole are called lines of **longitude.** Lines of longitude are not parallel. They touch at the poles and are farthest apart at the equator. Lines of longitude are also called *meridians.*

Lines of latitude begin at the equator and are measured from 0° (the equator) to 90° (the poles). The equator divides the earth into two equal halves, the Northern Hemisphere and the Southern Hemisphere. Therefore, latitude readings must indicate direction from the equator as well as degrees; for example, 33°N or 20°S.

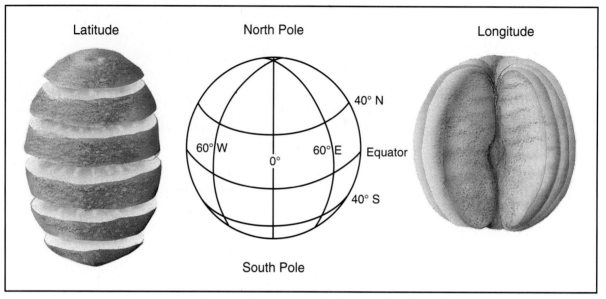

Figure 4–6. Lines of latitude are like slices of an orange. Lines of longitude are like sections of an orange.

In 1884 a line representing 0° longitude was established by international agreement. This line, the *prime meridian*, passes through Greenwich, England. Lines of longitude are measured in degrees east or west from the prime meridian. These measurements meet 180° from the prime meridian. The prime meridian and the 180° meridian, also called the *international date line,* divide the earth into two equal halves—the Eastern Hemisphere and the Western Hemisphere. As with latitude, all longitude readings must include both degrees and direction from the prime meridian; 29° E or 41° W, for example.

Many people find it hard to distinguish between latitude and longitude. You may find it helpful to use this memory device: lines of latitude go across the map like the rungs of a ladder. You move up and down them to find locations as if you were climbing up or down a ladder. Once you recall that latitude and ladder go together, you can remember that the lines of longitude go the opposite way. Study the diagram on the next page, and then try the activity that follows.

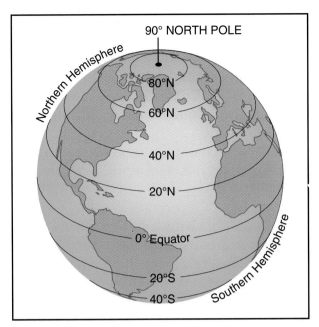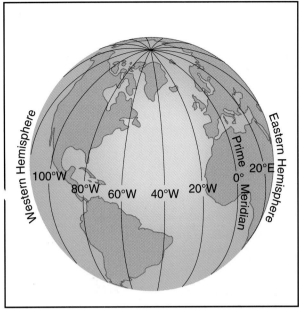

Figure 4–7. Latitude and longitude can be used to divide the earth into hemispheres.

ACTIVITY

How can you use latitude and longitude to locate places on a map?

MATERIALS

a world map

PROCEDURE

1. Review the definitions of latitude and longitude, and study the diagram above.
2. Familiarize yourself with the world map by locating the equator and the prime meridian. Determine the number of degrees between lines of latitude and lines of longitude.
3. Locate the 30° N parallel on the map. List the continents this parallel crosses.
4. Locate the 80° W meridian on the map. List the continents this meridian crosses.

APPLICATION

1. What city is located at 42° N, 87° W?
2. What city is located at 34° S, 18° E?
3. What city is located at 22° N, 88° E?
4. What is the latitude and longitude of Houston, Texas?
5. What is the latitude and longitude of Hong Kong?
6. What is the latitude and longitude of the major city nearest to you?
7. What is the latitude and longitude of Sydney, Australia?
8. What are the largest latitude and longitude readings possible?

 ASK YOURSELF

In which two hemispheres is the United States?

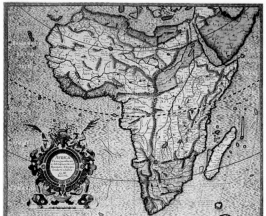

Figure 4–8. Mercator's maps were useful for navigation.

Map Projections

As handy as maps are, they have many short-comings as models of the earth. The main shortcoming is that maps are two-dimensional models of a three-dimensional object. Making a flat representation of the spherical earth causes distortions in the appearance of many of the earth's features. As a result, maps do not show totally accurate shapes of the continents and oceans.

In the 1500s, Flemish cartographer Gerardus Mercator made a breakthrough in the science of map making. Mercator's maps showed all parallels and meridians at right angles to each other. This type of map is known as a *Mercator projection.*

Mercator created his maps during the era of exploration, and the maps work perfectly for navigational purposes. A sailing course between any two points can be shown as a straight line. For example, if you were sailing from Rome, Italy, to Marseille, France, you would simply draw a straight line between the two cities. Then you would determine the angle of the line from any meridian and sail at that angle until you reached your destination. Of course, you would have to detour around the island of Corsica, but Mercator's maps show the shapes of islands and harbors fairly accurately, so this would not be a problem.

Mercator projections do have a major drawback, however. They widen and lengthen the size of areas at high latitudes. For example, on the Mercator projection below, Greenland appears almost as large as Africa. In fact, Africa is 15 times larger than Greenland. The map also distorts Antarctica, which does not fill nearly the area that it appears to on the map.

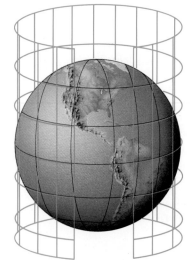

Figure 4–9. This illustration shows the development of a Mercator projection. Notice the distortion of size at high latitudes.

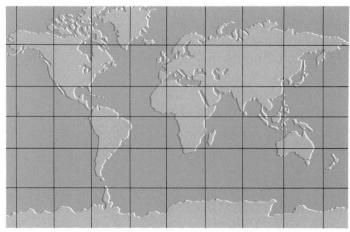

Different projections can be used to study specific parts of the world without the distortions of the Mercator maps. If you wanted to study the countries around the Arctic Ocean, you might use a *polar projection,* such as the one shown. Polar projections are made as if the observer were looking down on the world from the North or South Pole. A polar projection is circular rather than rectangular.

Note in the diagram that the North Pole is at the center of the map and the equator forms the outer boundary. Land areas near the pole are shown in true proportions, but areas near the equator are greatly distorted. Polar projections are valuable to people who live and work in the Arctic and Antarctic. Try the following activity to see how you would have to adjust your perspective of the world in order to use a polar projection.

Figure 4–10. Polar projections such as this show areas at high latitudes without much distortion.

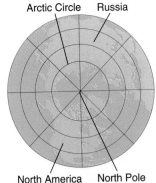

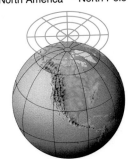

DISCOVER BY *Doing*

Use the map of Antarctica below to answer the following questions. Notice that the projection does not extend all the way to the Equator. You may need to use a globe to help you answer some of the questions.

1. In what direction do lines of longitude lead away from the South Pole?
2. What is the closest continent to Antarctica? What part of Antarctica is closest to this continent?
3. Through which ocean would you travel if you sailed north along the 120° W meridian?
4. If you sailed north from Wilkes Land between meridians 120° W and 150° W, in which direction would your compass point? ✐

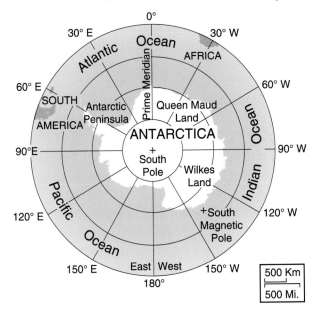

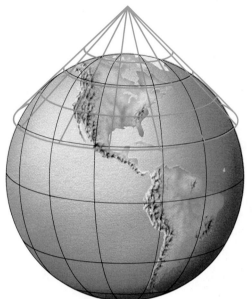

Figure 4–11. Conic projections are often used to make small area maps because they do not distort the size or shape of the land.

A *conic projection* is used to show relatively small areas of the world without much distortion of any landmasses. As you can see, however, these maps would not be useful for certain navigation tasks because they do not cover a large area. Many different maps would have to be used. Conic projections are often used for road maps.

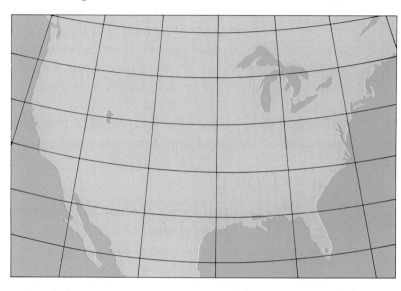

▼ **ASK YOURSELF**

What problem do map projections share?

Modern Maps

Figure 4–12. A bench mark shows the exact location and elevation of a surveyed spot.

Artists and explorers made maps in ancient times. But map making became a science, and scientists continue to make maps today. Since 1807 the United States Coast and Geodetic Survey and later the United States Geological Survey have been surveying much of the surface of the earth to improve the accuracy of maps.

Within the United States, each place that has been precisely surveyed is marked with a bronze marker called a *bench mark*. Perhaps you have noticed such markers while walking along a trail, in a field, or even in a city. Engraved on each bench mark are the elevation and location. Cartographers include bench marks on their maps. They also help land surveyors set property boundaries and engineers plan bridges and highways. In the next activity you can use a map to learn some things about the area in which you live.

DISCOVER BY *Researching*

Your teacher will provide you with a road map of your area. Use the map to find out the following: What specific area does the map cover? What kind of projection was used? How up-to-date is the map? How does the map show some familiar points, such as your city and the surrounding area? With a classmate, use the map to plan a trip to a nearby park or city. ✎

In recent years, greater accuracy and speed in updating maps has been made possible through the use of remote sensing satellites, aerial photographs, and computers. However, it will still take years to finish the job of accurately mapping the world.

Figure 4–13. Satellite photographs help cartographers make accurate, up-to-date maps.

▶ ASK YOURSELF

To whom would accurate mapping of a coastline be important? Why?

SECTION 1   *REVIEW AND APPLICATION*

Reading Critically

1. Why were Ptolemy's contributions to cartography important?
2. How did Mercator's map projection contribute to world exploration?

Thinking Critically

3. An old treasure map shows the X at 95° N and 190° E. Explain why this is impossible.
4. A straight line between New York and Moscow passes through different cities on different map projections. Explain how this could be possible.

Maps in Earth Science

Objectives

Describe the common types of map scale.

Explain the use of a map legend.

Compare and contrast topographical and geological maps.

You may think that there is little use for maps, except for road maps and those you use in social studies. But maps are used everywhere. When you go on a treasure hunt at a party or look for a certain store at a new mall, you probably use a map. Some maps are particularly useful for studying the earth.

Map Scales and Legends

If you were planning a trip from your home to your state capital, a world map probably wouldn't be much help. People often need to see more details than could possibly fit on a world map. To show details, cartographers could make larger maps or change the scale of a map. **Map scale** is the relationship between a distance on the map and a distance on the earth. For example, a scale of 1 cm = 100 km means that 1 cm on the map represents 100 km on the earth's surface. With a scale of 1 cm = 1 km, what would 2 cm on the map represent on the earth's surface?

Figure 4–14. Finding Main Street in Capital City on this map might be a problem.

"But where is 139 Main Street?"

The previous example of 1 cm = 1 km is called a *verbal scale*. This type of scale often equates two different units, in this case centimeters and kilometers. A verbal scale is often found on road maps, because it is easy for people to understand.

Map scales are also written as ratios. This kind of scale, called a *fractional scale,* shows the ratio of a represented distance on a map to the actual distance on the earth's surface. For this scale, the two numbers must be in the same units. For example, the scale on the map below is 1:5 000 000, which means any measurement on the map represents 5 million of the same unit of measurement on the earth's surface. So 1 cm on the map equals 5 million cm (50 km) on the earth.

This kind of scale, though accurate, is not practical for ordinary map use. Therefore, many maps include a *graphic scale*. A graphic scale consists of a measured line marked off in specific distances. You can learn how to use a graphic scale in the next activity.

DISCOVER BY Calculating

Look at the map of Hawaii, and examine the graphic map scale. To use the scale, lay a strip of paper next to the scale of miles to make yourself a paper ruler. Do the same thing with the kilometer scale. Then measure the length of the Hawaiian Island chain with each paper ruler. Measure along a diagonal, starting at the top right corner of the island of Niihau and ending at the bottom-left corner of the island of Hawaii. How many miles long is the Hawaiian chain? How many kilometers long is it? ✏

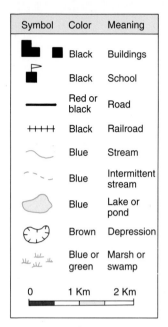

Symbol	Color	Meaning
◼ ◼	Black	Buildings
⚑	Black	School
———	Red or black	Road
+++++	Black	Railroad
~	Blue	Stream
- - -	Blue	Intermittent stream
⬭	Blue	Lake or pond
⬯	Brown	Depression
⫶⫶	Blue or green	Marsh or swamp
0 1 Km 2 Km		

Figure 4–15. Some common map symbols

In addition to the scale and the names of places, you probably noticed other information on the map of Hawaii. Symbols on the map show the location of airports, volcanoes, and the state capital. Some additional symbols are shown in the box on the left of this page. This box is a map legend. A **legend** lists all the symbols used on a map and tells what they represent. Not all maps use the same symbols. Some maps, for instance, use lines and colors to show different elevations, rock formations, or landforms. This legend shows some of the symbols you might find for buildings, roads, and streams.

▼ **ASK YOURSELF**

How does a map scale help model the earth more accurately?

Topographical Maps

Maps are basic tools of an earth scientist. One kind of map earth scientists use is the topographical map. The word *topos* is Greek for "place," and *graphe* means "drawing." **Topographical maps** show the shape of the land with lines. These lines connect points of equal elevation. A photo and part of a topographical map of the area around Mount St. Helens are shown here. Notice how the volcano is represented on the map.

Figure 4–16. Photo and topographical map of Mount St. Helens

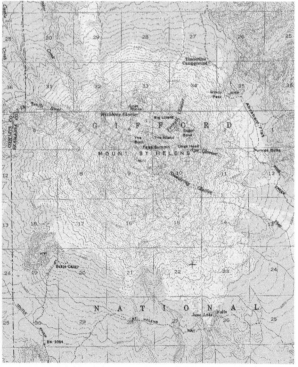

Who might need such a map? If you were an engineer planning a new highway, you would need to know about differences in elevation at your highway site. If you were planning a hike through mountainous country, you would want to know how steep certain trails were or where streams intersect your path.

On a topographical map, a pattern of lines is used to show elevation. The lines are called **contour lines,** and they run around a landform and connect all the points that are the same height above sea level. Streams, lakes, and other natural features are also represented on topographical maps, as well as political boundaries, towns, roads, and other important features. Examine the map shown here. The summary below will help you interpret contour lines.

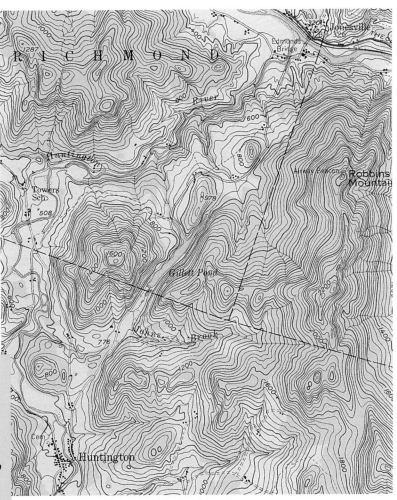

Figure 4–17. Contour lines are used to represent differences in elevation.

Interpreting Contour Lines

1. A contour line connects points of equal elevation.
2. A contour line is an endless line; it closes upon itself either on the map or at some point outside the map area.
3. Contour lines never branch or fork.
4. Contour lines never cross one another.
5. Closely spaced contour lines represent steep slopes; widely spaced contour lines represent gentle slopes.
6. Circular depressions in the earth's surface are shown by closed contours that have marks on the inside.
7. Contour lines point upstream when crossing a stream, river, or valley.

You may have noticed on the topographical map that contour lines are brown, and that some of them are labeled with the elevation. Some topographical maps have so many contour lines that the elevation of a single line is difficult to determine. As an aid to reading the map, every fifth contour line is drawn darker and thicker, so it stands out. It is also labeled with the elevation. The labeled contour lines are called *index contours*. Use the topographical map on page 117 to do the next activity.

DISCOVER BY *Observing*

Notice that the index contours on this map increase or decrease by 200 ft. What can you infer about the lines between two index contours? Notice also that certain sites have specific elevations labeled, such as the peak near the top-left corner of the map. Locate some others. How does the elevation change between Gillett Pond and the small peak to the left of it? Now locate Airway Beacon, just to the left of the map's right edge. Notice that it has not been given a specific elevation label. How could you estimate its elevation? What would it be? ✐

Now that you have learned something about topographical maps, consider this. Many developments in contour mapping were aided by the use of computers. What information does a computer provide that cartographers would need?

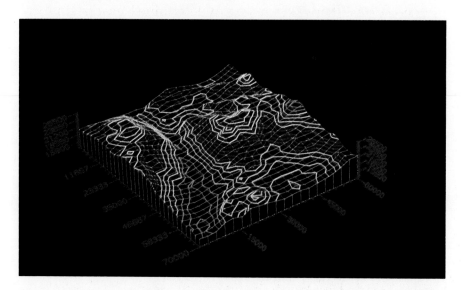

Figure 4–18. Computers help cartographers visualize the shape of the land.

 **ASK YOURSELF**

How do topographical maps show differences in elevation?

Hydrological and Geological Maps

In addition to topographical maps, earth scientists use other types of maps, such as hydrological maps and geological maps. **Hydrological maps** show where surface and underground water is located. They also show drainage patterns and areas where water has produced unique landforms.

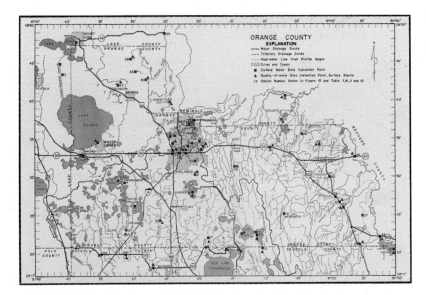

Figure 4–19. Hydrological map of Orange County, Florida

Geological maps use patterns or colors and symbols to show layers of rock at or just below the earth's surface. By studying geological maps, scientists can determine how certain rock formations have been folded and faulted.

Figure 4–20. Geological maps show rock formations at or just below the surface.

Most rock formations are formed in flat sheets, one layer piled upon another like the pages in a book. If each page were a different color, a book could serve as a model for making a geological map. However, a book lying flat on a desk would make a map of only one color—the color of the top page. If you stood this multicolored book on its binding, you could see all the colors of the now vertical pages. This is how vertical layers of rock look on an geological map. How do you think folded rock formations might look on a geological map? The next activity can give you an idea.

DISCOVER BY Doing

You can show how a folded rock formation would look on a geological map using modeling clay. Flatten three different colors of modeling clay into thin sheets, and stack the layers on top of each other. Then form an anticline by pushing up the center of the clay block. Using a butter knife, carefully slice off the top of the fold, forming a flat surface. If you now draw your formation, you will have a geological map. What would a geological map of a syncline look like? ✏

Since the pattern of synclines and anticlines can look very confusing on geological maps, scientists have developed a better system to describe how rock formations are tilted, folded, and faulted. This system is called *strike and dip*. **Dip** is the angle at which a rock layer is tilted from a horizontal plane. Imagine a flat sponge being plunged into a bucket of water. The water surface is the horizontal plane. If you hold the sponge straight up and down, its angle with the water is 90°. If you tilt it a little, the angle might dip to 60°, or even to 30° if you tilt the sponge a little more.

Figure 4–21. The orientation of a rock layer is described by its strike and dip.

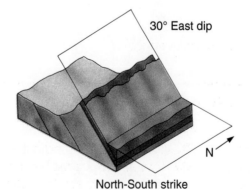

30° East dip

N

North-South strike

Just as the sponge might dip 30° to the left or to the right, rock layers can also tilt in different directions. Therefore, dip also includes the direction in which the layer is tilted. The rock layer shown in the diagram, for example, is dipping 30° east.

Think again of the sponge in the water. Try to visualize the line where the plane of the water touches the surface of the sponge. In geological terms, this line would be the *strike* of the sponge. **Strike** is the line formed where a rock layer intersects the horizontal plane. Strike is always at right angles to dip. If the dip of a rock formation is east, then the strike must be north-south. On a geological map, a T-shaped symbol is used to indicate strike and dip. The bottom of the T points in the direction of dip. Other symbols on geological maps show synclines and anticlines, horizontal and vertical rock layers, and specific rock types. Sometimes rock formations are named as well.

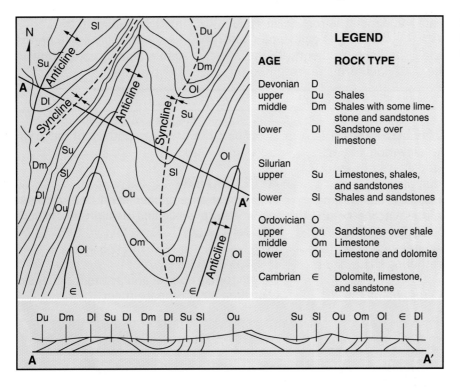

Figure 4–22. How are synclines and anticlines shown on this map?

 ASK YOURSELF

How do geological maps represent strike and dip?

SECTION 2 *REVIEW AND APPLICATION*

Reading Critically
1. Give an example of a verbal scale and a fractional scale.
2. What does a hydrological map show?

Thinking Critically
3. What would be indicated by a zero index contour?
4. How are topographical and geological maps similar? How are they different?

INVESTIGATION

Comparing Features on Topographical and Geological Maps

▶ MATERIALS
- topographical and geological maps of Jack Mountain ● hand lens

▼ PROCEDURE

1. Locate streams, rivers, mountains, and other landforms on the topographical map of Jack Mountain. Locate as many of these same features as possible on the geological map of Jack Mountain.

2. Study the rock formations on the geological map, and compare them with the landforms represented on the topographical map.

3. Determine which rock formations make up the valleys and which make up the mountains by comparing the two maps.

▶ ANALYSES AND CONCLUSIONS

1. Which map shows the elevation of Jack Mountain? Which map shows the rock formations of Jack Mountain?

2. Is the valley between Back Creek Mountain and Jack Mountain a syncline or an anticline? How can you tell?

3. Which rock formations seem to be the most resistant to weathering? How can you tell?

4. Which map would be most helpful in describing the drainage pattern of the area? Why?

▶ APPLICATION
Why would an engineer need both of these maps before designing a highway through the area?

✳ Discover More
Obtain topographical and geological maps of your area and compare them.

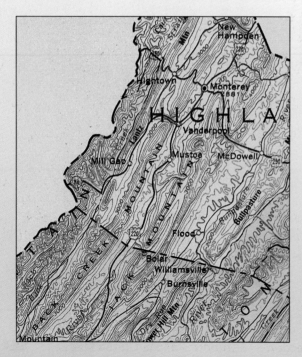

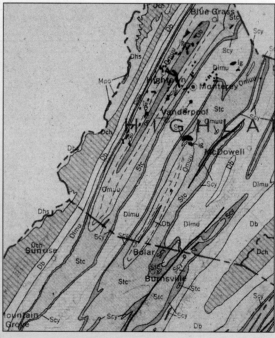

Landforms of the United States

Objectives

Identify *factors that create specific landforms.*

Describe *the major geomorphic provinces of the United States.*

Compare and contrast *the landforms of each geomorphic province.*

Can you identify the three places shown below by looking at photographs of each? Maybe identifying the Grand Canyon was easy, but you might make an educated guess about the others based on the landforms shown.

Geomorphology and Geomorphic Provinces

The study of landforms and the processes that produce them is the basis of geomorphology. The word *geomorphology* sounds like a tongue twister, but it has three simple parts. *Geo* means "the earth," as in *geology*, and *morph* means "form." The ending *-ology* is attached to many words and means "the study of." For example, *biology* is "the study of living things." *Geomorphology* then is "the study of the earth's forms, or landforms." You can practice using these word parts in the following activity.

Figure 4–23. Can you identify each of these landforms?

See how many words you can write that contain the word part _geo_. Think of school subjects, for instance. Now see how many words you can write that contain the word part _morph_. You may have to use the dictionary to give you some ideas. Finally, see how many words can you write with the suffix _-ology_. Then compare your lists with those of your classmates. ✐

If you look at a photograph of the area near your home, you may recognize certain landforms. Landforms are the result of interactions between processes that build up the crust—faulting and folding—with processes that wear away the crust—weathering and erosion. Each region of the earth has recognizable landforms that set it apart from all other regions.

North America can be divided into several regions, called geomorphic provinces. **Geomorphic provinces** are areas with distinctive landforms and topography. A study of all the geomorphic provinces in the United States will require some traveling. So pack a bag; we're going on quite a field trip.

Figure 4–24. The United States includes many different geomorphic provinces.

▼ **ASK YOURSELF**

What is geomorphology, and how does it relate to geomorphic provinces?

The Western Provinces and Hawaii

The first region we will visit includes the western provinces and the islands of Hawaii. The western provinces are primarily mountainous. In general, the mountain-building events started along the eastern edge of the region and moved westward. Part of the region is affected by the boundary of two major plates—the North American plate and the Pacific plate. Hawaii, in the Pacific Ocean, is a separate province.

Figure 4–25. The western provinces

The Rocky Mountain Province   Our first stop is Denver, Colorado, at the foot of the Rocky Mountains. The Rocky Mountains are about 4500 km long and are part of a mountain chain extending from Alaska to the tip of South America. The summits of most of the mountains are more than 1800 m above their bases. The rugged topography throughout the Rocky Mountains indicates that it is a young mountain range. Between the mountain peaks are U-shaped valleys, carved by glaciers.

Figure 4–26. Some Rocky Mountain landforms were created by glacial erosion.

The Colorado Plateau   From Denver we travel to Flagstaff, Arizona, on the vast Colorado Plateau. The Colorado Plateau is a dry area characterized by deep canyons, buttes, and mesas. The upheavals that created the mountainous areas surrounding the plateau had little effect on this province, except that the entire area was raised more than 1 km.

Figure 4–27. The Colorado Plateau (top), was created by uplifting. Death Valley National Monument (bottom) is located in the Basin and Range Province.

There are many national parks on the Colorado Plateau. Among them are Zion, Bryce Canyon, and Grand Canyon. Most show stunning erosional features where wind and water have carved the colorful sedimentary rocks. The Grand Canyon is a virtual historical tour of the earth as told in its rock layers. You will be taking that tour in a later chapter. Right now we're heading for Palm Springs, California, in the Basin and Range Province.

The Basin and Range Province

This province makes up a large area extending from southern Oregon into Mexico. The landforms of the Basin and Range Province are mostly the result of block faulting. The movement of huge blocks of rock along faults created the mountains of the province. Located at the bottom of a fault block is Death Valley, the lowest, hottest, and driest place in the country.

The Columbia-Snake River Plateau

Next stop, Spokane, Washington, and the Columbia-Snake River Plateau. This province, in Oregon and Washington, contains a variety of plains, hills, mountains, and plateaus. The predominant feature, however, is the extensive lava flow that blankets the province. In some places the lava deposits are more than 3000 m thick. In Craters of the Moon National Monument in Idaho, a small cinder cone stands starkly on a rugged blanket of lava.

Figure 4–28. Craters of the Moon National Monument is located on the Columbia-Snake River Plateau.

The Pacific Mountain System We're heading west again, to Portland, Oregon, and the Pacific Mountain System, which includes a continuous line of young mountains paralleling the western coast of the United States. The mountains known as the Cascades are volcanic in origin. Some are still active—Mount St. Helens erupted as recently as 1980, blanketing much of the Pacific Northwest with volcanic ash. The Sierra Nevada, also in this province, formed from a single fault block.

Beautiful landscapes of the Pacific Mountain System can be seen in several national parks in this province, including Yosemite in California, Crater Lake in Oregon, and Mount Rainier in Washington. Certain national parks are identified with specific people. You can learn about some of them in the following activity.

Figure 4–29. Yosemite National Park is located in the Sierra Nevada.

DISCOVER BY *Researching*

Use your school or community library to learn about John Muir, John Wesley Powell, Gifford Pinchot, and Marjory Stoneman Douglas. Find out what area each is identified with. Also find out what each person did to explore, publicize, or preserve the unique environments, wildlife, or landforms of the United States. ✐

The Pacific Border Province From the mountains it is only a short drive to the Pacific Border Province. This province extends the length of the West Coast. The southern part of the province includes the coastal ranges and lowlands to the east. The lowlands, such as the Great Valley of California, are formed by faults—great cracks at or just beneath the earth's surface. Earthquakes are frequent throughout this province as the western edge of California continues to slide along the San Andreas fault. California's Redwood National Park, home of the world's tallest trees, is located in this province.

Figure 4–30. Giant redwood trees are protected in Redwood National Park.

The Alaskan Provinces We really have some distances to cover now—in Alaska. Alaska is so large that it actually contains several geomorphic provinces within its borders. The southeastern sliver of Alaska is a mountainous coastal region with many islands and fiords. The Alaska Peninsula and the Aleutian Islands form a volcanic island arc, and the Yukon Basin is a plateau and lowland area drained by the Yukon River.

Figure 4–31. The Arctic Slope (left) and the Alaskan coast near Sitka National Park (right)

South central Alaska has a coastal strip leading to a band of mountains known as the Alaska Range. The Brooks Range in northern Alaska is roughly parallel to the Alaska Range. The most northern area is a gently sloping plain called the Arctic Slope, which borders the Arctic Ocean. Mount McKinley, in the Alaska Range, is the highest mountain in North America. Athabaskan Indians call the peak *Denali,* which means "the high one." Located in Denali National Park, the peak is nearly 6200 m high and has several alpine glaciers.

Figure 4–32. Alaska has several geomorphic provinces, including the Alaska Range with its magnificent Denali peak.

Figure 4–33. The Hawaiian Islands comprise a unique province. Their volcanic structure can be studied at Hawaii Volcanoes National Park.

The Hawaiian Province

The flying time from Anchorage, Alaska, to Honolulu, Hawaii, is relatively short, but the difference in geomorphology is great. The Hawaiian Islands, which are the tops of underwater volcanic mountains, have developed over a hot spot below the Pacific plate. The oldest volcanoes are to the northwest of Honolulu. Some have been weathered down to *atolls,* or round islands. The southeastern volcanoes are actively building up the "big island." Southeast of the island of Hawaii are submerged volcanoes that will eventually reach the ocean surface. The islands are all shield volcanoes, with gentle slopes and relatively quiet eruptions.

 **ASK YOURSELF**

Why are there so many mountains in the western provinces?

The Plains and Appalachian Provinces

After a little rest and relaxation on the beaches of Hawaii, it's time to head back east to the geomorphic provinces of America's heartland and eastern coast.

In the center of the continent are great prairies and plains that stretch from the Arctic Circle to the Tropic of Cancer. The coastal plains are primarily located along the Atlantic and Gulf coasts. Although these regions seem to be similar, there are many subtle differences that make each unique. The Appalachian Provinces include the oldest mountains on the continent.

Figure 4–34. The region of great prairies, coastal plains, and ancient mountains

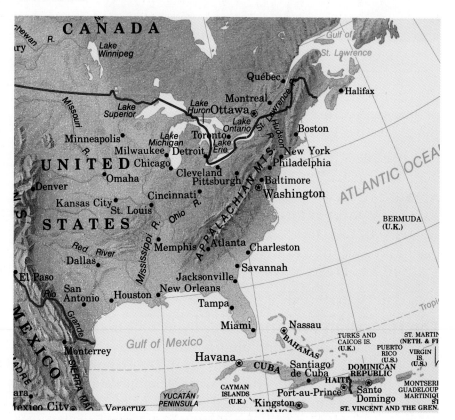

Figure 4–35. The famous Wisconsin Dells are located in a glaciated area of the Great Plains.

The Great Plains and Central Lowlands There are many cities that could be used as jumping off points for a study of the Great Plains—Chicago, St. Louis, Dallas. The Great Plains and the Central Lowlands extend from the Rocky Mountains to the eastern mountains and from Canada into Mexico. Located in the center of the North American plate, they form a vast plain composed mostly of rocks that lie relatively undisturbed by geological events. The five Great Lakes, carved out by Ice Age glaciers, are located on the eastern edge of the province.

The Coastal Plains

Our trip now takes us from the prairies to the oceans. The Coastal Plains, formed of gently sloping layers of rock, are located along the Atlantic Ocean and the Gulf of Mexico. Here young layers of rock blanket the older formations of the interior, and barrier islands line the shore.

In this region is Florida, with its examples of *karst topography*—sinkholes and caverns with beautiful deposits caused by the action of underground water on limestone. Florida has unique topography because much of the state was underwater about 100 000 years ago. For this reason, Florida is often called "the land from the sea."

Figure 4–36. Many karst features, such as these caverns, occur in the Blue Ridge Mountains.

The Blue Ridge Mountains

Next we travel north, into the Blue Ridge Mountains, which extend from Tennessee into Virginia. Although this area must have once been very high, it has since eroded extensively. There are many caverns and two beautiful national parks in the Blue Ridge Mountains: Shenandoah National Park in Virginia, and Great Smoky Mountains National Park in North Carolina and Tennessee. Both provide dramatic landscapes as well as important historical details about the westward growth of the United States.

The Piedmont Plateau

Our next stop lies east of the Blue Ridge Mountains. The city of Charlotte, North Carolina, is in the center of the Piedmont Plateau. *Piedmont* means literally "the foot of the mountain." The rocks of the Piedmont are a complex mixture, showing both folding and faulting. But the area has eroded to such an extent that the underlying rock structure is not reflected in the plateau.

Figure 4–37. Small farms dot the valleys of the Piedmont Plateau.

The Valley and Ridge Province

We continue traveling generally northeast, through northern Virginia and Maryland to Harrisburg, Pennsylvania. This is the Valley and Ridge Province, which provides a classic example of folded mountains that have been extensively eroded. The ridges and valleys run in long, narrow lines from northeast to southwest. The ridges are composed of the hardest rocks; the valleys, the most easily eroded rocks. Shallow rivers, such as the Susquehanna, flow through gaps in the mountains, on their way to the coast.

Figure 4–38. Hardwood forests dominate the hillsides in the Valley and Ridge Province.

The New England Province

From eastern Pennsylvania, we continue northeast, into New York and New England. The rock structure of the New England Province is similar to that of the Piedmont; however, its topography is very different. It includes mountains, such as the Green Mountains of Vermont.

Ice Age glaciers carved and recarved the New England Province, rounding the highest ridges and gouging out river valleys. Acadia National Park in Maine is located in this province. The park offers an excellent opportunity to view landforms shaped by ice and waves.

Figure 4–39. Glaciers and ocean waves helped shape the coast of the New England Province.

The Appalachian Plateau

Next we head southwest again, to West Virginia and the Appalachian Plateau. The plateau forms a large province to the west of the Valley and Ridge Province. The area covered by the Appalachian Plateau is equal in size to the Piedmont Plateau, the Blue Ridge Mountains, and the Valley and Ridge Province combined. Its rock layers are nearly horizontal or only slightly folded, but the entire plateau dips westward, forming a cliff along its eastern edge.

The Ouachita and Ozark Mountains

Our final stop is in Little Rock, Arkansas. Near here, an extension of the Appalachian Plateau pushes into the central lowlands of Arkansas and

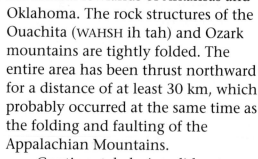

Oklahoma. The rock structures of the Ouachita (WAHSH ih tah) and Ozark mountains are tightly folded. The entire area has been thrust northward for a distance of at least 30 km, which probably occurred at the same time as the folding and faulting of the Appalachian Mountains.

Continental glaciers did not reach as far south as these mountains. Instead, the area's topography is due mainly to stream erosion. Hot Springs National Park is located in the

Figure 4–40. Hills and winding rivers (left) are typical landforms of the Appalachian Plateau. Hot Springs (right) is located in the Ouachita Mountains of Arkansas.

Ouachita Mountains. The spring water in this area is unique because it does not have the offensive odor and taste of sulfur that is associated with most hot springs.

It has been a long trip, but you should have a sense of how forces within the earth act to create the varied landforms of our country. In the next activity you can share some of what you have learned with those who might be thinking of taking a similar trip.

ACTIVITY

What's involved in making an informational brochure?

MATERIALS
paper, colored pencils, rulers, topographical and geological maps

PROCEDURE
1. Imagine that you have been hired to develop a brochure for a park, either in your region or somewhere else. Use your textbook and additional resources to obtain the information you need for your brochure.
2. Design a booklet or brochure to inform people about the geology, topography, and natural history of the area. Your booklet should have at least six pages, including graphics.

APPLICATION
1. How would this brochure be useful to park visitors?
2. In addition to the required information, what else might you include in the brochure to make it more informative and interesting?

 ASK YOURSELF

Why are the rock formations of the Appalachian Provinces so eroded?

SECTION 3 REVIEW AND APPLICATION

Reading Critically
1. Compare and contrast the Rocky Mountain Province to the Pacific Mountain System.
2. Compare and contrast the Columbia-Snake River Plateau to the Appalachian Plateau.

Thinking Critically
3. If you were to draw a cross section of the United States from San Francisco, California, to Norfolk, Virginia, how would the elevations change?
4. What is unusual about the erosion in Acadia National Park?

SKILL Drawing Profiles

▶ **MATERIALS**
• ruler • topographical map • graph paper

▼ **PROCEDURE**

1. A profile is a representation of an object as seen from the side—it is like looking at hills outlined against the horizon. The first step in drawing a profile is to draw a line across the map to show where the profile is to be taken. On the map, the line marked AB crosses the peaks of two hills.

2. Draw a horizontal line on a piece of graph paper. Fold the paper along this line, lay the fold along line AB on the map, and mark the ends of the profile on your graph paper with dots labeled A and B.

3. Next you must determine the vertical scale of your profile. The horizontal scale is determined by the scale of the map. The scale of the map in this exercise is 1:20 000. The vertical scale is your choice. However, if you use a vertical scale that is the same as the horizontal scale, the slopes in your profile will be proportional to the actual slopes of the land, but nearly flat.

4. On your graph paper, draw a vertical line upward from dot A. Make each line along the vertical a 20 m contour interval starting with 0.

5. To mark the positions of the contour lines, place your line AB along line AB on the map. Mark on the horizontal where each contour line crosses line AB. Now you are ready to draw the profile.

6. Find the mark that represents the place where the 20-meter contour line crossed your line AB. Directly above this mark, make a dot on the 20-meter line of your vertical scale. Mark the elevations of other contour lines in the same fashion. Connect the points on your graph paper with a smooth, curving line. You should now have a profile like the one below.

▶ **APPLICATION**
Compare your profile to the topographical map. Does it look as you expected? Explain.

✳ *Using What You Have Learned*
Make a profile of the area between points A and C.

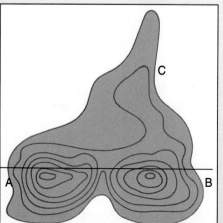

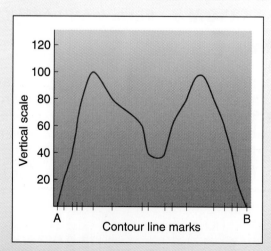

The Big Idea

How we find our way around the world has evolved over time. Through the use of maps—models of the earth's surface—we can represent any spot on the earth fairly well. Topographical and geological maps reveal a wealth of information about the geomorphology of an area. They help us understand how various landforms were created.

Jan Vermeer, The Astronomer, 1668, Louvre, Paris, France.

For Your Journal

Look again at the questions you answered at the beginning of this chapter. How have your ideas about maps changed? Write your new ideas in your journal.

Connecting Ideas

Copy the concept map into your journal and fill in the names of the projections.

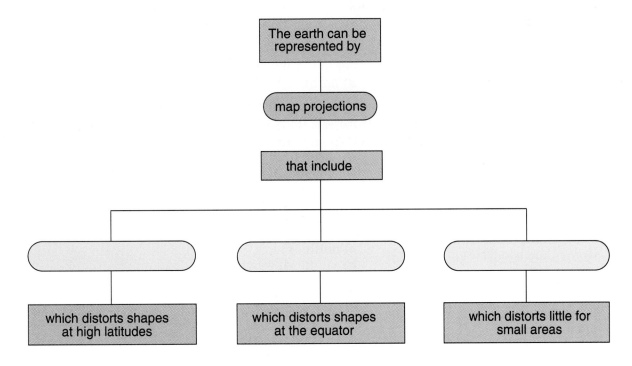

The earth can be represented by

map projections

that include

| which distorts shapes at high latitudes | which distorts shapes at the equator | which distorts little for small areas |

CHAPTER 4 REVIEW

Understanding Vocabulary

Explain how the terms in each set are related.

1. **a)** latitude (107), longitude (107)
   b) map scale (114), legend (116), contour lines (117)
   c) topographical maps (116), geological maps (119), hydrological maps (119)

2. Define the term *cartography* and explain why it was considered an art, rather than a science, in ancient times.

Understanding Concepts

MULTIPLE CHOICE

3. The Mercator projection distorts landforms in the area around the
   a) poles.   **c)** coasts.
   b) equator.   **d)** oceans.

4. Which type of map shows rock structure?
   a) topographical
   b) geological
   c) hydrological
   d) contour

5. On a topographical map, exact elevation is shown by
   a) contour lines.
   b) legends.
   c) bench marks.
   d) scale.

6. Which of the following geomorphic provinces has little in common with the others?
   a) the Basin and Range Province
   b) the Rocky Mountain Province
   c) the Pacific Mountain System
   d) the Colorado Plateau

7. Geomorphology is the study of
   a) caves.
   b) islands.
   c) mountains.
   d) landforms.

SHORT ANSWER

8. Explain why the earliest civilizations had maps.

9. What do the Great Valley and Death Valley have in common?

Interpreting Graphics

Use a hand lens and this topographical map of Clark's Falls to answer the questions that follow.

10. In which direction does Green Fall River flow?
    a) north   **c)** east
    b) south   **d)** west

11. What is the contour interval of this map?
    a) 5 feet   **c)** 15 feet
    b) 10 feet   **d)** 20 feet

12. Which side of the hill northwest of Clark's Falls is the steepest?
    a) north   **c)** east
    b) south   **d)** west

13. What are the elevations of the two bench marks?
    a) 90 and 98   **c)** 199 and 255
    b) 150 and 90   **d)** 90 and 199

Reviewing Themes

14. *Systems and Structures*
Using the world map on page 674, determine what date and time it would be in Atlanta, Georgia, if it is 5:00 P.M., October 15, in Sydney, Australia. Explain how time zones are related to longitude.

15. *Technology*
Draw a simple topographical map of a hill using six contour lines. Make one side of the hill steeper than the other sides.

Thinking Critically

16. These photographs show a very complex system of levees, locks, and dams that protect much of the country of Holland. The components of these structures have been tested on a working scale model. Of what value is a working model for predicting the consequences of various happenings?

17. Explain why the eastern mountains of the United States are more eroded than the western mountains.

18. If you visited Los Angeles, you would be as far from the equator as what major city in the Southern Hemisphere?

19. Why can contour lines never cross?

Discovery Through Reading

Brown, Stuart F. "A New View of America." *Popular Science* 241 (November 1992): 86–89. This article explains how the U.S. Geological Survey has used digital image-generation technology to produce the most accurate topographical map in existence. The map was based on data representing over 12 million elevation measurements!

Glacier National Park

Glacier National Park is a majestic beauty of many contrasts. Here one can see glacier-carved mountain peaks that rise over 3000 m to form horns and long, pointed ridges. Along upper valley walls are hanging valleys, where tributary glaciers once joined the main glaciers of the valley. Beautiful waterfalls plunge hundreds of meters from these upper walls into the U-shaped valleys.

Waterfall in a hanging valley

Grinnell Glacier

Contrasting Sights

Waterfalls spill down from lakes in the hanging valleys into the cold, blue lakes of the valley floors. These lakes with their sapphire-blue color stand in sharp contrast to the glistening white of the 50 glaciers found throughout the park. These glaciers still erode the valleys, but their effects can not compare to those of the massive bodies of ice that were here during the last ice age.

Today, the largest glacier in Glacier National Park is Grinnell Glacier. Each winter this glacier moves down the slope about 10 m and then melts back about the same distance in the summer. The piles of rubble and loose rock at its base contrast with the massive rock walls that form towering peaks and ridges above the glacier. Many of its valleys are lined with moraines of long, rolling mounds.

Contrasting Climates

The drainage basins of the Pacific Ocean and Gulf of Mexico begin at the Continental Divide, which separates Glacier National Park. The high mountains of the divide also create distinctly contrasting climates on the east and west sides of the park. The western side receives warm, moist winds, and vegetation thrives in the moderate temperatures. As the air here rises to

cross the mountains, it cools and loses much of its water content. This results in the eastern side of the park having a drier, colder climate. Vegetation there struggles to survive the cold winter winds and the hot, dry winds of summer.

Rocks of Time

The history of Glacier National Park can be seen in its rocks. They show alternating periods of calm sedimentation and times of violent upheaval. The rocks show evidence of extremely old deposits as well as relatively recent changes.

Most of Glacier National Park's rocks are about a billion years old. Ancient red and gray-green mudstones and siltstones reveal amazing details about the environment in which they formed. Some have ripple marks made by wave motion, while others show mudcracks and impressions made by raindrops. There are rocks that contain stromatolites. The stromatolites indicate that there was a time when this part of the continent was a shallow mudflat at the edge of a sea. The mudcracks indicate that the sea must have dried out on

Purcell sill lava formation

occasion. There are also thick limestone deposits in this area, indicating periods when it was a deep-water marine environment.

Even evidence of flowing lava is found in the park. The Purcell sill, a thin black layer of igneous rock, appears between layers of white limestone high above Grinnell Glacier in the valley wall. Iron-rich magma once forced its way into cracks in the limestone, leaving these dark vertical dikes and horizontal layers.

Isolated Block

As the Rocky Mountains were forming about 70 million years ago, a large block of land slid eastward as a low-angle thrust fault. For years scientists were confused by this formation. They could not figure out why the mass of billion-year-old rock called Chief Mountain was on top of rocks only 100 million years old. Chief Mountain was actually an isolated mountain; the rest of the thrust block it was a part of had eroded away. This block moved eastward about 55 km and came to rest on top of layers of much younger rocks.

These contrasting features make Glacier National Park a beautiful and interesting place to visit. The conservationist John Muir said, "Give a month at least to this precious reserve. The time will not be taken from the sum of your life. Instead of shortening it, it will indefinitely lengthen it, and make you truly immortal." ◆

Chief Mountain thrust

VOLCANO!

by Caroline Wakeman Evans from *Ranger Rick*

It's 5:00 A.M. and I'm already wide awake. I live on a volcano—and when my windows start to rattle, I know it's time to get up.

I'm a ranger in Hawaii Volcanoes National Park, on the big island of Hawaii. That's why I live and work on a volcano. My home is on top of the most active volcano in the world: Kilauea.

Quickly I dress and go out on my porch. The clouds are glowing red. And I hear a deep roar like the noise of a jet engine. Kilauea's erupting again!

LOTS OF LAVA

The roaring and shaking are all part of an eruption. When Kilauea erupts, super-hot melted rock called lava comes out of the volcano from deep inside the earth. The lava may flow faster than a river. It also may shoot into the sky, a fiery fountain up to 10 times higher than the Statue of Liberty. And sometimes, like today, it does both. At other times the lava just oozes, slowly stretching its dark fingers over the land.

Lava can burn or bury everything in its path: trees, roads, even houses. But as it cools and hardens, lava becomes new land. In fact, the whole island of Hawaii is made of lava from Kilauea and four other volcanoes.

In the last 200 years, Kilauea has erupted over 60 times. But it's one of the

Kilauea's erupting again!

safest active volcanoes in the world. Scientists can forecast where and when the lava is likely to flow. That's why people can live on the volcano, and visitors can come to watch eruptions!

FLYING OVER KILAUEA

This morning I report to work dressed in heavy boots, fireproof clothes, and a flight helmet. My job is to take a helicopter ride over the eruption to see the action firsthand. Park visitors won't be able to see the eruption up close today, so I want to be able to describe to them what's going on.

The lava is shooting out of a vent, or opening, on the east side of the volcano. That's 10 miles (16 km) from the visitors' center, and no roads go near it. But people will be able to see the fiery fountain in the distance as they drive back down the mountain.

From the helicopter, I'll look for places people can see lava that flows down the mountain. If it's flowing slowly near a road, visitors can walk close to the lava and feel its heat.

As the pilot flies the helicopter near the fountain of lava, I get my first whiff of the rotten-egg smell of sulfur dioxide. This gas comes from inside the volcano. It makes my throat scratchy, and I start to cough. The ride gets rougher as the helicopter bounces around on the currents of hot air. Suddenly

the helicopter is lifted up, then just as fast it drops down. My stomach somersaults and I'm a little scared. But soon we fly away. The air is now calmer and so am I.

I hang on for a hot and scary ride.

WORKING ON THE EDGE

I can see scientists working down below, near lava that's flowing away from the vent. One scientist is wearing thick leather gloves and holding a rock pick. She dashes up, dips her pick into the taffylike lava, and drops a red-hot blob into a can of water. When the sample has cooled, she'll find out what chemicals are in it. The chemicals will help her tell how far

The lava burns or buries everything in its path.

A scientist grabs a red-hot sample.

ger. And I wonder whether the kids are in school or at home helping their parents move their belongings to a safer place.

In the last few months lava has burned several houses that were in its path. I remember the kids I met who had lost their homes. They were brave. They knew their homes were

inside the earth the melted rock came from.

Another scientist is almost hidden in a cloud of gas near the vent. He's collecting samples of the gas. This can be dangerous work! Even though he has a gas mask on, the gases could become so thick that he couldn't breathe. But the gas samples are important: They help scientists estimate how much lava is still in the volcano.

I can see a friend of mine, Tina, standing on cool, hard lava at the edge of the lava river. She's holding a long metal rod that looks like a fishing rod. She dips it into the hot, flowing lava to take the lava's temperature.

I hold my breath as I remember a close call that Tina had recently: She thought she was standing on a bank of solid lava. But suddenly it cracked and started to split apart. As Tina leaped away, the place where she had been standing melted into the super-hot lava river. If she

hadn't moved so quickly, she could have fallen in and been badly burned or even killed. Today, though, she takes the river's temperature and calmly walks away.

The scientists also estimate the lava river's depth, its width, and its speed. They radio their findings back to the lab on top of Kilauea. On the helicopter radio I hear a scientist at the lab call them back: The measurements mean that this lava river might cause big trouble. It might flow over houses farther down the mountain.

LIVING WITH LAVA

The pilot circles the helicopter around the fountain one more time, and then we fly down the mountain. The lava river is flowing slowly and seems a long way from houses and the highway. Still, I'm sure that people who live in its path are being warned now of the dan-

I have a lot to tell the visitors.

gone forever, and they would be living with relatives for a while. But they love the island that Kilauea and the other four volcanoes have created. And they love to see Kilauea's beautiful eruptions.

The pilot heads the helicopter back to the top of the volcano, and we land smoothly near the visitors' center. I can see the parking lot is already filled with cars. People, curious about what Kilauea's been up to, are waiting for my report. So I quickly take off my flight helmet and head over to them. I have a lot to tell them about that fountain of fiery lava. ◆

143

James Hutton (1726–1797)

In his book *Theory of the Earth*, James Hutton suggested that the earth was much older than most scientists thought. He hypothesized that the same rock-forming processes that can be observed today were working in the past and will continue to work in the future.

James Hutton lived most of his life in Edinburgh, Scotland. His interest in science led him to investigate many parts of his environment. Hutton was a careful observer. He studied the land and drew conclusions about the origin of the rocks and the processes that changed them. Eventually he

constructed his "theory of the earth." He believed that earth processes were continuous cycles. To support this theory he described the processes that had occurred. As a result, he made major contributions to the fields of geology, agriculture, chemistry, physics, and philosophy.

Hutton's theory has been so thoroughly tested and supported by experimental data that it forms the basis for the science of geology. His theory proved to be so important to the development of the field that Hutton is now called the founder of modern geology. ◆

Lisa Rossbacher (1952–)

Lisa Rossbacher's knowledge of erosion and landforms here on Earth is helping us identify patterns that show that erosion takes place on other planets.

Lisa Ann Rossbacher was born in Fredericksburg, Virginia. After graduating from Dickinson College with a bachelor's degree in science, Rossbacher attended the State University of New York at Binghamton, where she earned her master's degree. Rossbacher continued her studies at Princeton University, where she earned her doctorate in geology in 1983.

Rossbacher was recently a professor of geology at California State Polytechnic Uni-

versity. Her work involved the study of landforms and geological processes. Using maps, photographs, and on-site inspections, Rossbacher recognized the forces of nature acting on landforms. Her expertise has aided NASA in identifying possible examples of erosion on the planet Mars.

The work Rossbacher does helps identify areas prone to flooding, landslides, earthquakes, sinkholes, and pollution. This information alerts communities to geological hazards and provides the knowledge we need to use land wisely. ◆

Karen McNally, Seismologist

Every thirty seconds or so there is an earthquake somewhere in the world. Scientists record about one million earthquakes a year. Fortunately, most of them occur beneath the ocean and are so slight that it takes special instruments to detect them. However, a great earthquake occurs once every two or three years, and can result in a huge loss of life and property. Through the years, our knowledge about earthquakes has increased, and so have our questions.

Karen C. McNally is an earth scientist. She has a doctorate in geophysics, the science that deals with the physics of the earth, including weather, winds, tides, volcanoes, and earthquakes. She is also a seismologist, a scientist who studies seismic activity, or earthquakes.

7:00 A.M. Few of McNally's days are routine. She arrives at her office

early and examines data supplied by the National Earthquake Information Service in Golden, Colorado. All earthquake stations in the world send data here. McNally notices a sudden drop in seismic activity around the Cocos Plate in Central America. This drop in activity is a signal that an earthquake might be about to happen in the area.

8:00 A.M. After analyzing computer data from earthquake stations throughout South America, McNally puts the finishing touches on the lecture she will give to her geophysics classes. She plans to tell them about research being conducted on hidden, or "fold," earthquakes. These earthquakes do not rupture the earth's surface, and take place under folds where the earth is layered or buckled upward. "All earthquakes are different," McNally says. "Some result in landslides, others may cause tidal waves called tsunami. Seismic tremors can move horizontally or vertically

across the earth. They are not all alike; therefore, analyzing and predicting each one is different."

8:30 A.M. A call comes from an architect who is designing a new building in Los Angeles. He is concerned that the building be structurally safe, since it will be located along the San Andreas fault. There is also a call from an oil company. Seismographs help in the search for oil and natural gas. McNally also takes time to accept an invitation to speak at a conference. She is often asked by governments and groups from around the world to give talks about earthquakes.

9:00 A.M. McNally is in the classroom teaching geophysics. "Seismologists do a lot of writing," she explains. "It would be good if students had a good background in writing."

1:00 P.M. McNally contacts a scientist with the U.S. Geological Survey, who has the responsibility of conducting and coordinating earthquake predictions. She explains, "By studying historical information both in geology and in reports of measurements of quakes, we develop our predictions." The two scientists talk about the latest activity in Central America. There could be an earthquake in that area in anywhere from a few days to a few hours.

Sharing information as soon as possible is critical.

1:45 P.M. McNally meets with a group of scientists to plan ways to get grants for earthquake research. Making sure there is enough money to continue research involves many hours of paperwork, telephone calls, and speaking engagements.

4:00 P.M. A call comes from the government in Mexico. Their seismographs have picked up a change in seismic activity. Based on all the information she has received in the past 24 hours, McNally makes the decision to take a team of scientists to Mexico. Within a day and a half, McNally and 30 other scientists will be in Mexico, ready to set up the seismographs and make detailed measurements.

6:30 P.M. McNally meets with one of her graduate students to discuss a project he is doing on the March 1986 earthquake in California. She is busy preparing for the trip to Mexico, but she is very excited about this project and does not want her student to fall behind in his research.

9:30 P.M. McNally's work day has ended. At this time tomorrow, she will be in Mexico. There is always the chance that the earthquake she is "chasing" will be a minor one, or that it will not be an earthquake at all. Whether it is or not, the efforts of her team and of scientists from around the globe could save many lives. They could also lead to a better understanding of our earth ◆

Discover More

For more information about careers in seismology, write to the

National Earthquake Center
M.S. 967
Box 25046
Denver, CO 80225

SEISMIC RECORDS

For years people have tried to find ways to record the strength of earthquakes. Over time, scientists developed the seismograph. Not all seismographs are alike. There have been different types of seismographs used over the years.

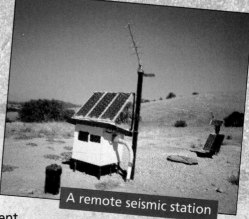
A remote seismic station

Types of Seismographs

One type of seismograph uses a suspended pen, which rests lightly on a rotating, paper-covered drum. The drum is anchored in rock, so that any movement of the earth causes the pen to make zigzag lines on the drum. This type of seismograph records vertical movement. A second type of seismograph uses a pen attached to an arm that swings freely. This seismograph records horizontal movements. Both types are necessary to accurately measure earthquake magnitude.

In the 1930s, Victor Hugo Benioff designed a seismic instrument that was similar to a telephone. A continuous electrical current was passed through the instrument. When the earth vibrated in an earthquake, this instrument showed tiny variations in current, which could be recorded. Modern seismographs are based on this same principle.

Improved Methods

Scientists' ability to interpret seismograms has been greatly improved by advances in technology over the past 20 years. The orientation of a fault plane, the type of movement, and even the direction of movement along the fault can now be determined. The speed and accuracy of recording seismographic data has also improved with the use of computers.

Scientists at the National Earthquake Center

Seismic recording stations are located around the world. These stations report their data to the National Earthquake Information Center in Golden, Colorado.

Approximately 60 000 seismic records are fed into the computers at the center each month. This information is analyzed by scientists to obtain a better understanding of earthquakes and to aid in the development of new and better methods of predicting earthquakes. ◆

A computer-enhanced seismogram

147

EARTH'S RESOURCES

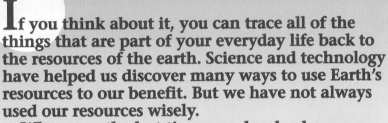

If you think about it, you can trace all of the things that are part of your everyday life back to the resources of the earth. Science and technology have helped us discover many ways to use Earth's resources to our benefit. But we have not always used our resources wisely.

When was the last time you drank a beverage from a glass bottle or an aluminum can? What did you do with that bottle or can when you were finished? And where did that bottle or can end up?

There is still a lot we need to learn. We must all use our resources wisely.

Science PARADE

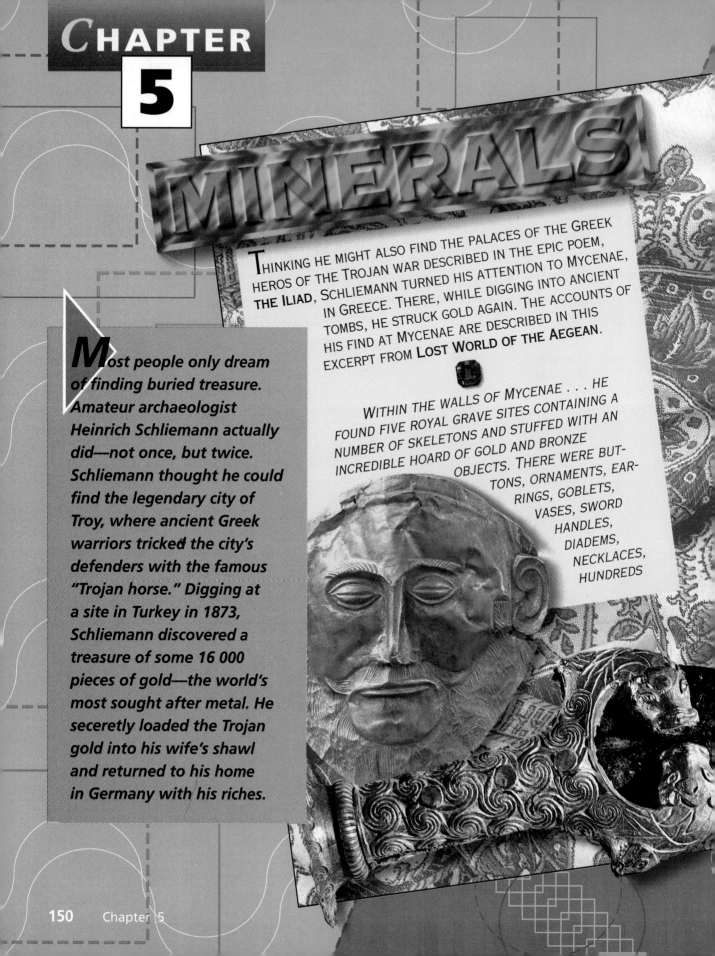

CHAPTER 5

MINERALS

THINKING HE MIGHT ALSO FIND THE PALACES OF THE GREEK HEROS OF THE TROJAN WAR DESCRIBED IN THE EPIC POEM, THE ILIAD, SCHLIEMANN TURNED HIS ATTENTION TO MYCENAE, IN GREECE. THERE, WHILE DIGGING INTO ANCIENT TOMBS, HE STRUCK GOLD AGAIN. THE ACCOUNTS OF HIS FIND AT MYCENAE ARE DESCRIBED IN THIS EXCERPT FROM **LOST WORLD OF THE AEGEAN**.

WITHIN THE WALLS OF MYCENAE . . . HE FOUND FIVE ROYAL GRAVE SITES CONTAINING A NUMBER OF SKELETONS AND STUFFED WITH AN INCREDIBLE HOARD OF GOLD AND BRONZE OBJECTS. THERE WERE BUT-TONS, ORNAMENTS, EAR-RINGS, GOBLETS, VASES, SWORD HANDLES, DIADEMS, NECKLACES, HUNDREDS

Most people only dream of finding buried treasure. Amateur archaeologist Heinrich Schliemann actually did—not once, but twice. Schliemann thought he could find the legendary city of Troy, where ancient Greek warriors tricked the city's defenders with the famous "Trojan horse." Digging at a site in Turkey in 1873, Schliemann discovered a treasure of some 16 000 pieces of gold—the world's most sought after metal. He secretly loaded the Trojan gold into his wife's shawl and returned to his home in Germany with his riches.

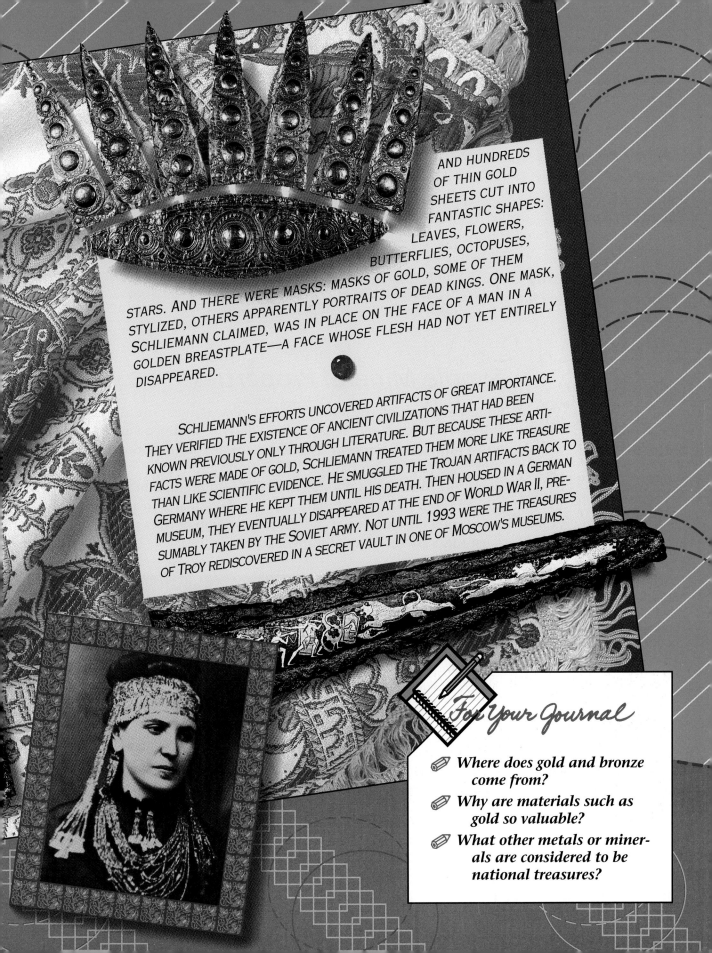

AND HUNDREDS OF THIN GOLD SHEETS CUT INTO FANTASTIC SHAPES: LEAVES, FLOWERS, BUTTERFLIES, OCTOPUSES, STARS. AND THERE WERE MASKS: MASKS OF GOLD, SOME OF THEM STYLIZED, OTHERS APPARENTLY PORTRAITS OF DEAD KINGS. ONE MASK, SCHLIEMANN CLAIMED, WAS IN PLACE ON THE FACE OF A MAN IN A GOLDEN BREASTPLATE—A FACE WHOSE FLESH HAD NOT YET ENTIRELY DISAPPEARED.

SCHLIEMANN'S EFFORTS UNCOVERED ARTIFACTS OF GREAT IMPORTANCE. THEY VERIFIED THE EXISTENCE OF ANCIENT CIVILIZATIONS THAT HAD BEEN KNOWN PREVIOUSLY ONLY THROUGH LITERATURE. BUT BECAUSE THESE ARTIFACTS WERE MADE OF GOLD, SCHLIEMANN TREATED THEM MORE LIKE TREASURE THAN LIKE SCIENTIFIC EVIDENCE. HE SMUGGLED THE TROJAN ARTIFACTS BACK TO GERMANY WHERE HE KEPT THEM UNTIL HIS DEATH. THEN HOUSED IN A GERMAN MUSEUM, THEY EVENTUALLY DISAPPEARED AT THE END OF WORLD WAR II, PRESUMABLY TAKEN BY THE SOVIET ARMY. NOT UNTIL 1993 WERE THE TREASURES OF TROY REDISCOVERED IN A SECRET VAULT IN ONE OF MOSCOW'S MUSEUMS.

For Your Journal

✐ Where does gold and bronze come from?

✐ Why are materials such as gold so valuable?

✐ What other metals or minerals are considered to be national treasures?

Mineral Properties

Materials in the earth's crust are similar, in many ways, to materials in a kitchen—each substance has its own set of properties. In a kitchen, these properties might be taste and color. Some substances are the same in all kitchens. Even if you have two different brands of table sugar, for example, you can't tell one from another by its taste or color. All table sugar is the same and has the same properties. Similarly, there are materials in the earth's crust with properties that are the same no matter where they are found. These substances are called *minerals*.

Simple Mineral Properties

Some people think the terms *rock* and *mineral* mean the same thing—but they don't. What *is* the difference between a rock and a mineral? A **mineral** is a solid, naturally formed inorganic substance that always has the same composition and the same properties. A **rock** is a naturally formed combination of minerals and other natural materials. A mineral compares to a rock as a tree compares to a forest. A forest is made up of trees and most rocks are made up of minerals.

Figure 5–1. This sample of quartz has the same composition and properties as all quartz samples.

Just as trees come in different sizes and shapes, minerals also differ in size and shape. In fact, each mineral has its own definite set of properties. And just as sugar can be identified by the properties of taste and color, a mineral can be identified by its unique properties.

A detailed analysis of a mineral is usually done in a laboratory. However, there are some simple physical properties that can be determined easily and used to identify a mineral sample. In the next activity you can observe the most obvious property of minerals.

DISCOVER BY *Observing*

Your teacher will give you several samples of the mineral quartz. What obvious differences do you notice in the samples? If they are all quartz, why are there differences in the samples? ✎

Color   Probably the most obvious property you will notice about a mineral is its *color.* Unfortunately, this is also one of the least dependable properties you can use to identify a mineral. If you have ever used food coloring in cake frosting, you know how little coloring it takes to make a brightly decorated cake. Natural coloring agents, such as iron and manganese, when combined with a mineral, can change its color as food coloring changes the color of frosting.

Figure 5–2. These minerals are all samples of quartz. The color differences are due to various coloring agents.

Quartz is usually clear or white, but with various coloring agents, it can be pink, tan, red, black, yellow, or purple! How can you identify quartz? Color alone is obviously not reliable, so you must use other properties. In the next activity, you can learn a simple test for a more dependable mineral property.

Your teacher will give you two mineral samples and a rough, white tile. One mineral, pyrite, looks like gold. The other mineral, hematite, looks like silver. Rub each mineral sample across the white tile. How does the color of each mineral's mark on the tile compare to its color? ✐

Streak   If you rub a piece of chalk on a sidewalk, it leaves a mark. The color of the powder that is left when a mineral is rubbed on a rough surface is called the mineral's **streak.** Although the color of the streak may be different from the color of the mineral sample, all samples of a specific mineral have the same streak. Therefore, streak is often a better property for identifying minerals than color is.

Figure 5–3. The girl is preparing a mineral streak. Notice that these two differently colored mineral samples produce a similar streak.

Luster   Have you ever admired the appearance of a shiny new car? What you are admiring is the luster of the car's surface finish. The **luster** of a mineral refers to the appearance of its surface in light. There are two types of luster: metallic and nonmetallic. If a mineral shines like gold, copper, or silver, then its luster is *metallic.* A nonmetallic luster can be described using terms such as *waxy, pearly, glassy, dull,* or *brilliant.* What common objects can you think of that might be examples of these lusters?

Figure 5–4. Luster is the way a mineral's surface shines in reflected light.

ASK YOURSELF

What is the difference between color and streak?

Additional Properties

The mineral properties described so far can be observed without laboratory equipment. Additional properties can be determined by using simple tools.

Density   If someone handed you two cups of equal size, one filled with salt and the other filled with iron, you could probably guess which cup contained the iron, even with your eyes closed. The cup of iron is heavier than the cup of salt because iron is more dense than salt. **Density** is a measure of how much matter, or mass, an object contains for its size, or volume. Density can be calculated by dividing the mass of the object by its volume. A chunk of gold, for example, has eight times as much mass as a chunk of salt of the same size because its density is much greater. Density is a characteristic property of all matter; it doesn't change with the size of the sample.

Hardness If you were to list 10 objects in order from softest to hardest, chances are you would place a diamond at the bottom of your list. A diamond is the hardest substance on the earth. At the other end of the list, at least in terms of minerals, would be talc—the mineral from which talcum powder is made. All other minerals fall somewhere between talc and diamond.

Table 5-1

Mohs' Scale of Mineral Hardness

Hardness—mineral	Common objects—hardness
1—Talc	
2—Gypsum	
	Fingernail—2.5
3—Calcite	
	Penny—3.5
4—Fluorite	
	Common nail—4.5
5—Apatite	
	Glass plate—5.5
6—Orthoclase	
	Steel file—6.5
7—Quartz	
8—Topaz	
9—Corundum	
10—Diamond	

Hardness is one of the most important properties used to identify minerals. A set of ten standard minerals is used to make a scale to measure the hardness of all minerals. This scale, called *Mohs' Scale of Mineral Hardness,* is shown in Table 5–1. Talc, the softest mineral, is assigned number 1, and diamond, the hardest mineral, is assigned number 10. Other common minerals are assigned numbers 2 through 9. A range of numbers is sometimes used to describe the hardness of a mineral that does not fit one of the standards exactly. For example, a mineral that can scratch calcite (number 3) but cannot scratch fluorite (number 4) would have a hardness between 3 and 4.

Frequently, the minerals of the Mohs' scale are not available for comparison, so geologists have created a list of common objects that can be used to test a mineral's hardness. If a certain mineral can scratch a penny, for instance, its hardness is greater than 3.5. If a common nail can scratch that same mineral, its hardness is less than 4.5. Therefore, its hardness is somewhere between 3.5 and 4.5. In the next activity, you can make and use a modified scale of hardness.

ACTIVITY

How can you make and use a scale of mineral hardness?

MATERIALS
pencil, chalk, penny

PROCEDURE
1. Try scratching the wood of the pencil with your fingernail. Which is harder, the wood or your fingernail?
2. Try the same test with the piece of chalk. Which is harder, the chalk or your fingernail?

3. List, in order of increasing hardness, your fingernail, the wood of the pencil, the lead of the pencil, the piece of chalk, and the penny.

APPLICATION
Suppose you are given two minerals. Both look like calcite, but one is not. Using what you know about mineral hardness, how can you determine which mineral is really calcite?

Cleavage and Fracture   Another useful property for identification is the way a mineral breaks. Some minerals, such as quartz, always break with irregular, uneven *fractures.* Others break along flat planes. Minerals that break along flat planes are said to have **cleavage.** Minerals cleave in one, two, three, or more directions. Mica cleaves in one direction, as the photograph on the next page shows. It breaks into thin, flat sheets.

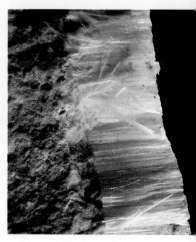

Figure 5–5. The cleavage of calcite (left) is along three planes, while mica (center) forms thin, flat sheets. Asbestos (right) fractures, producing long, thin splinters.

Minerals can fracture in different ways also, and some ways can be helpful in identification. Some fractures have a curved surface, similar to the inside of an eggshell. Other fractures produce long splinters or fibers like the asbestos in the photograph. A mineral with cleavage will break in such a way that the broken pieces have the same general shape as the original piece. Mineral grains that fracture, such as quartz, break into irregular pieces that do not look like the original piece. In the next activity, you can determine whether a common mineral cleaves or fractures.

DISCOVER BY *Observing*

Pour some halite (table salt) onto a dark piece of paper. Using a hand lens, observe the shape of the mineral grains. In your journal, draw some of the halite. How many sides does each halite grain have? With the point of a sharp pencil, break some of the halite into smaller pieces. Is the shape of the broken halite grains the same as the original piece? Does halite cleave? If so, how many cleavage directions does it show? ✑

Crystals Cleavage should not be confused with crystal shape. Cleavage is a property of the way a mineral breaks, while crystal shape is a property of the way a mineral grows. When minerals have plenty of space to grow, they form *crystals*.

Each mineral has a characteristic crystal shape that is useful in its identification. However, large and perfectly formed mineral crystals are not often seen, because most of the time the crystals grow in limited space. Some examples of mineral crystal shapes are shown in Table 5–2.

Mineral	Geometry	Mineral
Galena	< Cubic	Chalcopyrite
	Tetragonal >	
Quartz	< Hexagonal	Olivine
	Orthorhombic >	
Gypsum	< Monoclinic	Microcline
	Triclinic >	

Table 5-2 **Crystal Shapes**

Unique Properties Some minerals show unique properties that can be useful for identification. A form of magnetite called *lodestone,* for example, is magnetic—a piece of lodestone will attract iron objects just as a magnet does.

Figure 5–6. Lodestone will attract a variety of iron objects.

Calcite can be identified by its chemical reaction with weak acids. If a weak solution of acid is dropped onto calcite, the acid begins to bubble. Other minerals give off light when exposed to ultraviolet light or X-rays. But properties such as these are useful for identifying only a small number of the more than 3300 kinds of minerals.

 ASK YOURSELF

What simple tools are needed to determine the hardness of many minerals?

SECTION 1 *REVIEW AND APPLICATION*

Reading Critically

1. Describe the properties of color and luster.
2. How is the hardness of a mineral determined?

Thinking Critically

3. How can you distinguish between crystal shape and cleavage?
4. Why is color alone not a good property for identifying minerals?

SKILL Using a Classification Key

▶ MATERIALS

- galena ● Mineral Classification Key, page 670 ● streak plate
- penny ● common nail ● glass plate ● steel file ● samples of pyrite, chalcopyrite, sphalerite, magnetite, quartz, hematite, talc, and calcite

Classifying objects into groups as a means of identification is a technique that scientists often use. To speed up the identification process, information is often organized into a classification key. At each step in a mineral classification key, one property is identified and minerals not meeting the criteria of that property are eliminated. The process continues until the sample is identified.

▼ PROCEDURE

1. Obtain a sample of galena from your teacher. Open your book to the Mineral Classification Key on page 670. Using the galena sample, follow each step in the classification process.
2. Note the mineral's luster. Galena belongs in Category I because of its metallic luster.
3. Determine the mineral's streak. Galena's black streak puts it into Subcategory A.

4. Test the mineral's hardness. Galena may or may not be scratched by your fingernail because its hardness is close to 2.5. Galena belongs in Group 1.
5. Determine the cleavage and any other important characteristics of the mineral sample. Because of its cubic cleavage, galena can be selected from among the other minerals in Group 1. Record the properties of galena in Table 1.

TABLE 1: MINERAL PROPERTIES				
Sample	**Luster**	**Streak**	**Hardness**	**Cleavage**
1. Galena				
2.				
3.				
4.				
5.				

▶ APPLICATION

1. Why is color an unreliable characteristic for classification?
2. What are the advantages and disadvantages of this type of classification key?
3. Which properties are most useful in classifying minerals with this key?

※ Using What You Have Learned

Obtain other samples of minerals from your teacher, and classify them using the key. Extend the table of mineral properties, and fill it in as you classify your samples. (Hint: Since you know what samples you need to identify, you can eliminate minerals that are not on your list.)

Atoms, Elements, and Compounds

Objectives

Describe the parts of an atom.

Explain the relationship between atoms and elements and between compounds and molecules.

Justify the placement of elements in certain groups on the Periodic Table.

Why are diamonds so hard? Why are sapphires blue and rubies red? Why is copper a good conductor of electricity? What makes each mineral unique? To find the answers to these questions, you need to peer inside a mineral to see how it is put together. You must study the building blocks from which minerals are made.

Atoms and Elements

Every mineral, as well as every person, every table, every chair, every book, and every chalkboard, is made of very small particles called *atoms*. So, too, is all the water on the earth, all the air around the earth, all the other planets, the sun, the moon, and the stars. In fact, all matter, everywhere, is made of atoms. Atoms are the building blocks of matter.

Figure 5–7. Everything that you see in this picture is made of atoms.

What do atoms look like? How big are they? Atoms are so small that the tiniest speck of dust you see floating in the light of a window contains millions of atoms. Imagine the smallest thing you have ever seen—maybe even something you've seen through a microscope. Whatever you are imagining, it is made of thousands and thousands of atoms. Now imagine how many atoms there must be on the earth! The next activity can help you to do this.

ᴄᴏᴠᴇʀ ʙʏ *Doing*

Take a piece of scrap paper, and tear it in half. Now tear the half in half. Continue this until you cannot tear the paper anymore. Of course this is still far bigger than an atom. Now imagine taking the tiniest piece of paper and tearing it in half 1000 more times. If you could, you would now be approaching something close to the size of an atom. But these are the atoms from *one* sheet of paper only! ✐

There are many different kinds of atoms. Each kind of atom forms an element. **Elements** are substances made of only one kind of atom. There are 91 different elements that occur naturally on Earth. In addition to the natural elements, 18 other elements are known. If they do not occur naturally, where do you think they come from?

In nature, some minerals, such as gold, silver, and copper, often occur as uncombined elements. Most minerals, however, are combinations of two or more elements. The photograph shows a purple mineral, called *fluorite*. It is made of two elements—fluorine and calcium. Combined, these two elements form the mineral fluorite. Uncombined, fluorine is a yellow-green gas, and calcium is a silvery metal.

Even a single atom of fluorine or calcium has all the characteristics of its element. But if an atom of fluorine or calcium is broken down, it is no longer an element. And atoms can be broken down. Even though atoms are small, they are not the smallest particles that scientists know about. Atoms are made up of subatomic particles called *protons, electrons,* and *neutrons.*

Subatomic particles do not have the characteristics of the elements from which they come. Think of an atom as being like a stool. If someone offered you a stool leg to sit on, would you consider this a stool? Not likely! A stool must have several legs and a seat. Likewise, an atom must have a proton and an electron. There are often neutrons in atoms as well, just as there

Figure 5–8. Calcium, a solid, and fluorine, a gas, combine to form the mineral fluorite.

are often backs on stools. The protons, neutrons, and electrons of fluorine are the same as the protons, neutrons, and electrons of calcium. Calcium and fluorine are different from each other because they contain different numbers of subatomic particles. In fact, all 109 elements contain different numbers of these particles.

Protons have a positive electrical charge. **Electrons** have a negative electrical charge. You may already understand something about the terms *positive* and *negative* from working with magnets. Opposite magnetic poles attract each other. In atoms, protons and electrons attract each other. This helps to hold atoms together. **Neutrons** have no electrical charge.

Figure 5–9. A computer-generated model of an atom, with protons and neutrons in the center, surrounded by electrons in a cloud

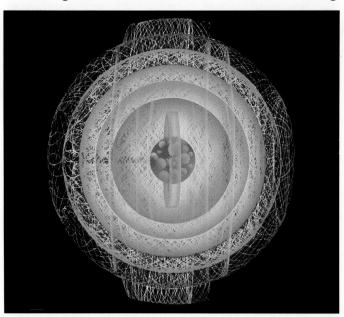

Protons and neutrons are similar in mass. Together they form the core of an atom. This core is called the *nucleus.* Most of the mass of an atom is in the nucleus. Neutrons help separate protons. Remember what happens when like magnetic poles come together. Neutrons keep protons from tearing an atom apart. Electrons have very little mass. They are in constant motion around the nucleus, forming a negatively charged cloud.

Atoms of different elements contain different numbers of subatomic particles. But all the atoms of any particular element have the same number of protons and electrons. And within any individual atom, the number of protons is the same as the number of electrons. Since protons are positively charged and electrons are negatively charged, equal numbers of protons and electrons make for electrically balanced atoms. In the next activity, you can try making models of some simple atoms.

Use colored pencils to draw flat models of atoms with from 1 to 15 protons. An atom of the element sulfur, with 16 protons and 16 neutrons, is shown in the diagram. As a rule of thumb, you will need the same number of neutrons as the number of protons in your atom. Notice the orbitals around the sulfur nucleus. The orbital nearest the nucleus has only 2 electrons, while the next orbital can have up to 8 electrons. All your drawings of atoms with more than 2 electrons should have at least 2 orbitals. Which drawings will have a third orbital? Why do you think the number of electrons in an orbital is limited? ✎

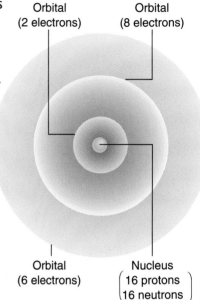

Orbital (2 electrons)
Orbital (8 electrons)
Orbital (6 electrons)
Nucleus (16 protons 16 neutrons)

▼ **ASK YOURSELF**

Which subatomic particles must be present in all atoms?

Compounds

Many atoms combine to form compounds. A **compound** is a combination of two or more different elements. Most minerals are compounds. Compounds exist in any of three states—solid, liquid, or gas. Water, a simple compound, can exist in any of the three states, but minerals are always solids. The mineral fluorite, for example, is a compound called *calcium fluoride*. But how does a compound form?

Compounds form as atoms bond together. There are two basic types of bonds: ionic and covalent. An *ionic bond* forms when electrons move from one atom to another. The bonds in fluorite, for example, are ionic bonds. Atoms of calcium give electrons to atoms of fluorine. The bonds in many common minerals are ionic. Halite, or table salt, forms ionic bonds—sodium atoms give electrons to chlorine atoms.

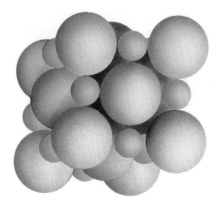

Figure 5–10. Halite contains atoms of sodium and chlorine, held together with ionic bonds.

A *covalent bond* is formed when atoms share electrons. Many common compounds, such as water and carbon dioxide, have covalent bonds. When atoms share electrons, they form molecules. A **molecule** is a group of two or more atoms bonded together by covalent bonds. In water molecules, two hydrogen atoms and one oxygen atom share electrons. In the carbon dioxide molecule shown below, which elements and how many atoms of each share electrons?

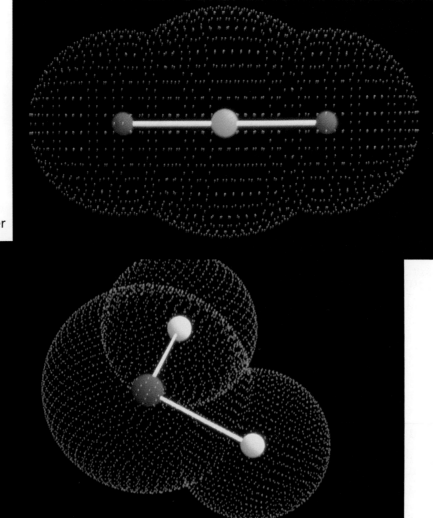

Figure 5–11.
Computer-generated models of carbon dioxide (top) and water (bottom)

Many common minerals have covalent bonds. In quartz, for example, the elements silicon and oxygen share electrons. Sharing electrons produces strong bonds between atoms. In a diamond, atoms of carbon alone are covalently bonded.

 **ASK YOURSELF**

What is the difference between ionic and covalent bonds?

The Periodic Table

In addition to the 91 naturally occurring elements, scientists have produced 18 other elements in the laboratory. Trying to remember the characteristics of all these elements would be very difficult. To help organize information about each element, all of the elements have been placed in a table. This table, called the *Periodic Table,* is shown on the next two pages.

Figure 5–12. This key will help to explain the Periodic Table.

Each element on the Periodic Table has a symbol. In the diagram above, you can see that the symbol for uranium is U. The atomic number represents the number of protons in the element. Since the atomic number of uranium is 92, it has 92 protons. The atomic mass is the average mass of the atoms found in nature. Remember, most of the mass of an atom is in its nucleus. Uranium's atomic mass of 238.03 represents the average mass of all naturally occurring forms of uranium.

Elements found in the same vertical column in the Periodic Table make up a group. Elements in the same group have similar properties. Knowing this makes it easier to predict the characteristics of an element. The next activity can help you discover some other things about groups of elements.

DISCOVER BY Problem Solving

Look at the Periodic Table on the next two pages. All the elements in Group 1 can lose one electron, and all the elements in Group 17 can gain one electron. Therefore, elements from these two columns often form compounds with ionic bonds. See how many different compounds you can create by combining elements from these two groups. Using *sodium chloride* as an example, try naming the compounds as you form them. ✐

The Periodic Table

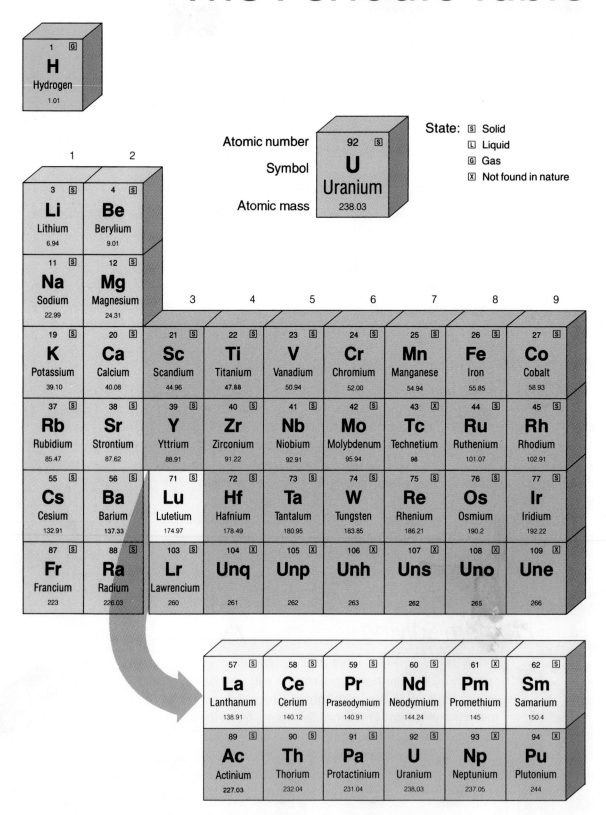

H
1 G
Hydrogen
1.01

Atomic number — 92 S
Symbol — U
Uranium
Atomic mass — 238.03

State: S Solid
L Liquid
G Gas
X Not found in nature

1	2		3	4	5	6	7	8	9
3 S **Li** Lithium 6.94	4 S **Be** Berylium 9.01								
11 S **Na** Sodium 22.99	12 S **Mg** Magnesium 24.31								
19 S **K** Potassium 39.10	20 S **Ca** Calcium 40.08		21 S **Sc** Scandium 44.96	22 S **Ti** Titanium 47.88	23 S **V** Vanadium 50.94	24 S **Cr** Chromium 52.00	25 S **Mn** Manganese 54.94	26 S **Fe** Iron 55.85	27 S **Co** Cobalt 58.93
37 S **Rb** Rubidium 85.47	38 S **Sr** Strontium 87.62		39 S **Y** Yttrium 88.91	40 S **Zr** Zirconium 91.22	41 S **Nb** Niobium 92.91	42 S **Mo** Molybdenum 95.94	43 X **Tc** Technetium 98	44 S **Ru** Ruthenium 101.07	45 S **Rh** Rhodium 102.91
55 S **Cs** Cesium 132.91	56 S **Ba** Barium 137.33		71 S **Lu** Lutetium 174.97	72 S **Hf** Hafnium 178.49	73 S **Ta** Tantalum 180.95	74 S **W** Tungsten 183.85	75 S **Re** Rhenium 186.21	76 S **Os** Osmium 190.2	77 S **Ir** Iridium 192.22
87 S **Fr** Francium 223	88 S **Ra** Radium 226.03		103 S **Lr** Lawrencium 260	104 X **Unq** 261	105 X **Unp** 262	106 X **Unh** 263	107 X **Uns** 262	108 X **Uno** 265	109 X **Une** 266

57 S **La** Lanthanum 138.91	58 S **Ce** Cerium 140.12	59 S **Pr** Praseodymium 140.91	60 S **Nd** Neodymium 144.24	61 X **Pm** Promethium 145	62 S **Sm** Samarium 150.4
89 S **Ac** Actinium 227.03	90 S **Th** Thorium 232.04	91 S **Pa** Protactinium 231.04	92 S **U** Uranium 238.03	93 X **Np** Neptunium 237.05	94 X **Pu** Plutonium 244

Metals
Transition Metals
Nonmetals
Noble gases
Lanthanide series
Actinide series

					13	14	15	16	17	18
										2 [G] **He** Helium 4.00
					5 [S] **B** Boron 10.81	6 [S] **C** Carbon 12.01	7 [G] **N** Nitrogen 14.01	8 [G] **O** Oxygen 16.00	9 [G] **F** Fluorine 19.00	10 [G] **Ne** Neon 20.18
	10	11	12		13 [S] **Al** Aluminum 26.98	14 [S] **Si** Silicon 28.09	15 [S] **P** Phosphorus 30.97	16 [S] **S** Sulfur 32.07	17 [G] **Cl** Chlorine 35.45	18 [G] **Ar** Argon 39.95
	28 [S] **Ni** Nickel 58.69	29 [S] **Cu** Copper 63.55	30 [S] **Zn** Zinc 65.39	31 [S] **Ga** Gallium 69.72	32 [S] **Ge** Germanium 72.61	33 [S] **As** Arsenic 74.92	34 [S] **Se** Selenium 78.96	35 [L] **Br** Bromine 79.90	36 [G] **Kr** Krypton 83.80
	46 [S] **Pd** Palladium 106.42	47 [S] **Ag** Silver 107.87	48 [S] **Cd** Cadmium 112.41	49 [S] **In** Indium 114.82	50 [S] **Sn** Tin 118.71	51 [S] **Sb** Antimony 121.75	52 [S] **Te** Tellurium 127.60	53 [S] **I** Iodine 126.90	54 [G] **Xe** Xenon 131.29
	78 [S] **Pt** Platinum 195.08	79 [S] **Au** Gold 196.97	80 [L] **Hg** Mercury 200.59	81 [S] **Tl** Thallium 204.38	82 [S] **Pb** Lead 207.2	83 [S] **Bi** Bismuth 208.98	84 [S] **Po** Polonium 209	85 [S] **At** Astatine 210	86 [G] **Rn** Radon 222

63 [S] **Eu** Europium 151.96	64 [S] **Gd** Gadolinium 157.25	65 [S] **Tb** Terbium 158.93	66 [S] **Dy** Dysprosium 162.50	67 [S] **Ho** Holmium 164.93	68 [S] **Er** Erbium 167.26	69 [S] **Tm** Thulium 168.93	70 [S] **Yb** Ytterbium 173.04
95 [X] **Am** Americium 243	96 [X] **Cm** Curium 247	97 [X] **Bk** Berkelium 247	98 [X] **Cf** Californium 251	99 [X] **Es** Einsteinium 252	100 [X] **Fm** Fermium 257	101 [X] **Md** Mendelevium 258	102 [X] **No** Nobelium 259

Elements in the same or adjacent groups are similar enough that they sometimes substitute for each other in compounds. This is similar to the substitution of walnuts for pecans in chocolate chip cookies. The texture and appearance of the cookies stays the same—the taste is just a little different. Feldspar is a good example of this in minerals. A feldspar may contain potassium, sodium, or calcium.

Figure 5–13. The minerals shown are all feldspars.

▼ ASK YOURSELF

What does the mass number of an atom represent?

SECTION 2 REVIEW AND APPLICATION

Reading Critically

1. A silicon atom has 14 protons. What is its atomic number? How many electrons does silicon have?

2. Which groups of elements most easily form compounds with ionic bonds?

Thinking Critically

3. If two sodium atoms were to lose one electron each, would they be attracted to each other? Explain your answer.

4. What elements could probably replace silicon in compounds?

Metals and Minerals—
National Treasures

Since human societies began using the natural materials around them, certain metals and minerals have been thought of as national treasures. As civilizations progressed, humans began refining metals to make coins, tools, weapons, and the machines of modern society. Over the years, many different metals and minerals have been considered important or valuable, and wars have been fought to obtain or secure these materials.

Figure 5–14. Gold, silver, and copper have long been considered valuable metals.

Strategic Metals and Minerals

Everyone knows the value of gold and silver, but about 5500 years ago, the Egyptians considered copper a strategic metal. *Strategic* means "essential for something, often war, politics, or business." Copper is a soft metal that the Egyptians shaped into tools, jewelry, and armor.

About 5000 years ago, people discovered that if tin is added to copper, a harder and stronger metal alloy, called *bronze,* results. An **alloy** is a mixture of two or more metals or a mixture of metals and nonmetals. Bronze made stronger tools and weapons than copper. In fact, bronze was so important that the time in history when it was commonly used became known as the *Bronze Age.* Tin and copper for making bronze were in high demand during the Bronze Age and both were considered strategic metals. Early explorers searched for these metals across Europe, the Middle East, and Asia.

Figure 5–15. Tools of the Bronze Age and Iron Age

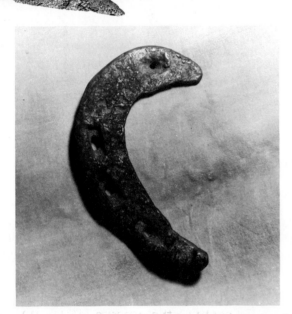

As civilization advanced, different metals and minerals became strategic. About 3000 years ago, a process for hardening iron was developed in Persia and Eastern Europe. Blacksmiths, as early iron workers were called, discovered that hot iron plunged into cold water hardened rapidly. The hardened iron could be sharpened and would stay sharp longer than any known metal, even longer than bronze. The Bronze Age was replaced by the Iron Age.

Figure 5–16. This may be the age of specialty alloys, such as those used in modern aircraft.

Today most iron is made into steel, an alloy of iron, carbon, and other materials. The United States classifies tin, beryl, cadmium, chromite, cobalt, niobium, diamond, fluorite, graphite, manganese, mercury, nickel, antimony, titanium, and platinum as strategic metals and minerals. Many of these are used to make specialty steels and superalloys, such as those used in lightweight aircraft like the "stealth" fighter.

What metals and minerals would you consider "strategic" for your lifestyle? The battery in your radio or tape player is made with manganese dioxide, graphite, and other metals and minerals. Cobalt is used in the production of high-strength magnets for modern sound equipment. Chromium is used in stainless steel and gives a shine to the automobile in your older brother's or sister's life. Diamonds are everyone's "best friend," since diamond drill bits and saw blades cut through the hardest substances, shaping many manufactured goods. And if your lifestyle includes a superlight bike, titanium makes it strong and lightweight. In the next activity, you can try your hand at a simple method of obtaining a metal that has long been considered necessary for maintaining a quality lifestyle—gold.

Figure 5–17. The stainless steel body of this car should last indefinitely.

ACTIVITY

How can gold be obtained by panning?

MATERIALS
apron; mixture of sand, gravel, and fool's gold; shallow dish or pan; water; large container

PROCEDURE

1. Place a handful of the mixture containing the "gold" into the shallow pan. Fill the pan half full of water.
2. Hold the pan with both hands, and, using a circular motion, gently rock the contents back and forth over the large container.
3. When the water becomes cloudy, gently tip the shallow pan and allow some of the cloudy water to run into the large container.
4. Add more water to the pan, and continue the process, tilting the pan to allow some of the excess sand and water to flow into the large container. Pick out any "gold" you see.
5. Repeat steps 2-4 until you have recovered all of the "gold."

APPLICATION

1. Why did the "gold" stay in the pan, while the sand and dirt were sloshed out?
2. What could you do to increase the amount of "gold" recovered each time you panned?

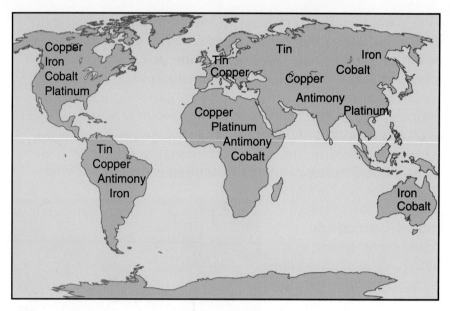

Figure 5–18. A map of the world's major deposits of strategic metals and minerals

Unfortunately, many of the metals and minerals most essential for manufacturing the products of our present lifestyle are also the most rare. Sometimes a metal or mineral may not be rare on a worldwide basis but may be abundant only in one or two countries. This means that there are countries of haves and countries of have-nots, and the countries that have the strategic metals or minerals can control the markets and set whatever prices they want for these materials. In the next activity, you can find out what metals and minerals are important to the industries in your area.

DISCOVER BY Researching

Contact an industry in your area, and inquire about metals and minerals used to manufacture their products. Some metals or minerals may be used as raw materials in the manufacturing process. Others may play a role in cleaning equipment or in making a manufacturing process safer or faster. Find out where they buy the resources they need. Record your information in your journal and report back to your class.

◀ ASK YOURSELF

Why do some metals and minerals have a strategic value?

A Different Kind of Treasure

In addition to strategic metals and minerals, some nations possess another kind of natural treasure—gemstones. **Gemstones** are mostly minerals with exceptional brilliance, color, durability, and rarity. A few gemstones, such as opals, are not true minerals. Although not of strategic importance, throughout history gemstones have been fought over, bartered with, and used for social prestige because of their beauty and value.

Amethyst

Figure 5–19. The world's major gemstones are unevenly distributed across the continents.

Diamond Ruby Sapphire Emerald

Turquoise

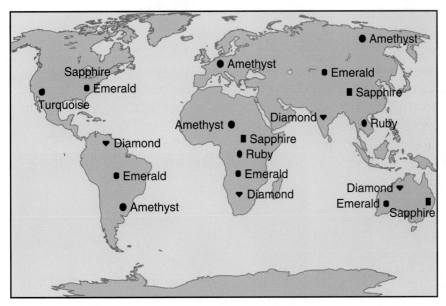

Figure 5–20. No other opals can compare in quality to the fiery beauty of Australian opals.

The best-quality gemstones are rare indeed. As you can see on the map, gemstones are not equally distributed on each continent. In addition, deposits of a gemstone may exist in a certain location, but the quality may not be comparable with deposits from another location. For example, an opal with large areas of iridescent reds and greens is almost certainly an Australian opal. Although opals are mined in other countries, including the United States, they are usually milky white with only small spots of bright colors.

Figure 5–21. The natural crystal shapes (right) reflect the internal structure of the minerals. A jeweler cuts the gemstones to form the shapes found in jewelry-quality gems (left).

In its natural form, a gemstone may look quite different from a gem. Remember, as a mineral grows, it takes a certain crystal form. The crystal form provides information about the internal structure of the mineral.

Did you ever wonder what determines the value of a gem? The next activity can help you find out.

DISCOVER BY Researching

Visit a local jewelry store, and ask the jeweler about the criteria used to classify and price different gems. Find out which gemstones are easily fractured and the relative hardness of the various minerals. ✎

You can learn a lot about gemstones by talking with a jeweler. Many department stores have fine jewelry with diamonds, sapphires, rubies, aquamarines, opals, and topazes on display. The shapes of the gems you see in rings and other jewelry are the result of a jeweler's knowledge of the cleavage, crystal structure, and other properties of various minerals.

Each gemstone differs from all others in two ways: the elements from which it is made and the arrangement of those elements. You can see many beautiful mineral crystals by visiting a museum or attending a gem and mineral show. In the next activity, you can try your hand at designing a piece of jewelry from a natural gemstone.

DISCOVER BY Problem Solving

Imagine that you are a jewelry designer. Make a scale drawing of an idea you have for a ring or a pin that shows off some of the natural crystal forms or cut stones you have seen. ✎

 **ASK YOURSELF**

What characteristics distinguish gemstones from other minerals?

SECTION 3   *REVIEW AND APPLICATION*

Reading Critically

1. Why do nations compete for minerals used in making certain types of alloys?
2. Where can deposits of iron ore and diamonds be found?

Thinking Critically

3. Compare and contrast the discovery of bronze and the resultant Bronze Age with the discovery of specialty alloys in our age.
4. Why do you think that durability and hardness are important qualities in gemstones?

INVESTIGATION

Growing Crystals

▶ **MATERIALS**
- cotton string, 30 cm ● Petri dishes (3) ● goggles ● apron ● water ● beakers, 100 mL (3)
- hot plate ● copper sulfate ● glass stirring rod ● ferrous sulfate ● sodium chloride
- ice-filled tray ● hand lens

▼ **PROCEDURE**

1. Cut the string into three 10-cm lengths, and lay each piece across a Petri dish.
2. **CAUTION: Goggles and aprons must be worn when heating the copper sulfate.** Prepare a saturated solution of copper sulfate by heating water in a beaker on the hot plate and slowly adding copper sulfate while stirring with the glass rod until no more copper sulfate will dissolve.

3. Soak the string from the first petri dish with the copper sulfate solution. Then return the string to the dish.
4. Prepare saturated solutions of ferrous sulfate and sodium chloride, and soak the other strings in these solutions.

5. Observe the strings as the water evaporates.
6. When the strings are nearly dry, place the beakers containing the saturated solutions into the ice-filled tray. Dip each string into the beaker containing its original solution. Observe the results with the hand lens.

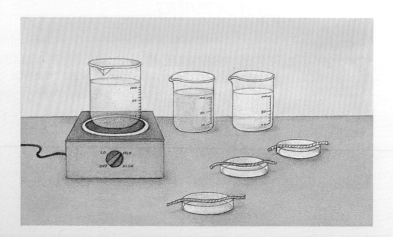

▶ **ANALYSES AND CONCLUSIONS**

What happens when the nearly dry strings are placed into the cooling solutions? Why does this happen?

▶ **APPLICATION**

How might a knowledge of the way crystals grow help jewelers produce large, pure crystals of gemstones?

✳ *Discover More*

Devise a method for making rock candy using a procedure similar to the one in the Investigation.

HIGHLIGHTS

The Big Idea

Everything in the earth's crust is made of elements or combinations of elements. Some of these combinations—minerals—have structures and physical properties that enable them to be identified wherever they occur.

As with all matter, minerals are composed of small particles called atoms. Specific characteristics make some minerals and metals strategic or valuable as gemstones.

For Your Journal

Look again at the way you answered the questions at the beginning of the chapter. How have your ideas changed about why some metals and minerals are so valuable that they are considered national treasures? Write your new ideas in your journal.

Connecting Ideas

Copy the concept map into your journal. Complete the map using what you have learned about minerals.

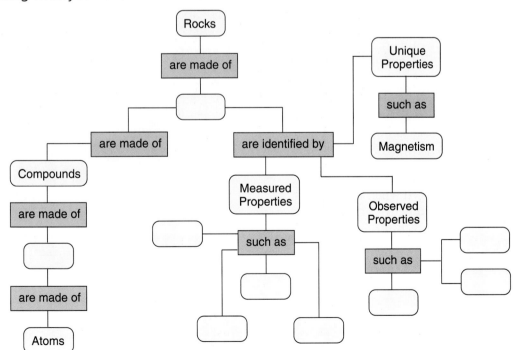

▶ ## Understanding Vocabulary

1. For each set of terms, explain the similarities and differences in their meanings.
   a) elements (163), compound (165), molecule (166), atoms (162)
   b) protons (164), electrons (164), neutrons (164)
   c) streak (154), luster (155), density (155), cleavage (157)
   d) mineral (152), rock (152)
   e) gemstones (175), crystals (158)

▶ ## Understanding Concepts

MULTIPLE CHOICE

2. The smallest building block of an element is a(n)
   a) neutron.
   b) atom.
   c) electron.
   d) proton.

3. All the atoms of a particular element have the same number of
   a) protons and neutrons.
   b) protons and electrons.
   c) neutrons and electrons.
   d) covalent bonds.

4. Which of the following is not a mineral?
   a) opal
   b) magnetite
   c) diamond
   d) quartz

5. Granite is not a mineral because it
   a) has an irregular shape.
   b) is too hard.
   c) is not useful.
   d) does not have definite properties.

6. What alloy is produced when tin and copper are combined?
   a) iron
   b) steel
   c) bronze
   d) titanium

SHORT ANSWER

7. Compare and contrast the properties of fracture and cleavage.

8. Explain why some minerals are considered to be gemstones.

Interpreting Graphics

Use the mineral key below to identify these three minerals.

9.

10.

11.

	Silver	Antimonite	Enstatite
Color–Streak	gray–silver/white	lead/gray–gray/black	gray–white
Hardness	2.5-3	2.5	5.5
Crystal shape	moss-like	needle-like	column-like

12. Which properties helped you identify each mineral? Explain why some properties were not testable by you.

Reviewing Themes

13. *Systems and Structures*
Why is it possible to have a single atom of some minerals, while for most minerals a molecule is the smallest unit possible?

14. *Environmental Interactions*
Why do some atoms form ionic bonds while others form covalent bonds?

19. This person is standing inside a special mineral formation known as a geode. Most geodes are not this large, but they do represent some of the earth's largest mineral structures. A geode is a hollow ball with mineral crystals growing on the inside. How do you think a geode might form?

Thinking Critically

15. How would you estimate the hardness of a mineral if the Mohs' Scale of Mineral Hardness were not available?

16. Consider the definition of a mineral. Decide whether ice should or should not be considered a mineral. Explain.

17. Which two mineral identification tests do you think are the most useful? Explain your choices.

18. How would the Periodic Table help you find elements that could replace calcium in a compound?

Discovery Through Reading

von Baeyer, Hans Christian. "Atom Chasing." *Discover* 13 (July 1992): 42-49. Imagine being able to count all the atoms in your desk! By using the universal atom-counting method he developed, Sam Hurst can determine not only which elements are present in an unidentified substance, but also how many atoms of each element the substance contains. Read this article to find out how he does it.

CHAPTER 6

ROCKS

Whenever people have wanted to remember, or to be remembered, they have built memorials of rock. A few places in the world—Mount Pentelicus in Greece, Carrara in Italy—are famous for the beauty of the rock quarried there, the raw material for timeless works of art.

Ancient Athens built its soaring temples of Pentelic marble, which has weathered to a reddish hue after 2500 years. The titanic artist, Michelangelo, spent five years of his life at Carrara, seeing that workers properly quarried the gleaming marble he would fashion into his famous statue of David. Carrara marble was once thought to have been flawed, but Michelangelo sculpted the rock into a work of art that many consider perfect.

Visitors to Washington, D.C., find the heroes of American history preserved in glistening white rock. The Tomb of the Unknowns, in Arlington National Cemetary, is built with the largest piece of marble ever quarried. It weighed more than 50 tons and took a year to cut from a Colorado Mountain. Rock from the same quarry was used in the Lincoln Memorial, which sits just across the Memorial Bridge from Arlington, like a Greek temple at the entrance to the capital city.

Rock monuments stand against time, keeping their beauty in spite of the centuries. Shaped and hewn by artists, rock bears the hopes and sorrows of each generation, so that later generations may remember.

For Your Journal

- Why are statues and memorials made of rock?
- How many different kinds of rock are there?
- What other uses are there for rock?

Studying Rocks

Objectives

List some of the uses of rocks.

Describe the criteria used to classify rocks.

Correlate the three rock groups with the rock cycle.

If you have ever traveled through New England, you may have seen dozens of rock walls crossing the countryside. The New England landscape was shaped by massive glaciers during the Ice Age. When the ice melted, sand, pebbles, and large rocks were left behind, some pushed all the way from Canada. Before colonial farmers could begin growing crops, they had to remove the rocks. Naturally, they chose to carry these heavy rocks as short a distance as possible, so they formed mounds and walls at the edges of their fields.

Figure 6–1. There are many of these rock walls in New England.

"Reading" Rocks

Have you ever stopped to pick up an interesting rock as you walked across a field or along a road? What was there about the rock that drew your attention? A geologist encountering a rock wall would certainly be drawn to examine each boulder. Rock walls are like libraries to geologists. Each rock is a history book, with its story locked inside. What do you think a geologist might learn by studying a rock?

In the case of a rock from a wall, the most recent step in its history is the most obvious: A farmer picked up the rock and added it to the wall. Earlier stages of a rock's history can also be inferred, once you understand how a rock is formed.

Figure 6–2. Unraveling the history of rocks such as these challenges geologists. What can you tell about these rocks from the photographs?

Just as cells are the building blocks of the earth's living things, rocks are the building blocks of the earth itself. The structure of a cell reveals information about an organism. In a similar way, the structure of a rock reveals information about the earth. It tells of ancient environments, movements and stresses within the crust, and the evolution of life.

Rocks contain minerals that provide metals and mineral products for society. Early humans lived in rock caves and ground their food in rock bowls. As technology advanced, humans learned to take advantage of the minerals found in rocks. Look around you. Many of the objects you see were manufactured from rocks and minerals. Buildings are constructed of stone, brick, or concrete made from rocks. The plaster in your classroom walls, the metal in the chairs, and the glass in the windows come from minerals in various rocks.

Rocks and minerals have influenced the growth and development of entire nations. Some countries have rich deposits of the rocks and minerals their industries need. This gives them an economic advantage—they don't have to import these materials. The availability of rocks and minerals has even influenced the location of certain industries, which often choose to locate near needed materials rather than incur the expense of transportation. The early pottery industry in America, for example, was centered in Georgia, near the rich kaolin clay found there. Perhaps you can explain the location of industries near you on the basis of available materials. The next activity can help you.

Even though you use materials made from rocks every day, you may not have thought about how rocks are formed or why one rock is different from another. Some rocks, like coal, consist of organic particles. Other rocks consist of mineral particles. A few of these rocks consist of just one type of mineral. However, most rocks contain two or more different minerals. Although thousands of different kinds of minerals occur, only about 20 are common to most rocks.

Figure 6–3. Most rocks are combinations of only a few minerals.

Mica

Quartz

Feldspar

Granite

If most rocks are made of the same few minerals, what makes one rock different from another? Examine the rock shown in the picture. Notice how it can be divided into two different parts. The lighter part is made of large grains of tan, white, and black minerals. These minerals are interlocking, making a very dense rock. The second part is a band of fine black minerals, cutting across the light-colored mineral mixture.

These things you can observe, but what is the history of this rock? The first step in determining a rock's history is to identify the minerals in the rock—testing for hardness, cleavage, or one of the other properties you learned about previously. Then you can focus on the rock's origin—how it was formed—deciding which part formed first and how the two parts came to be connected.

Figure 6–4. The history of this rock is a mystery, waiting to be solved.

 ASK YOURSELF

What is the relationship between minerals and rocks?

Rock Groups

When you hear a new song on the radio, you can often identify it with a certain group you're already familiar with. In a similar way, geologists identify, or classify, rocks as belonging to certain groups. In geology, as in music, classification is important because it provides ways to understand similarities and differences among rock groups. A well-designed classification system, based on the characteristics of the objects being grouped, also provides an easy way to communicate ideas about the objects.

Rock classification is based on certain characteristics. From these characteristics, a rock's origin can be inferred. All rocks can be divided into one of three groups, based on origin. These groups are *igneous, sedimentary,* and *metamorphic.* Within each group, other criteria are used to identify individual rocks. These criteria include the types of minerals present, the shape and size of the mineral grains, and the way the minerals are arranged.

It's not usually possible to watch rocks forming in nature. You can, however, observe similar processes by using simple materials. The next activity will show you how one kind of rock forms.

🔥 Your teacher will take some old candles and melt them in a pan, and then pour the melted wax into a mold. Observe the wax as it melts and again as it cools in the mold. Carefully touch the wax at various stages with the eraser end of a pencil. Record your observations in your journal. ✎

Figure 6–5. Cooling lava hardens to form igneous rock.

When you allowed the wax to cool completely, it hardened—you had a solid object formed from liquid, or molten, material. Rocks are formed in a similar manner inside the earth. Where the temperature is high, rocks melt, forming magma. Many pockets of magma are found in the earth's crust and in the mantle.

Recall that magma that reaches the earth's surface is known as lava. As magma or lava cools, it forms **igneous** (IHG nee uhs) **rock.** The word *igneous* comes from the Latin word *ignis*, meaning "fire." In the next activity, you can learn how another kind of rock forms.

Take your hardened wax, and carefully scrape off some of the wax with the edge of a plastic knife. When you have made a large pile of wax shavings, cover it with a piece of paper and a heavy book and press it as hard as you can. What happens to the pile of wax shavings? ✎

Figure 6–6. The mud of this drying lake bed may eventually become sedimentary rock called mudstone.

A second type of rock forms from rock fragments in a way that is similar to the wax in the activity. Rocks exposed to weather break into small pieces. An accumulation of such pieces is called *sediment*. Sediments may be transported by wind or water to new locations. Eventually the sediments are deposited in layers.

As layers of sediments are compressed and naturally cemented, they stick together to form **sedimentary rock.** The word *sedimentary* also comes from the Latin word *sedimentum,* meaning "that which has settled." In the next activity, you can observe the process that forms the third kind of rock.

Discover by Doing

Hold a piece of either your "igneous rock" or your "sedimentary rock." Warm it for a while in your hands and then squeeze it. What happens to the "rock"? ✏

Heat and pressure change rock. Unlike the activity, however, where only the shape of the "rock" changed, heat and pressure within the earth cause changes in a rock's texture or mineral content. These changes produce **metamorphic** (meht uh MOR fihk) **rock.** The word *metamorphic* comes from the Greek word *metamorphosis,* which means "change."

You may already know about the metamorphosis of certain insects or amphibians, in which the adult form of the organism does not resemble the young. In a similar way, a metamorphic rock does not resemble the rock from which it formed.

Metamorphosis of a Rock Group

Then

Now

 ASK YOURSELF

What are the three groups of rocks, and on what basis are rocks classified?

The Rock Cycle

In biology, scientists often study food webs, tracing the path of energy from plants that produce food, to animals that eat plants, to predators that eat the plant eaters, to decomposers that return nutrients to the soil. There are many such examples of interactions in the natural world. The processes that form rocks, for instance, interact through the **rock cycle.**

As magma cools, it forms an igneous rock—say granite. If the granite is near the sea, it may be weathered by waves into sand and clay—sediments. The sand may be transported, deposited, and cemented to become sandstone, and the clay may be deposited as mud and harden into mudstone—both sedimentary rocks. With enough heat and pressure, the sandstone becomes quartzite and the mudstone becomes slate, or with even more heat and pressure, schist. Quartzite, slate, and schist are metamorphic rocks. The schist might be forced deep within the earth, where it would melt, forming magma. What kind of rock would form when the magma cooled?

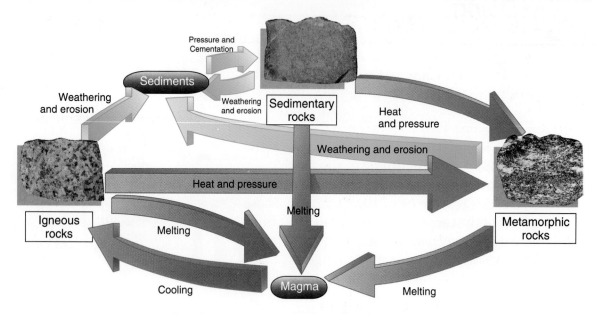

Pressure and
Cementation

Sediments

Weathering
and erosion

Weathering
and erosion

Sedimentary
rocks

Heat
and pressure

Weathering and erosion

Heat and pressure

Melting

Igneous
rocks

Melting

Metamorphic
rocks

Cooling

Magma

Melting

Figure 6–7. Rocks may enter the rock cycle at any place and be changed into any other type of rock.

Unlike the flow of energy in an ecosystem, the rock cycle is not a one-way process. It is possible for igneous rock to be changed into sedimentary rock or metamorphic rock, and for metamorphic rock to be changed into sedimentary rock or igneous rock. A sedimentary rock can become another kind of sedimentary rock, an igneous rock, or a metamorphic rock. In fact, any rock may change from one type to any of the others.

▼ **ASK YOURSELF**

Why is the rock cycle not a one-way process?

SECTION 1 *REVIEW AND APPLICATION*

Reading Critically

1. Explain why rocks are to a geologist as cells are to a biologist.
2. How has the location of rocks and minerals influenced industrial development?

Thinking Critically

3. How is the process of describing the history of a rock similar to solving a mystery?
4. How is the rock cycle similar to the flow of nutrients in a food web? How is it different?

Igneous Rocks

All kinds of rocks are used in the construction of buildings and monuments, and America's capital city, Washington, D.C., provides an ideal place to see some examples of each of the three rock groups. Let's imagine we're on a walking tour of Washington. We'll begin our walk on the National Mall, near the Washington Monument. Between the monument and the White House is a memorial to the soldiers who fought in the Civil War. Although the statues are made of bronze, the base on which they stand is granite.

Objectives

Discuss the process that forms igneous rocks.

Classify common igneous rocks by their characteristics.

Describe some uses for igneous rocks.

Figure 6–8. The base of this memorial is granite.

Formation of Igneous Rocks

If you look closely at the granite of the memorial, you will see that the rock is composed of interlocking minerals of several different colors. This interlocking texture is typical of most igneous rocks. Remember that igneous rock begins as magma inside the earth, where the temperature and pressure are high enough to melt solid rock.

What does magma look like? How does it behave? At more than 1000°C, magma is much too hot to work with in a classroom, but we can observe materials that behave in similar ways. You can see this for yourself in the next activity.

DISCOVER BY *Observing*

Your teacher will demonstrate the formation of igneous rock by heating granular sugar in a pan on a hot plate. As soon as it the sugar melts, he or she will remove it from the heat and carefully pour it down an inclined board. Observe how it moves and how it cools. Why do bubbles form on the surface of the cooling "magma"? Record your observations in your journal. ✎

 **ASK YOURSELF**

What causes solid rock to melt?

Classifying Igneous Rocks

If you were given the job of designing a classification system for a box of buttons, you might group the buttons by characteristics such as color, size, and shape. Similarly, geologists use characteristics to group igneous rocks. The two characteristics that are the basis for the classification of igneous rocks are texture—the size of the mineral grains—and mineral content.

Texture As magma cools, it eventually reaches a temperature at which the elements within the magma join to form minerals. If the cooling process is very slow, the elements have time to move around, combine with similar elements, and form large mineral grains. As these grains grow next to each other, they lock together, like pieces in a jigsaw puzzle.

If the mineral grains are large enough to be distinguished from each other, the rock is said to have a coarse texture. The granite of the Civil War monument has such a texture. Which other rock(s) pictured here has/have a coarse texture?

Granite

Gabbro

Basalt

Rhyolite

Figure 6–9. Although these rocks are all igneous in origin, they have different textures.

You've probably seen pictures of glowing lava, streaming from a volcano. Because it is on the surface, lava cools more quickly than magma does. This rapid cooling causes the minerals in lava to become locked in place before large grains have time to form. Igneous rocks with very small mineral grains are said to have a fine texture. Rhyolite and basalt are examples of fine-textured igneous rocks.

Sometimes magma begins to cool slowly. Then, as the magma nears the surface, it cools more quickly, resulting in a rock with both large and small mineral grains. A rock with this mixed texture is called a *porphyry.*

Mineral Content   In addition to texture, mineral content is also used to classify igneous rocks. The kinds of minerals found in any igneous rock depend on the chemical composition of the magma from which the rock formed. Magmas of the oceanic crust are rich in magnesium, iron, and calcium that form dark-colored minerals like hornblende and olivine. These minerals form dense, dark-colored rock. The less dense rocks of the continental crust contain light-colored minerals like feldspar and quartz that are formed from magmas rich in sodium, potassium, and silicon.

 ASK YOURSELF

Granite and rhyolite are light colored, while gabbro and basalt are dark colored. Which rocks probably contain quartz?

Glassy Igneous Rocks

Some igneous rocks have no visible crystals at all; they either cooled too quickly or the cooling lava had such a thick consistency that the atoms and molecules could not move together to form mineral crystals. The volcanic glass, or obsidian, in the photograph always fascinates people because it looks so much like manufactured glass. It is, however, an igneous rock.

Glassy rocks have properties that make them uniquely useful. For example, when obsidian breaks, it forms curved, sharp-edged surfaces. Early Native Americans took advantage of this property, using obsidian to make spearheads and arrowheads.

Pumice and scoria are two other glassy igneous rocks with unusual properties. In the next activity you can find out how they differ from most other rocks.

Figure 6–10. Is this manufactured glass, or is it a glassy rock?

Doing

Your teacher will give you two rocks that look very similar. How are the rocks alike? How are they different? Fill two deep dishes with water, and place one rock in one dish and the other rock in the second dish. What do you observe? What can you infer from your observations? ✐

Figure 6–11. Pumice (top) and scoria cool so quickly that trapped gases in the magma leave many holes in the rocks.

Pumice floats because it is less dense than water. As you might observe, it is very lightweight for its size, and it is full of holes, like a cork. The lava that cooled to form pumice had such a thick consistency that gas bubbles were trapped within it. This process is similar to the one that forms holes in bread. Gases given off by the yeast in bread dough are trapped in the thick dough and remain there even after the bread is baked.

The other rock with holes is scoria. Scoria forms in basically the same way as pumice, but it does not have enough bubbles to lower its density to the point that it can float in water.

◤ ASK YOURSELF

How can you distinguish between pumice and scoria?

Identifying Igneous Rocks

Using the characteristics of texture and mineral content, geologists assign names to igneous rocks—light-colored, coarse-textured rock is granite. Table 6–1 shows several other examples.

Table 6-1

Common Igneous Rocks

	Rock name	Identifying characteristics
	Granite	Light colored with coarse texture
	Gabbro	Dark colored with coarse texture
	Rhyolite	Light colored with fine texture
	Basalt	Dark colored with fine texture
	Pumice	Light colored, low-density, glassy
	Scoria	Dark colored, medium-density, glassy
	Obsidian	Dark colored, high-density, glassy

Identifying igneous rocks can help you reconstruct the history of a rock. Look at the rock pictured here. It is the same one you saw on page 187. The magma containing the light-colored minerals cooled slowly, forming granite. At some later time, magma containing darker minerals forced its way into the granite and cooled quickly, forming basalt. How do you know the basalt didn't form before the granite? To answer this question, think about the conditions necessary for granite and basalt to form.

 ASK YOURSELF

What does classifying an igneous rock as fine- or coarse-textured tell you about its formation?

Figure 6–12. The mystery is solved!

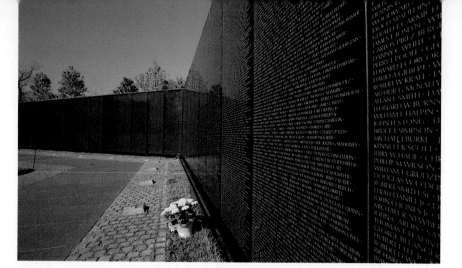

Figure 6–13. The "black granite," or gabbro, used to make this memorial will last for centuries.

Uses of Igneous Rocks

Many igneous rocks are used commercially. Pumice, for example, is not only interesting, but it is also useful as a polishing material, because of its sharp-edged, glassy fragments. Sometimes it is ground up and used as a scouring powder or in hand soap. Igneous rocks are also valuable sources for certain minerals and metals. These rocks are often mined for these materials.

If we continue walking down the National Mall, past the Washington Monument, and near the Lincoln Memorial, we see a low wall, dedicated to the soldiers who died in a more recent war—Vietnam. The Vietnam Veteran's Memorial is made of 150 panels of gabbro, or "black granite." Although its mineral content is different from the granite of the Civil War memorial, its texture is similar. Many igneous rocks will take a high polish, and stay that way for a long time. Carvings in granite can withstand the effects of weather for centuries, so future generations can remember fallen American heros.

 ASK YOURSELF

Why are igneous rocks often used to construct monuments?

SECTION 2 REVIEW AND APPLICATION

Reading Critically

1. What are the two characteristics used to classify igneous rocks?
2. Why did the granite have to form first in the rock shown on page 187?

Thinking Critically

3. What kinds of igneous rocks would you be likely to find on volcanic islands, such as the Hawaiian islands?
4. Describe the formation of a rock containing large feldspar grains and very fine quartz grains.

SKILL Designing Tables

▶ **MATERIALS**
- paper ● pencil

▼ **PROCEDURE**

1. When you design a table, you must first give the table a title. Then list all the information or relationships you wish to record in your table. These are your headings. Headings may be listed either horizontally or vertically in the table.

2. Table 1 is set up to show the relationship between color and type of igneous rock. Copy the table, and fill in the missing information.

TABLE 1: IGNEOUS ROCKS	
Color	**Rock types**
light	granite, _____, and _____
_____	gabbro, basalt, and scoria

▶ **APPLICATION**

Design a table that shows the relationship between the texture of igneous rocks and the ways in which they form.

✳ *Using What You Have Learned*

After completing the sections on sedimentary and metamorphic rocks, design tables to show various relationships within these rock groups.

TABLE 2: SEDIMENTARY ROCKS

TABLE 3: METAMORPHIC ROCKS

Sedimentary Rocks

Describe *three types of sedimentary rocks.*

Compare and Contrast *the formation of various limestones.*

Identify *some common sedimentary rocks by their characteristics.*

On your walk down the National Mall toward the Vietnam Memorial, you have passed many of the museums of the Smithsonian Institution, sometimes called "America's attic." The various buildings contain everything from paintings and sculptures by the great masters to the flag that inspired Francis Scott Key to write the "Star-Spangled Banner." You can also admire the airplanes that the Wright brothers and Charles Lindbergh flew, moon rocks, and the Hope diamond. But no less interesting are the buildings themselves and the materials from which they are constructed. The original Smithsonian building, often called "the castle" because of its medieval architecture, is made of red sandstone, while the interiors of many of the other buildings of the Smithsonian are made of limestone. Sandstone and limestone are sedimentary rocks.

Figure 6–14. These buildings were both constructed with sedimentary rocks.

Formation of Sedimentary Rocks

You may not have noticed it, but rocks are forming all around you. Have you ever stood in a lake and felt the mud ooze between your toes? Have you ever picked up a flat stone to skip across a pond? As you searched for just the right stone, you probably saw layers of fine sand, clay, and pebbles—the building materials of sedimentary rocks.

Cements   Maybe you have taken clay from a stream bed and pressed it together to make a mud ball. This process of pressing clay sediments until they stick together also occurs naturally, and eventually forms a sedimentary rock. Could you also press sand together and make it into a rock? Try the next activity.

DISCOVER BY Doing

Take a handful of dry sand and try pressing the sand together to form a rock. What happens? Try wetting the sand before forming your rock. What happens when the sand dries? Now make a solution of water and school glue, and pour it over the sand. Allow it to stand for several hours; then observe. How is this process similar to the formation of real sedimentary rocks? How is it different? ✑

In nature, water soaking through the ground carries with it natural cements. Some of these natural cements are silica, from quartz; calcium carbonate, from calcite; and iron oxides. They coat sediments, such as sand, and bind them together. This process is called **cementation.**

Natural cementation is similar to the process of making concrete. To make concrete for a driveway or sidewalk, you need three things—sand, water, and cement. The cement you use in making concrete has the same function as the natural cement in sedimentary rocks—it holds the sand together. In fact, cement is powdered limestone, a sedimentary rock. The water in the concrete ensures even distribution of the sand and the cement.

Figure 6–15. Three sizes of clastic sediments: clay (top), sand (middle), and pebbles (bottom).

Sediments   There are three types of rock-forming sediments: clastic sediments, chemical sediments, and organic sediments. Sediments made of fragments of other rocks and minerals are called *clastic sediments*. These fragments are formed when their parent rock or mineral is weathered. After the fragments are deposited, they become cemented together into a new rock. Clastic sediments are classified by size. Figure 6-15 shows three common sizes. The degree of rounding of fragments and the type of cementation involved are also a part of the history of clastic sedimentary rocks.

Figure 6–16. You can still identify the shell fragments in this rock called *coquina.*

Chemical sediments come from minerals once dissolved in water. The salts in seawater, for example, become chemical sediments as they collect in thick layers on the bottom of the sea. The weight of overlying layers compacts these sediments into rock.

The remains of once-living organisms, such as shells or plants, form *organic sediments.* In some places these remains accumulate in thick layers and form rock. Many of these rocks contain organic remains, called *fossils,* that help geologists to determine how the rock formed.

▼ **ASK YOURSELF**

List the characteristics used to classify sediments.

Classifying Sedimentary Rocks

You might think of sedimentary rocks as you would cookies. Some cookies contain coarse materials, such as nuts, chocolate chips, or oatmeal. Others are uniformly smooth and fine grained, like sugar cookies. Sedimentary rocks have characteristics that are similar to cookies—they just don't taste as good.

Clastic Sedimentary Rocks Clastic sedimentary rocks are formed from clastic sediments. Remember, some clastic sediments are large, while others are very small. So the size of fragments, or texture, is a useful property for the classification of clastic sedimentary rocks.

Coarse fragments form conglomerates and breccias. Both of these rocks contain pebble-sized fragments with diameters of 2 mm or more. The difference between conglomerates and breccias lies in the shape of the fragments. Conglomerates have rounded fragments, and breccias have sharp-edged fragments. They may be of a single mineral, or they may be fragments of different rocks or minerals, each with its own characteristics. Most commonly, the single mineral fragments are quartz. In the next activity, you can make models of conglomerates and breccias.

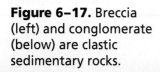

Figure 6–17. Breccia (left) and conglomerate (below) are clastic sedimentary rocks.

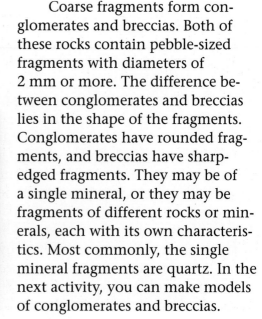

ACTIVITY

How can you make models of clastic sedimentary rocks?

MATERIALS

20 pieces of assorted hard candies, plastic bag, mallet or hammer

PROCEDURE

1. **CAUTION: Do not eat any of the candy.** Place 20 pieces of assorted hard candies into a plastic bag.
2. Carefully break the candy into pebble-sized fragments with a mallet or hammer.
3. Take half of the fragments out, warm them in your hands, and press them together into a ball.
4. Place the remaining candy in a beaker of water, and stir until the fragments loose their sharp-edged appearance. Remove the rounded fragments, and form a second rock in the same manner as you did the first one.

APPLICATION

1. Which candy "rock" is similar to a breccia? Which is similar to a conglomerate?
2. How do pebbles become rounded in nature?

Conglomerates and breccias are composed of fairly large fragments. Smaller-sized fragments form other types of clastic sedimentary rocks. For example, sandstone is formed of sand-sized particles, usually grains of quartz. The surface of a sandstone feels like sandpaper. Clastic rocks formed from fragments smaller than sand are mudstones.

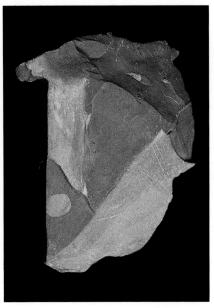

Figure 6–18. Sandstone (left) is made of sand-sized grains. Mudstone (right) is composed of nearly microscopic fragments.

Clastic materials are deposited over long periods as sediments are stacked one layer upon another. There are many histories to be read in clastic sedimentary rocks. Sometimes different layers are different colors, indicating changes in the environment. Color changes can easily be seen in the rock layers in the Badlands of South Dakota (left). At other times, fossils of living organisms are preserved between layers, as shown in the photograph below.

Figure 6–19. Layers of different sediments provide a history of environmental change.

Chemical sedimentary rocks

Water can be a source of sediments. Most water contains dissolved minerals. You can observe some of these minerals by doing the next activity.

DISCOVER BY *Doing*

Half fill a shallow pot or baking pan with tap water. Allow the pan to sit uncovered for several days until all the water evaporates. You can speed up the evaporation by gently heating the pan on a burner or in a warm oven. After the water has evaporated, look at the bottom and sides of the pan. What do you observe? ✏

Figure 6–20. Rock salt (top) and rock gypsum are common evaporites.

There are two processes that form chemical sedimentary rocks: evaporation and precipitation. *Evaporites* form when substances dissolved in water are left behind as the water evaporates. This is the process you saw in the activity. The water evaporated, leaving a film of minerals that were once dissolved in the water.

The most common evaporite rocks are rock salt and rock gypsum. Thick layers of these rocks are found in many places in the United States. Most were formed when large quantities of sea water became isolated from the oceans. A continuing example of the formation of evaporites occurs along the Great Salt Lake in Utah. As water in the lake evaporates, the lake shrinks and the salts are left behind, eventually becoming rock salt.

Figure 6–21. As the Great Salt Lake evaporates, deposits of salt are left behind. These deposits eventually become rock salt.

Precipitates are chemical sediments that form when dissolved substances precipitate, or fall out of solution. A similar process causes rain or snow to fall. When air cannot hold any more water vapor, some of it condenses out into droplets. The water droplets combine until they become so heavy that they can no longer be held by the air, so they fall to the ground. Similarly, sea water can only hold a certain amount of minerals in solution. As microscopic organisms change the composition of the water around them, some of the minerals come out of solution and precipitate, or "rain" down, to the ocean floor.

Figure 6–22. Compact limestone (left) and chert (right) form thick layers on the sea floor.

The most common precipitate rock is compact limestone, which forms as calcite precipitates out of sea water. Some silica also precipitates out of sea water, forming chert, a hard, dense rock. Like quartz and obsidian, chert fractures into sharp, curved pieces. Dark-colored chert, called flint, was used for making cutting tools by Native Americans because of the way it fractures.

Organic Sedimentary Rocks

Organic sedimentary rocks form from the remains of once-living organisms. Although compact limestone is a chemical sedimentary rock, other types of limestone are organic sedimentary rocks. Organic limestone forms from the hard remains of animals, such

Figure 6–23. Animal remains are often preserved in limestone.

as corals. Organic limestone often contains fossils of animals trapped in the sediments, as shown in the picture. These fossils provide clues about life in the past.

Organic sedimentary rocks also form from compressed plant material. In a swamp, plants are quickly buried and decay very slowly to form peat. Over thousands of years the buried peat becomes lignite, sometimes called *brown coal.* Brown coal is compacted, but it is not as hard as other coals. With more pressure and time, a harder coal, called *bituminous coal,* forms. Millions of years are required for bituminous coal to be produced from peat. Bituminous coal, once used to heat homes, is now used primarily in the production of electricity.

Figure 6–24. Coal deposits form where thick layers of plant material have been buried and then compressed.

 **ASK YOURSELF**

What are the three ways that sedimentary rocks can be formed?

Uses of Sedimentary Rocks

Many sedimentary rocks are used in industrial processes and construction projects. Shale and siltstone, for example, are used for manufacturing sewer pipe, drain tile, and chimney linings.

Rock gypsum is used by the building industry in the production of drywall and plaster. The quartz grains in sandstone are the basic mineral component of glass. Porcelain—used in fine china, bathtubs, sinks, and toilets—is a combination of clay, feldspar, and quartz sand. Fiberglass insulation and silicon computer chips are also made from sand.

Many buildings in Washington, D.C., are constructed of sandstone. Huge blocks are sawed out of thick sandstone beds. These blocks are then cut into smaller rectangular building blocks. The main building of the Library of Congress is made of gray sandstone, and the White House is built mostly of sandstone, painted white to hide the damage of a fire set by British soldiers during the War of 1812.

Figure 6–25. The White House in Washington, D.C., is built of sandstone that is painted white.

 ASK YOURSELF

What is porcelain, and what is it used for?

SECTION 3 *REVIEW AND APPLICATION*

Reading Critically
1. What kind of rocks form from rounded, pebble-sized sediments?
2. Why is rock salt considered to be an evaporite?

Thinking Critically
3. Why might you expect to find fossils in organic limestone but not in compact limestone?
4. Describe the history of a sedimentary rock containing the fossil remains of a fish. Include steps of erosion, transportation, deposition, and cementation.

Metamorphic Rocks

Describe *how metamorphic rocks form.*

Define *the terms* foliated *and* non-foliated *as they apply to metamorphic rocks.*

Justify *the use of metamorphic rocks in so many buildings and monuments.*

There is a line in "America the Beautiful" that goes "Thine alabaster cities gleam, undimmed by human tears." The poet must surely have been inspired by Washington, D.C., when writing this line. Nowhere else in America is there such a collection of brilliant white buildings and monuments. The Lincoln and Jefferson memorials, the Washington Monument, the Kennedy Center, the Supreme Court Building, the United States Capitol, and many others are made with gleaming white marble—a metamorphic rock.

Figure 6–26. The Lincoln Memorial is made of white marble.

Formation of Metamorphic Rocks

You probably know that many animals go through a metamorphosis, or change, in their lives. Frogs, for example, start life as fishlike tadpoles. They gradually change their body form, growing legs and lungs. Their lifestyle changes also as they become adults. Instead of living in water as tadpoles do, frogs live on land. What other things can you think of that undergo great changes in form?

Although we tend to think of rocks as strong, permanent objects, under certain conditions they too can change. Metamorphic rocks form when heat and pressure cause changes in existing rocks. During metamorphosis, heat and pressure can change the mineral composition of a rock or its texture. You can model some of these changes in the next activity.

DISCOVER BY Doing

You will need two different colors of modeling clay. Form about 10 small spheres from each color of clay. These represent rock fragments of two different mineral types. Place all the spheres together in a loose pile. This represents the original sediment. Press the pile of clay spheres with the palm of your hand, but do not press hard enough to remove all the spaces between the spheres. What kind of rock does the pile represent? Now press the pile of spheres until their shapes are completely destroyed. What kind of rock have you now formed?

Figure 6–27. Quartzite is the metamorphic form of sandstone.

The quartzite shown here is a metamorphic rock. The original rock was a quartz sandstone. Sandstones, with their rounded fragments, have many open spaces between the mineral grains, just as there were open spaces between the clay spheres in the activity. When put under pressure, individual quartz grains squeeze together, change shape, and become interlocked with other grains of quartz, forming quartzite.

Where do rock-altering pressures occur? You might guess that the pressures involved with mountain building, for example, could alter minerals and make metamorphic rocks. Over time, great forces, such as those of moving plates, "put the squeeze" on mountainous areas. This causes rock layers to fold and the texture and mineral content of individual rocks to change. Sometimes these forces cause rocks to melt, and magma intrudes into existing rocks. More often the heat from nearby magma sources is not enough to melt surrounding rock, but it is enough to cause changes in rocks for many kilometers around the source.

ASK YOURSELF

What two kinds of changes can occur when metamorphic rocks are formed?

Classifying Metamorphic Rocks

Examine the metamorphic rocks shown in the photographs. If you were asked to separate them into two groups, you would probably make a pile of layered ones and a pile of nonlayered ones. When describing metamorphic rocks, the obviously layered rocks are called *foliated*. Metamorphic rocks without obvious layers are *nonfoliated*.

Figure 6–28. Two of these rocks are foliated, and two are not.

Marble

Phyllite

Schist

Quartzite

Foliated rocks are formed by pressure from one direction. The layers are obvious, due to new mineral grains that have formed. Foliated rocks often contain micas, which are very distinct. Phyllite and schist are examples of foliated rocks. Nonfoliated rocks have no visible layers. Examples of nonfoliated rocks are quartzite and marble.

Some foliated rocks show a pattern of bands. When adjacent layers of foliated rocks contain different minerals, a layer of quartz next to a layer of mica, for example, the rock is banded. Both of the rocks in this photograph are metamorphic, but only the gneiss is banded. Which one is the gneiss?

Figure 6–29. Which of these foliated rocks is banded?

 **ASK YOURSELF**

How is banding different from foliation?

Uses of Metamorphic Rocks

Metamorphic rocks have many common uses. We have seen that marble is used extensively as a building material in Washington, D.C. Its surface polishes beautifully, displaying many interesting patterns. Because marble is also very durable, it is often used in monuments and memorials. By doing the next activity, you can find out how marble is used in your community.

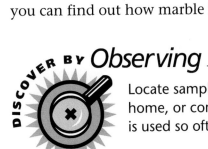

DISCOVER BY Observing

Locate samples of marble used in your school, home, or community. Why do you think marble is used so often in public buildings? ✏

Figure 6–30. Slate makes a good roofing material.

In addition to marble, other metamorphic rocks are used as building materials. Slate is used for roofing in some areas, because it breaks easily into thin slabs and it is waterproof and fireproof. Slate was once widely used for gravestones as well. The inscriptions on many of these stones, some over 200 years old, are still readable.

 ASK YOURSELF

Why are metamorphic rocks used for buildings and monuments?

SECTION 4 *REVIEW AND APPLICATION*

Reading Critically

1. What are the factors that help to form metamorphic rocks?
2. What are some of the uses for metamorphic rocks?

Thinking Critically

3. How does the presence of foliation provide information about the direction in which pressure was applied to a metamorphic rock?
4. Would you expect to find more metamorphic rocks near a plate boundary or in the middle of a plate? Explain your answer.

INVESTIGATION

*I*dentifying Rocks

▶ **MATERIALS**

- safety goggles ● laboratory apron ● rock samples ● Rock Classification Key, page 672 ● hydrochloric acid, 0.5M ● hand lens

▼ **PROCEDURE**

CAUTION: Put on your safety goggles and laboratory apron, and leave them on for the entire investigation.

1. Make a table like the one shown. Fill in the table as you complete the Investigation.
2. Select one of the rock samples. Try classifying the sample using the key on page 672. The key is organ-

ized to guide you, step by step, eliminating possibilities as you move through the key. For example, the first category is for sedimentary rocks. If your sample seems to be made of rock fragments cemented together, then you should continue through the subdivisions in Category I. If your sample does not look like a sedimentary rock, move to

Category II. Continue through the key until you find the category that best matches your sample.

3. Work through the category that best fits your sample until you find the name of the rock. When you decide on the name of the rock, fill in the information in the table.
4. Follow this procedure for each of your samples.

▶ **ANALYSES AND CONCLUSIONS**

1. Which rocks did you find to be the most difficult to identify? Explain the reasons for the difficulty.
2. Which rocks were the easiest to identify? Why?

▶ **APPLICATION**

If you owned some property with light-colored rocks on it, how would you determine whether the rocks were rhyolite, limestone, or marble?

✳ *Discover More*

Find several interesting rocks near your home, and try to identify them using the key. Add this information to your table.

TABLE 1: Rock Identification				
Rock	Name	Mineral Content	Rock Group	Key Classification Characteristics
1				
2				
3				
4				
5				
6				
7				
8				
9				
10				

The Big Idea

Rocks are the building blocks of the earth. By studying the structure of rocks, scientists can learn about ancient environments, movements and stresses within the earth's crust, and even the evolution of life. Rocks can be classified by their characteristics of texture and mineral content. But rocks are not static. They can and do change as they move through the rock cycle.

For Your Journal

Look again at the way you answered the questions at the beginning of this chapter. How have your ideas about the way rocks form and how they are used changed? Explain now why certain kinds of rocks are used for statues and memorials.

Connecting Ideas

Imagine a silicon atom—part of the granite rock created when the earth was forming. Today—4.6 billion years later—this atom is part of a sedimentary rock. Describe its travels through the rock cycle over the years.

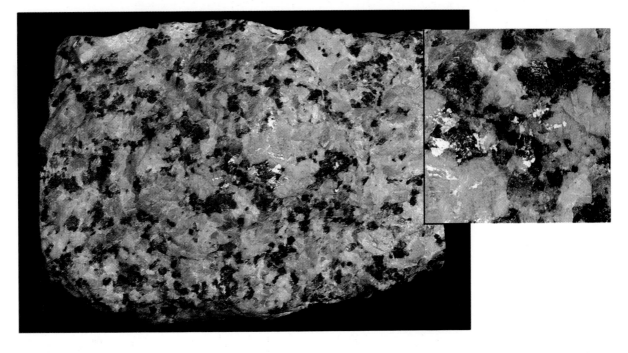

Understanding Vocabulary

1. Explain how the terms rock cycle (189), sedimentary rock (188), igneous rock (188), and metamorphic rock (189) are related.

Understanding Concepts

MULTIPLE CHOICE

2. Rocks are classified on the basis of texture and
   a) cooling period.
   b) mineral content.
   c) shape.
   d) weight.

3. Granite usually contains the minerals feldspar, mica, and
   a) quartz.
   b) olivine.
   c) pyrite.
   d) halite.

4. Which list contains only metamorphic rocks?
   a) basalt, schist, slate, marble
   b) slate, shale, granite, gneiss
   c) slate, schist, quartzite, marble
   d) slate, sandstone, schist, gneiss

5. A rock formed by the cementation of fine clay and silt-sized minerals is
   a) sandstone.
   b) mudstone.
   c) limestone.
   d) marble.

6. Which of the following is not a type of clastic sedimentary rock?
   a) limestone
   b) breccia
   c) sandstone
   d) conglomerate

SHORT ANSWER

7. Compare and contrast evaporites and precipitates.

8. What kind of sediments make up bituminous coal? How is bituminous coal formed?

9. Explain why the rock cycle is not a one-way process.

Interpreting Graphics

Use the drawing to answer the following questions.

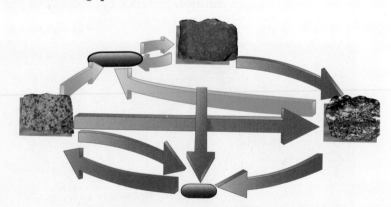

10. Which process must necessarily occur before the formation of sedimentary rocks? Explain why.

11. How can a sedimentary rock become an igneous rock?

Reviewing Themes

12. *Changes Over Time*
Explain how rocks from each of the three rock groups are formed.

13. *Systems and Structures*
How is the relationship between rocks and minerals similar to the relationship between a wall and bricks?

Thinking Critically

14. Would it be possible for fossils to be found in igneous rocks? Why or why not? Where are fossils most likely to be found?

15. Apollo astronauts brought back rocks from the surface of the moon. Scientists found the rocks to be almost exclusively igneous. What does this tell you about the moon's history?

16. Have you ever seen rocks that bend? Most people think of rocks as being very solid. The rock in the photograph is a type of sedimentary rock called *flexible sandstone*. The rock can bend without breaking. What characteristic of sedimentary rocks might allow them to bend in this way? Why are there no flexible igneous or metamorphic rocks?

Discovery Through Reading

Chadwick, Douglas H. "Arches National Park." *National Geographic Traveler* 9 (November/December, 1992): 46–57. Find out how the weathering of huge blocks of red sandstone have carved the unusual and almost impossible formations found in Arches National Park, on the Colorado Plateau.

CHAPTER 7

Conserving Resources

*E*very day you use products made from the earth's resources, such as glass bottles and aluminum cans. And every day you must decide what to do with these resources when you are finished with them. Whether you realize it or not, you already have an important role in managing resources. The decisions you make can affect the availability of resources that future generations will need. What **can** you do with discarded resources?

TOSS IT? Trash mars the landscape just about everywhere. Even the Pyramids of Egypt, symbols of an ancient civilization, are not immune from this symbol of modern civilization.

REDUCE IT! Every time you turn on a light, you use an energy resource. Energy resources can't be reused or recycled. But their use can be reduced.

RECYCLE IT! Many people recycle valuable resources. Your recycled aluminum can be "as good as gold" to future generations.

REUSE IT! This giant sculpture. called "Worlds Apart," was made from discarded metals.

BURY IT? Most waste today is buried in landfills. Future generations may have to "mine" the minerals they need from these sites.

Reusing, recycling, and reducing are good ways to save resources. By conserving resources, we can extend their availability. Conserving resources means using them wisely. Use the resources you need, but don't waste them. Wasted resources don't help anyone.

For Your Journal

- What other mineral resources do we use?
- What energy resources do we depend on most?
- Are there alternatives to the energy resources we use now?

Mineral Resources

Explain how all systems on the earth are interrelated.

Define the term natural resources.

Evaluate the need for conserving mineral resources.

Looking at the earth from space has truly changed the way we see our planet. Imagine how you would feel looking at the earth through the window of a planetary shuttle returning from Mars. Would you be awe-struck by the swirls of white clouds against the blue ocean? Would the experience change the way you feel about your home? For many people, this image of the earth has affected their behavior, leading them to recognize how fragile the earth and its interrelated systems are.

Natural Resources

Earth is a wonderous planet. It is often described as a spaceship, cruising the universe with us as its passengers, its interacting systems of atmosphere, oceans, and solid crust providing everything we need to survive.

Figure 7–1. Earth, as it appears from space

The atmosphere provides the air we need to breathe, distributes heat around the earth, and produces the weather that affects the way we live. The oceans and other waters of the earth are constantly in motion, eroding the land, modifying climates, and adding moisture to the atmosphere. The solid part of the earth is also in motion. You may recall that the crust changes through time, resulting in a replenishment of minerals and other resources from molten materials below the surface.

Many plants and animals have adapted to the various environments created by the interaction of the earth's systems. But people have found ways to use the earth's materials to change their environments. Any material from the earth that can be used in some way by people is called a **natural resource.**

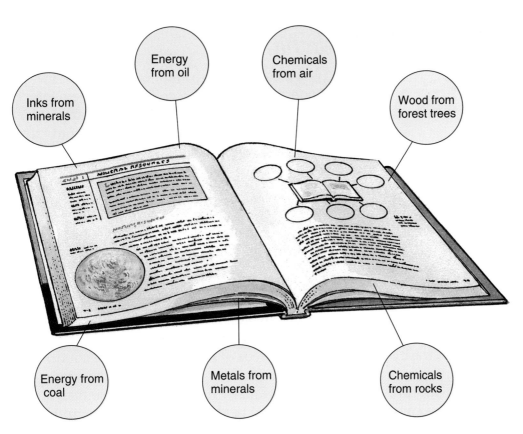

Figure 7–2. Many natural resources are needed for the manufactured products we use every day.

Inks from minerals

Energy from oil

Chemicals from air

Wood from forest trees

Energy from coal

Metals from minerals

Chemicals from rocks

Few of the earth's natural resources are used in their unaltered state. Most resources are made into products that people use to improve their lives. This book, for example, was made from many natural resources. The wood used to make the paper came from trees. Chemicals, made from rocks and air, helped to change the wood into paper. Minerals supplied the metals and the colored inks needed for printing on the paper. The energy for gathering and preparing the resources, manufacturing and transporting the materials, and printing and distributing the book came from coal and oil. With all our technology, it is easy to forget that the resources used for our basic needs and added comforts come from the earth.

There are two kinds of natural resources. Some resources, such as forests and water, are renewable. A **renewable resource** can be replaced, in time, after being used. But many of the earth's resources are not renewable.

Nonrenewable resources cannot be replaced or can only be replaced very slowly. It may take millions of years to replace minerals and fuels taken from the earth's crust. Most of the earth's iron-ore deposits, for example, were produced more than two billion years ago by bacteria that oxidized iron in the earth's oceans. In the following activity, you can learn more about nonrenewable resources.

DISCOVER BY *Researching*

Using pictures from magazines, make a collage of nonrenewable resources. Find out where they come from and how long they are expected to last, assuming the present rate of use. Record your information in your journal, and report your findings to your class. ✎

▶ ASK YOURSELF

In what ways do the earth's systems interact with each other?

Mineral Deposits

Americans are the world's greatest trash makers. In one year we produce enough trash to bury 1000 football fields under a pile of trash 90 m high! In hundreds of landfills across the country, piles of unsorted trash are built into mountains and covered with thin layers of soil. How many nonrenewable resources are you contributing to your local landfill every year? It might be more than you think. In the next activity you can find out.

DISCOVER BY *Calculating*

Figure 7–3. In addition to being sources of pollution, junked cars represent a tremendous waste of resources.

Analyze the mineral resources your family throws into the trash in one week. Catagorize the resources as glass, aluminum, steel, and so on. Calculate the amount of each by weight, and then estimate how much in each catagory is thrown away in a month and in a year. ✎

Landfills contain nearly every conceivable type of product, from old cars leaking toxic fluids to medical wastes containing radioactive materials. What additional concerns should we have about our growing landfills?

Look at the picture. In addition to the pollution, huge amounts of resources are wasted when cars, cans, bottles, plastics, and newspapers are buried in landfills. It is possible that, in the future, landfills will be our richest sources of minerals—perhaps our only sources. As you throw away that aluminum can, think about where the aluminum comes from, and what is required to turn it into a can.

Certain minerals are valuable because they contain economically important elements. For example, aluminum, copper, iron, silver, tin, and most other metals come from minerals. Any rock that contains a significant amount of these minerals is called a *mineral deposit*. A mineral deposit is considered an **ore** if it contains enough of the mineral to make mining and processing profitable. Ores can form in several ways.

Hot Water Have you ever been given an especially dirty dish or pan to wash? You probably used a lot of soap, muscle power, and hot water. Nearly all materials will eventually dissolve in hot water. This is true of many minerals, and is one way that mineral deposits form. In the first step of this so-called hydrothermal process, a mineral is dissolved from a rock by hot water. The term *hydrothermal* means "hot water." The hot water is usually ground water that is heated by a nearby source of magma. As the hot mineral solution flows through cracks in cooler rocks, the dissolved minerals precipitate, forming a deposit. This process often results in a *mineral vein* running through a rock. Mineral deposits containing copper, gold, silver, lead, and tin are often found in veins.

Weathering A second way that mineral deposits form is by weathering. Minerals can be separated from rock by physical weathering or by chemical weathering. For example, minerals containing aluminum are found in many kinds of rock. However, in most cases they are not concentrated enough to be ores. Sometimes chemical weathering breaks down aluminum-bearing rocks. Then rainwater removes quartz and other soluble minerals from the deposit, leaving an ore from which aluminum can be easily obtained.

Large deposits of aluminum ore are found in Jamaica. The weathering process is speeded up by the island's warm, humid climate, and the aluminum-containing minerals form deposits called *bauxite*.

Figure 7–4. This rock has several mineral veins running through it.

Figure 7–5. In the warm, humid climate of Jamaica, weathering concentrates minerals containing aluminum into bauxite, shown here being loaded onto a freighter.

Cooling Magma   Mineral deposits may form along with igneous rocks. Within a body of cooling magma, minerals tend to separate, the more dense ones sinking to the bottom. A mineral deposit rich in metals, such as chromium and nickel, can be formed this way. There are many of these metal-rich deposits in the province of Ontario, Canada.

Mineral deposits may also form from direct contact with an igneous source. Heat from nearby magma can even alter minerals within existing rocks, forming new deposits. Some of these contact deposits are found in the Iron Springs district of Utah.

 ASK YOURSELF

How do mineral deposits form?

Figure 7–6. Although the sea floor probably holds many mineral deposits, there is another way to meet the world's demands.

Conserving Mineral Resources

Mineral deposits form very slowly. It has taken millions of years to form the deposits that are in the earth's crust now, so minerals are considered nonrenewable resources. As the world's population grows, the need for mineral resources increases. Industrialized countries demand more and more resources as new technologies are developed. The need for minerals is growing in other parts of the world too. How can this demand be met?

One way to increase the supply of mineral resources is by finding new deposits. There are still large areas of the earth's surface, such as the ocean floor or under the Antarctic icecap, that have not been fully explored for minerals. Prospects for finding new deposits are good. However, the expenses of mining them will be high.

The increased need for mineral resources can partly be met by conservation. **Conservation** means using a material wisely. One method of conserving mineral resources is recycling. **Recycling** is the process by which a resource is recovered and used again. Certain metals, especially aluminum, are easily recycled. About half of all new aluminum cans are now made from recycled metal. It would be easy to increase that to 90 percent. It takes about 100 used cans to make 90 new ones, so very little new aluminum would be needed. Recycling aluminum also saves energy. It takes only about 5 percent as much energy to process recycled aluminum as it does to separate the metal from bauxite.

Figure 7–7. Many metals and other mineral resources can be recycled.

About one-third of the iron used today comes from recycled metal, such as junked cars. But we need to do much more. Mineral resources are limited. Conscientious citizens recognize that they must change their behavior to incorporate the three R's of conservation: *reuse, reduce,* and *recycle.* Think of creative ways to reuse materials rather than throwing them away. Try to reduce your use of certain resources. And when reuse and reduce are not options, recycle.

 ASK YOURSELF

What is saved when a mineral resource is recycled?

SECTION 1 *REVIEW AND APPLICATION*

Reading Critically

1. How are metal-rich mineral deposits formed?

2. What is the difference between renewable and nonrenewable resources? Give examples.

Thinking Critically

3. Explain why nonrenewable resources must be conserved.

4. Select a product made from a nonrenewable resource, and identify ways that the three R's of conservation can be applied to it.

SKILL Making Decisions

► MATERIALS

- paper and pencil

Nearly everyone agrees that steps must be taken to ensure that future generations will have the mineral resources they need. However, not everyone agrees on what these steps should be. Suppose you and your classmates are part of a task force that must decide how to solve the problem of wasted mineral resources in your community. You need to find out how the people of your community feel about wasting minerals and what they would be willing to do about it. To get the information your task force needs to make some decisions, you will need to take a survey.

▼ PROCEDURE

1. Select an issue that relates to the problem of wasted mineral resources. Likely issues might include community landfill policy, curbside recycling, programs to reduce mineral use, and so on.
2. Prepare 10 questions about the issue you have selected. You need to find out what people know about the issue and what they think should be done about it within the community.
3. Try your questionnaire on your classmates to make certain that the questions ask for the information you need. As your classmates respond to your questions, think about how you will interpret the answers you receive.
4. Decide on the characteristics of the population you wish to survey. The characteristics should reflect the community as a whole, since the answers of a few people will be used as a measure of the entire population.

► APPLICATION

1. Select 10 people in your community who match the population characteristics and ask them to fill out your questionnaire.
2. Summarize the data from your survey and present the results to your task force. Discuss the results, and decide what action the community should take.

✳ *Using What You Have Learned*

How is it possible to use information from a survey to decide what action to take on an issue?

Energy Resources

W hat do you think is the most economically important resource in the world? Do you think it might be gold or diamonds, or some exotic metal? Would it help you to know that in the 1970s the economy of the United States suffered severely as a result of less than a 20 percent reduction in the availability of this resource? You have probably guessed by now that the resource that is so important to our modern way of life is oil. No other resource is so directly involved in the way we live, work, and play.

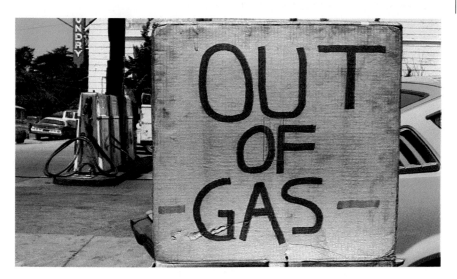

Figure 7–8. We never know how much we rely on some resources until they're gone.

Petroleum and Natural Gas

Oil belongs to a group of compounds know as *hydrocarbons*. Hydrocarbons are chemical compounds that contain the elements hydrogen and carbon. The simplest hydrocarbon, methane (CH_4), consists of four hydrogen atoms bonded to a single carbon atom.

Many hydrocarbons burn easily, so they are often used as sources of energy. Since they were made from once living organisms, these hydrocarbons are called *fossil fuels*. But hydrocarbons play a much larger role in our daily lives. Scientists have found ways of chemically altering hydrocarbon molecules to form many new products. The plastic chairs you sit on, the flexible soles of your sneakers, and the gum you chew all contain hydrocarbons.

Hydrocarbons exist as solids, liquids, or gases. A naturally occurring mixture of liquid hydrocarbons is called crude oil, or *petroleum.* Refining separates petroleum into gasoline, jet fuel, kerosene, diesel fuel, heating oil, fuel oil, motor oil, grease and other lubricants, and various petrochemicals. A naturally occurring mixture of gaseous hydrocarbons is called *natural gas.* Most natural gas is used for heating and for generating electricity. Solid hydrocarbons are called *asphalts.* Asphalts are used for paving and roofing materials.

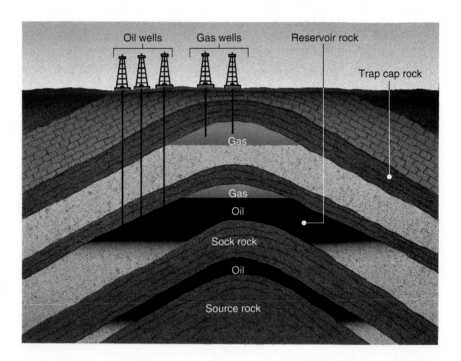

Figure 7–9. Hydrocarbons are produced in sedimentary deposits and trapped in porous rocks.

All the hydrocarbons we use today came from the decay of ancient forms of sea life hundreds of millions of years ago. The remains of these microscopic organisms settled on the sea floor, becoming part of the ocean sediments. With time, the sediments slowly changed into shale. It was within this shale, or *source rock,* that the chemicals of the decaying organisms were converted into oil and gas, often in as little as a million years. Under the pressure of overlying rocks and sediments, oil and gas were squeezed out of source rocks and into layers of more porous rocks. These rocks became *reservoirs* for petroleum and natural gas.

Nearly 60 percent of the world's petroleum is found in sandstone reservoirs. The remainder is stored in limestone and other porous rocks. In the next activity, you can calculate the capacity of various types of rocks to hold liquids.

MATERIALS

samples of sandstone, limestone, and shale; Petri dishes (3); medicine dropper; mineral oil

PROCEDURE

1. Place each rock sample in a separate Petri dish.
2. Fill the medicine dropper with mineral oil. Place 5 drops of oil on each rock sample.
3. Observe and record the time required for the oil to be soaked up by each of the rock samples.

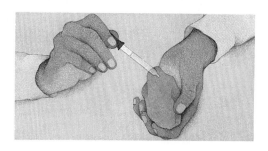

APPLICATION

1. Which rock sample soaked up the oil the fastest?
2. Which type of rock would probably make the best oil reservoir? Explain.

An oil-bearing formation is tapped by drilling a well into it. The oil in a new well is often under extreme pressure. If uncontrolled, the oil would gush violently from the well. To prevent this dangerous and wasteful situation, a special high-pressure valve is installed at the wellhead. After enough oil has been withdrawn, the pressure drops and the oil must be pumped out. To coax the last bit of oil from a reservoir, engineers may force steam or hot water into it to drive the remaining oil to the surface.

The United States has large deposits of petroleum and natural gas around the Gulf of Mexico; along the southern California coast; in Texas, Oklahoma, and Pennsylvania; and on the north slope of Alaska. There are also deposits of *oil shale* in Colorado and Utah. Oil shale contains unconverted hydrocarbons that can be extracted from the rock by heating. In spite of all these sources, we still import about 35 percent of our oil from the Middle East, South America, and Africa.

Figure 7–10. Oil must usually be pumped from a deposit, but occasionally a gusher will result if the oil is under pressure.

Figure 7–11. Obtaining oil deposited under the seas requires the construction of offshore drilling platforms.

Many petroleum and natural gas deposits are found off-shore, in sedimentary rocks under the sea. To drill in these locations, derricks are mounted on platforms, which are either set on the sea floor or floated on the surface. Large platforms, some more than 300 m tall, operate in the Gulf of Mexico and in the North Sea, between Great Britain and Norway.

 ASK YOURSELF

What are hydrocarbons, and how do they form?

Coal

Although the United States now imports about 35 percent of the petroleum it uses, it has vast supplies of another fossil fuel—coal. Coal used to be the leading source of energy in the United States. In the last half of the 19th century and the first half of the 20th century, most homes were heated with coal.

Figure 7–12. Most coal is used to generate electricity today.

Coal was also used in transportation—in the steam boilers of railroad locomotives and ships. Although coal is no longer used much as a heating or transportation fuel, it is very important in the production of electricity. Carbon, from coal, is also used in the steel industry.

Coal forms from decayed plants in areas that are, or were once, swamps. Fresh plant matter is about 80 percent water and 20 percent carbon. When plants die, they become waterlogged and sink to the bottom of the swamp. In the first stage of coal development, the activity of bacteria and fungi transforms the plants into *peat.* Peat is a dark brown, spongy substance that looks like rotted wood. It is about 60 percent carbon. In some places, peat is cut, stacked, dried, and used for fuel.

The second step in coal formation occurs as peat is buried under sediments and layers of additional peat. The pressure increases due to this additional weight, and the temperature rises. Water is expelled from the peat, and any oxygen in the peat combines with carbon to form carbon dioxide. Peat gradually becomes *lignite,* or brown coal, which is nearly 70 percent carbon. Lignite catches fire easily, but it burns with a smokey flame and produces relatively little heat.

With more heat and pressure, lignite becomes *bituminous coal.* Also called soft coal, bituminous coal is dark brown or black in color and breaks easily into blocks. Although it is the most abundant kind of coal and is about 80 percent carbon, bituminous coal burns with a smokey flame. The extensive use of bituminous coal as a heating fuel caused much of the air pollution that once darkened northern cities.

Figure 7–13. In some countries, peat is used as a fuel.

Figure 7–14. The burning of bituminous coal once darkened many cities.

Still more heat and pressure will change bituminous coal into *anthracite,* or hard coal. Anthracite is black, hard, and brittle. It burns with a blue, nearly smokeless flame and produces much heat, because it is more than 90 percent carbon. The diagrams and photos show the different stages of coal formation.

Peat   Lignite   Bituminous coal   Anthracite

Figure 7–15. Stages in coal formation

The major coal fields of the United States are in the Appalachian Mountains, extending from Pennsylvania to Alabama; under the prairies of the Midwest; and under the Great Plains and the eastern slopes of the Rocky Mountains. Coal beds range in thickness from a few centimeters to nearly 50 m.

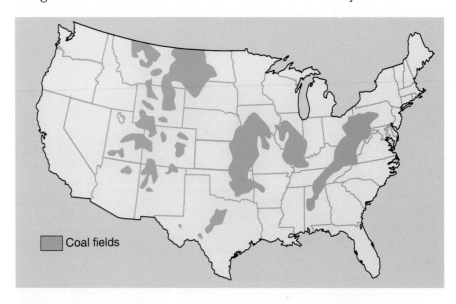

Coal fields

Figure 7–16. This map shows the locations of the major coal deposits within the contiguous United States.

ASK YOURSELF

Name the stages of coal formation.

Problems with Fossil Fuels

Although the presence of vast beds of coal is an economic security blanket for the United States, the use of coal is not without problems. The burning of coal can cause serious damage to the environment. Coal deposits often contain sulfur. When coal is burned, sulfur dioxide is released. Sulfur dioxide combines with moisture in the air to produce sulfuric acid, which is the acid in *acid rain*. Acid rain damages buildings and plants and is even responsible for fish kills in certain lakes. Nothing is immune to the effects of acid rain—the venerable Statue of Liberty, symbol of American freedom, was once in danger of being destroyed by acid rain. Only a costly renovation saved it.

Figure 7–17. Acid rain can damage statues and kill trees.

The mining of coal can create additional environmental problems. In some areas, coal is mined by stripping away the overlying soil and rock. If not properly repaired, strip mining permanently damages the land and destroys wildlife habitats.

Figure 7–18. Coal is obtained by strip mining and underground mining.

In other areas, coal is mined deep underground. This type of mining is costly and can be hazardous to the men and women working underground. Coal dust can damage the respiratory system, and it increases the always present danger of fire and explosion in the mines.

The biggest environmental problem with the burning of petroleum products is not acid rain; it is smog. **Smog** is a chemical fog, produced by the action of sunlight on automobile exhaust. Smog is particularly serious in places such as Denver and Los Angeles, where it is almost always sunny and where thousands of automobiles produce tons of pollution every day. Smog is also dangerous to human health. Since the 1940s, thousands of people have died from the effects of smog.

Figure 7–19. Automobile pollution is converted to smog by a chemical reaction started by the sun.

If we are not careful, we might find our smog problems solved the hard way—when we run out of fossil fuels. The supplies of fossil fuels in the earth's crust are limited. It has taken millions of years to make them, so they are considered to be nonrenewable. When the supplies are used up, there will be no quick way to replace them. This is especially true of petroleum, which is much less abundant in the earth's crust than coal. There is a great deal of disagreement as to just how long our petroleum will last at the current rate of usage. In the next activity, you can find out what the best estimates are.

Figure 7–20. Will there be enough petroleum for you to drive, or will you have to use alternate transportation?

DISCOVER BY *Researching*

Predictions of how much time we have before running out of oil and gas resources vary. In order to make such predictions, scientists must estimate how much will be used each year as well as how much remains in the ground, waiting to be recovered. Locate several reference books and articles about energy supplies and usage. Average any conflicting predictions about supplies, and decide whether alternative sources of energy will be necessary in your lifetime. ✎

▼ ASK YOURSELF

Why are coal, oil, and gas called fossil fuels?

SECTION 2 *REVIEW AND APPLICATION*

Reading Critically

1. How are hydrocarbons formed?

2. Why is coal an important energy source?

Thinking Critically

3. What are the dangers of dependence on fossil fuels?

4. What are the advantages and disadvantages of using the vast reserves of coal found in the United States?

Alternative Energy Resources

In the future, industrial, transportational, and residential energy needs will increasingly be met by electricity. However, most electricity is now produced from nonrenewable energy sources, such as fossil fuels. When these resources are used up, they will be gone forever. In the past, geologists have been able to discover new fuel reserves to balance those being consumed. For people to continue their present lifestyles, new sources of energy must become available before current supplies run out.

Nuclear Energy

You may already know something about nuclear energy. Many people have heard about the accidents involving nuclear energy plants at Three Mile Island, in Pennsylvania, and at Chernobyl, in Ukraine. But do you know that there are more than 400 nuclear energy plants still in operation around the world? Nearly 70 percent of France's electricity, for example, comes from nuclear energy. In the United States, nuclear energy produces less than 10 percent of our electricity.

Figure 7–21. There are dangers associated with nuclear energy, as the accident at Chernobyl showed.

The fuels used in nuclear energy plants, such as U-235, a form of uranium, are radioactive elements. Energy is released from the nuclei of radioactive atoms in a process called *fission*. When fission takes place, a large amount of heat is given off. The heat is used to produce steam to run electric generators.

Nuclear energy plants also produce wastes. The wastes are unsafe because they are radioactive. Radioactive wastes must be removed from the plant and stored until they lose their radioactivity. Nuclear wastes can remain dangerously radioactive for hundreds, even thousands, of years. A safe place must be found to put these wastes, a place where radioactivity cannot escape into the environment.

Nuclear energy plants are also potentially unsafe because of the way they are cooled—with water. If the cooling system were to stop working, the plant would overheat and could even melt. Then a large amount of radioactivity would escape into the environment, as it did at Chernobyl.

The engineers who build and operate nuclear energy plants assure the public that these problems can be solved. However, one problem with energy produced by fission may not have a solution. A piece of U-235 the size of a marble can produce as much heat as 100 metric tons of coal. Like coal, U-235 is nonrenewable, but unlike coal, there is not much U-235 left in the earth. All the U-235 could be used up within the next 50 years. Then nuclear energy plants would have to switch to more dangerous fuels, such as plutonium.

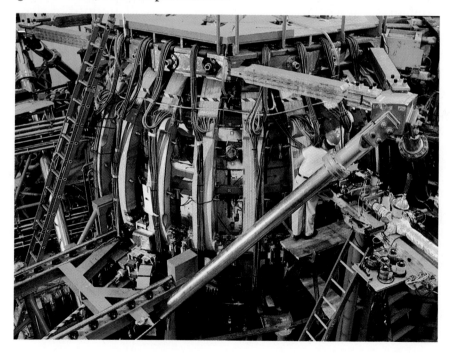

Figure 7–22. One day, fusion may provide a nearly inexhaustible supply of energy for generating electricity.

There is another type of nuclear energy that is potentially so abundant as to be considered inexhaustible. This energy source, called *fusion,* is produced when the nuclei of small atoms join to form new atoms. This is the same process that produces energy in the sun.

The main advantage of fusion compared with fission is that fusion produces few dangerous wastes. The main disadvantage of fusion is that very high temperatures are required for the reaction to take place. No known material can contain temperatures this high, so the reaction can occur only within an energy field. So far, fusion reactions have been limited to laboratory experiments.

▶ ASK YOURSELF

What are the advantages and disadvantages of nuclear energy?

Solar Energy

When sunlight falls on your skin, the warmth you feel is part of the energy the earth receives from the sun. Each year, the United States receives 500 times more solar energy than is currently provided by all other energy sources combined. And since we receive solar energy every day, it is a renewable resource.

Sunlight can be changed into electricity by the use of solar cells. You may have used a calculator powered by solar cells. The problem is, each solar cell produces only a small amount of electricity, so a large number of cells are needed to create a useful amount of electricity.

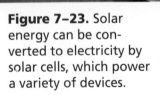

Figure 7–23. Solar energy can be converted to electricity by solar cells, which power a variety of devices.

While solar energy is free, solar cells are relatively expensive to make. In remote areas, where it is difficult and costly to run electric wires, solar panels positioned on roofs provide electricity for individual homes and businesses. There are even experimental solar-powered cars and airplanes.

Solar collectors are another way of using solar energy. Solar collectors are dark boxes with glass or plastic tops. Inside the boxes are liquid-filled tubes. After the liquid is heated by the sun, it is pumped into a tank, where it is stored until needed.

Figure 7–24. Solar energy can be collected to heat single rooms or an entire home.

In the photograph you can see a room with a different kind of solar collector. Here, drums of water are heated as the sun shines through a glass wall. After the sun sets, the water in the drums slowly releases heat into the room, keeping it warm during the night. This is similar to the way the oceans store the sun's heat and slowly release it, modifying climates.

In a solar-heated home, the stored energy must not be allowed to escape. One way to trap heat is with material called *insulation*. In the next activity, you can find out what kind of materials make good insulation.

Discover by Doing

You will need milk cartons, ice cubes, and insulating materials such as paper, cotton, and plastic. Wrap one ice cube in each of the materials to be tested. Then place each wrapped ice cube in a separate milk carton. The best insulator will keep the ice cube from melting for the longest time. Predict which material will insulate the best. Record your results in your journal. ✏

Figure 7–25. Solar energy can be used to produce electricity on a large scale in a solar thermal plant.

In southern California, electricity is being produced at a solar thermal plant in the Mojave Desert. This plant uses rows of mirrors that track the sun with the help of computers. The mirrors focus sunlight onto coated steel pipes filled with synthetic oil. The oil is heated to more than 400°C. The hot oil then heats water, which produces steam to turn electric generators. Solar technology is rapidly growing and holds great promise for areas that receive abundant sunshine.

ASK YOURSELF

Why is solar energy considered renewable?

Energy from Other Sources

The Wind

I saw you toss the kites on high
And blow the birds about the sky;
And all around I heard you pass,
Like ladies' skirts across the grass—
O wind, a-blowing all day long,
O wind, that sings so loud a song!

I saw the different things you did,
But always you yourself you hid.
I felt you push, I heard you call,
I could not see yourself at all—
O wind, a-blowing all day long,
O wind, that sings so loud a song!

O you that are so strong and cold,
O blower, are you young or old?
Are you a beast of field and tree,
Or just a stronger child than me?
O wind, a-blowing all day long,
O wind, that sings so loud a song!

Robert Louis Stevenson

Wind   Wind comes indirectly from solar energy. It is caused by the uneven heating of the air. There is a tremendous amount of energy in wind. You can see the effect of this energy, unleashed in a hurricane or tornado. Wind energy can also be used more productively. Some of the wind's energy can turn a windmill, pumping water or producing electricity. Small windmills have been generating electricity in the United States since the 1920s. In a few places, fields of modern wind turbines, technological updates of the old windmills, generate significant amounts of electricity.

Figure 7–26. Large wind farms can provide electricity for a small city.

Wind energy is renewable. However, there are not many places where the wind blows steadily enough to produce electricity on a large scale. In the next activity, you can determine whether there is enough wind where you live to generate electricity.

ACTIVITY

Is there enough wind to generate electricity?

MATERIALS
tape; colored thread, 20 cm; Ping-Pong® ball; protractor

PROCEDURE
1. Tape the thread to the ball and to the center of the protractor's curve.
2. Find a location outdoors where the wind is not blocked by trees or buildings. Hold one corner of the protractor so that the straight edge is parallel with the ground and the sharp edge of the curve faces into the wind.
3. When the wind blows the ball, the thread will line up with one of the marks on the protractor. Use Table 1 to estimate the wind speed.
4. Record the wind speed several times a day for a week.

APPLICATION
To produce electricity, the wind speed must be at least 13 km/h. The wind must also be constant. Would wind-generated electricity be practical where you live? Explain your answer.

TABLE 1: ESTIMATING WIND SPEED														
Mark on protractor	90	85	80	75	70	65	60	55	50	45	40	35	30	25	20
Wind speed (km/h)	0	9	13	16	19	21	24	26	29	31	34	37	41	46	52

Water The energy of falling water also comes from the sun. The sun's heat causes water to evaporate, mostly from the oceans. When the water returns to the earth as rain, it runs back to the oceans, releasing energy as it rushes downhill.

The energy of falling water has been used for thousands of years. Water wheels have been around since ancient times, turning giant millstones that grind grains into flour. Water wheels also provided energy for many factories in the early years of the Industrial Revolution. More recently, the energy of falling water has been used to generate electricity. Electricity produced by falling water is called *hydroelectric energy.*

Like wind energy, water energy is renewable. Water constantly cycles from the oceans to the air, to the earth, and back to the oceans. And like wind energy, water energy is not available everywhere. But the production of hydroelectric energy could be greatly increased. There are many large water resources in the world that are not currently being used to generate electricity.

Increased use of hydroelectric energy could reduce the demand for fossil fuels, but there are trade-offs. Construction of the large dams necessary for hydroelectric energy plants often destroys other resources, such as forests and wildlife habitats.

Figure 7–27. The sun starts a cycle of water that provides a renewable supply of energy.

Plants Plants are like green solar collectors, absorbing energy from the sun and storing it for later use. Leaves, wood, and other parts of plants contain the stored energy. This source of energy is called *biomass.*

Figure 7–28. Green plants can be turned into liquid fuel.

Biomass energy can be released in several ways. Wood can be burned as a fuel, or plant material can be changed into liquid fuel. Plants containing sugar or starch, for example, can be made into alcohol. The alcohol is burned as a fuel or mixed with gasoline to make a fuel mixture called *gasohol*. An acre of corn can produce more than 1000 L of alcohol.

Biomass is obviously a renewable source of energy, but producing biomass requires land that could be used for growing food. Most poor countries can't afford this kind of trade-off.

Geothermal Energy Imagine being able to tap into the energy of the earth itself. In a few places this is possible. This type of energy is called *geothermal energy*. Geothermal energy comes from heat within the earth's crust.

In some locations, rainwater penetrates porous rock near a source of magma. The magma turns the water to steam, which escapes through natural vents or through wells drilled into the rock. There is a place in California, called The Geysers, where escaping steam is used to generate electricity.

Geothermal energy can also be used as a direct source of heat. Buildings in some cities in Iceland are heated by hot water and steam from the country's many geothermal sites.

Figure 7–29. Geothermal energy can be used to generate electricity or provide heated water for swimming pools and homes.

 ASK YOURSELF

Why are wind, water, and biomass forms of solar energy?

SECTION 3 *REVIEW AND APPLICATION*

Reading Critically

1. Describe five energy alternatives to fossil fuels.
2. What are the safety concerns about the use of nuclear energy?

Thinking Critically

3. Where are people most likely to be able to take advantage of hydroelectric energy?
4. Why are so few alternative sources of energy in use in the United States today?

INVESTIGATION

Making a Solar Collector

▶ MATERIALS
- box ● black paper ● tape ● thermometers (2) ● transparent wrap
- light source

▼ PROCEDURE

1. Copy Table 1 into your journal.
2. Line the inside of the box with black paper.
3. Tape one thermometer inside the box. Be sure that the numbers on the thermometer are showing.
4. Cover the open side of the box with transparent wrap. Tape the edges to the sides of the box.

5. Place the box and the other thermometer under a light source. After five minutes, read both thermometers. Record the temperatures in your table.
6. Remove the light source. After five minutes, read both thermometers again. Record the temperatures in your table.

TABLE 1: SOLAR COLLECTOR DATA		
Condition	Inside Thermometer	Outside Thermometer
After 5 minutes with light		
After 5 minutes without light		

▶ ANALYSES AND CONCLUSIONS
1. Which thermometer had the higher temperature after the initial five minutes? Why?
2. Which thermometer had the higher temperature after the second five minutes? Why?

▶ APPLICATION
Based on your results, explain why solar collectors have a transparent cover and are painted black inside.

※ Discover More
Design a solar collector that will heat up more than the one in this Investigation. Test different ways to concentrate the light. Try varying the light angles or using aluminum foil to increase the heating.

The Big Idea

There are no isolated systems on the earth—all are interrelated. Natural resources are part of these interrelated systems. The world depends upon the availability of natural resources. Some of these resources will be available indefinitely because they are constantly renewed. Some may be available but will become unusable if the pollution they cause can't be controlled. Still others may be in such short supply that they will not be available for long. As responsible citizens, we must take action to preserve our natural resources and find alternatives to those in short supply.

For Your Journal

Look again at the way you answered the questions at the beginning of this chapter. Now that you have studied mineral and energy resources and methods of conserving them, how have your ideas changed?

Connecting Ideas

You have learned that natural resources are a major part of your life. You have also learned that some of them are in short supply. Develop a personal plan to help you reuse, reduce, or recycle resources important to you.

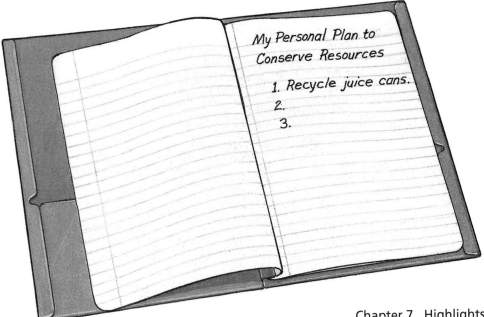

My Personal Plan to Conserve Resources

1. Recycle juice cans.
2.
3.

Understanding Vocabulary

1. Explain how the terms in each set are related.
   a) natural resource (216), renewable resource (217), nonrenewable resource (217)
   b) conservation (221), recycling (221)
   c) fossil fuels (223), acid rain (229), smog (230)

Understanding Concepts

MULTIPLE CHOICE

2. A mineral deposit that contains enough of a mineral to make the mining and processing of it profitable is called a(n)
   a) mineral vein.
   b) ore.
   c) natural resource.
   d) mineral resource.

3. Mineral deposits can be formed by
   a) hydrothermal processes.
   b) weathering processes.
   c) igneous processes.
   d) all of these processes.

4. Mineral deposits are
   a) natural resources.
   b) renewable.
   c) nonrenewable.
   d) both a and c.

5. The three Rs of conservation are
   a) reuse, reduce, and recycle.
   b) recover, reuse, and recycle.
   c) resources, reservoirs, and rocks.
   d) researching, refining, and recycling.

6. Peat, lignite, bitumin, and anthracite are all stages in the development of
   a) petroleum.
   b) natural gas.
   c) coal.

SHORT ANSWER

7. Why is coal a fossil fuel, but not a hydrocarbon?

8. Describe the process of obtaining oil from reservoir rocks.

Interpreting Graphics

Use this graph to answer the following questions.

9. What percentage of the world's total energy sources comes from coal? gas? hydroelectric?

10. Which two energy sources combined make up most of the world's total energy resources?

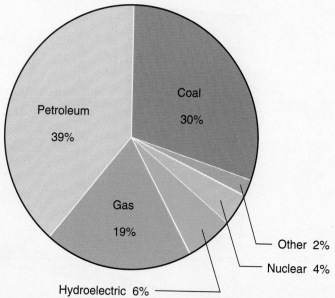

Coal 30%

Petroleum 39%

Gas 19%

Other 2%

Nuclear 4%

Hydroelectric 6%

Reviewing Themes

11. Systems and Structures
How would your life be different if all fossil fuels suddenly disappeared?

12. Environmental Interactions
Why is it important to conserve mineral and energy resources?

Thinking Critically

13. What risks are involved if the United States becomes more dependent on imported oil?

14. What are the environmental problems associated with the use of fossil fuels?

15. What are the advantages and disadvantages of using solar energy?

16. This is a photo of a nuclear breeder reactor in France. A breeder reactor makes electricity and at the same time makes plutonium fuel, which can also be used to produce more electricity. This eliminates somewhat the dependence on nonrenewable uranium as a source of fuel. However, plutonium is very dangerous to living organisms and produces wastes that take hundreds of thousands of years to decay. Discuss the advantages and disadvantages of using breeder reactors as an alternative way of generating electricity.

Discovery Through Reading

Rathje, William L. "Once and Future Landfills." *National Geographic* 179 (May 1991): 116-134. Digging deep into landfills, researchers seek to solve the problem of mounting waste, and uncover what's in our landfills and how long it lasts.

Science
PARADE

Recycling Aluminum Cans

When you toss your aluminum beverage can into a recycling bin, you are adding a page to an environmental success story. Recyclers of newsprint, plastics, and other materials face some tough problems, including too little demand for recycled goods. However, aluminum recyclers have usually been willing to pay hard cash for those shiny aluminum cans.

Aluminum is in demand.

EXPENSIVE AND COMPLICATED

The reasons for aluminum's recycling success lie in the technology of producing aluminum from bauxite ore. Producing the metal from bauxite is expensive and energy intensive. The ore must be dug out of the ground by a type of surface mining. Then the ore has to be purified. The purification process is expensive because it is complicated. Bauxite ore has to be crushed and mixed with sodium hydroxide. This mixture is then heated under pressure in a digester to dissolve the aluminum oxide, or *alumina,* out of the bauxite. This forms a solution of sodium aluminate. After further heating and mixing in a precipitator tank, the alumina is extracted as a fine white powder.

The alumina is now ready for smelting. In the smelting process, the alumina is dissolved at a temperature of about 950°C in a carbon-lined steel container called a *pot.* Carbon blocks are lowered into the dissolved alumina in the pot and an electric current is sent between the carbon blocks to release the oxygen still in the alumina. As a result, molten aluminum collects at the bottom of

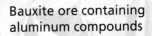

Bauxite ore containing aluminum compounds

the pot. The aluminum is siphoned into a large crucible and then poured out into molds to form blocks called *ingots* and *billets*. The ingots and billets are then shaped into the familiar forms that we associate aluminum with, such as cans, pots and pans, and aluminum foil.

RECYCLING SAVES

By using recycled aluminum, manufacturers can skip the expensive stage of purification. Recycled cans are almost pure aluminum. Cans are added to the smelting process, along with the new aluminum, melted down, and given whatever shape the manufacturer wants. Recycled aluminum can be back on the supermarket shelf in as little as six weeks.

The savings associated with recycling aluminum are enormous. One ton of recycled aluminum cans saves four tons of bauxite ore. In addition, a recycled can requires 70 percent less energy to produce than does a new aluminum can. Some states already recycle 70 percent of their aluminum cans.

It's not only possible that virtually all aluminum could be recycled, but perhaps even necessary. Recycling will become even more important as aluminum is used more widely. Auto manufacturers, for example, are now designing cars with aluminum frames and other aluminum parts. Aluminum is much lighter than steel and can increase fuel mileage, saving dwindling supplies of petroleum. And, of course, aluminum auto parts are more easily recycled. In twenty years, you may be leaving your old car at the recycling center, along with aluminum beverage cans. ◆

Aluminum, poured into molds (right), makes lightweight parts for cars (below).

THE GREAT AUSTRALIAN SUN RACE

By William Jordan from *Ranger Rick*

Australians line the streets to see the race.

Every few years, about two dozen weird-looking cars race from the top to the bottom of Australia. This is more than a race to see who can come in first. The racers also want to find out if sun-powered cars, or *solar cars*, might someday be able to take the place of cars that use gasoline.

The solar cars are built by inventors, engineers, and students. Some cars are even built by American car companies.

On top of most cars are panels that look like tables. And on these panels are glued hundreds of *solar cells*, the heart of sun-powered cars.

A solar cell contains special materials that turn sunlight into electricity. This electricity then powers the car's motor. The more cells that are used, the more energy is produced to move the car. The energy may be used right away, or it may be stored in a battery, so it can be used later.

This car's solar panel tilts toward the sun.

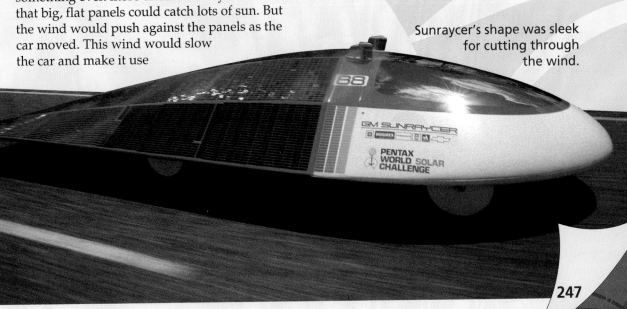

Solar cells work best when the sun is hitting their flat surfaces directly. But the sun doesn't stay in one place in the sky. So the inventors have to figure out how to arrange the cells to "catch" the direct rays of the sun all day long. That's why some of the cars look so strange.

Most of the inventors make big, flat panels that can be tilted to face the sun—no matter how low it is in the sky. A team once used a panel that curved up on both sides to catch the sun at any time of day. Another inventor used flaps that can be turned to face the sun each morning and afternoon.

Standouts in the Sun

Two of the recent designs are very unusual. A team from Hawaii designed a panel that looks a lot like a bent airplane wing. Like a wing, the solar panel uses the wind to help the car move.

The people from one car company tried something even more unusual. They knew that big, flat panels could catch lots of sun. But the wind would push against the panels as the car moved. This wind would slow the car and make it use more electricity than the solar panels could produce.

So the team built a sleek car that would move easily against the wind. And they bet that sleekness would be more important than catching as much solar energy as possible. They named their car "Sunraycer."

Ready, Set, Shine!

On the big race day, the hot sun rose into a clear sky. For two hours all the cars soaked up the sunlight. Then at 9:00 A.M. the race began. Sunraycer sped into the lead, with the Hawaiian car just behind. The cars headed south into the high mountains they had to cross. Then they drove into the desert in the center of Australia.

Sunraycer's shape was sleek for cutting through the wind.

Disaster struck the Hawaiian car that first afternoon. Its solar cells couldn't make enough power for it to climb one of the mountains, and there wasn't enough wind to give it a boost. So the driver had to use energy stored in the batteries, and that extra power ran out too. Then the car had to stop for the rest of the day. It never caught up, and it finally had to drop out of the race.

On the second day, a huge storm blew in from the Indian Ocean. Sunraycer kept just ahead of it, but other cars weren't so lucky. One car had to stop when a flash flood roared across the highway. Another car had to quit when its solar cells came unglued, and several cars lost wheels. By the end of the day, Sunraycer was 150 miles (240 km) ahead of the other cars.

Road Trains Roaring By

Storms were just one danger the cars faced. Another problem was the huge Australian trucks known as *road trains*. Road trains travel 80 miles (128 km) per hour, and everything has to stay out of their way. All of the solar cars had to be designed so wind from these gigantic trucks wouldn't blow them off the road.

When Sunraycer's driver saw the first road train of the race, he gripped the wheel tightly. As the truck roared by, the tiny solar car didn't even wobble. Sunraycer had passed one of its biggest tests.

Teaming Up to Win

Like the other cars, Sunraycer and its driver were never alone. A team of engineers, replacement drivers, and helpers followed it during the race. The drivers took turns at the wheel of the solar car. And the engineers were there to help if anything went wrong. Luckily, the only troubles Sunraycer had during the race were three flat tires!

By the rules of the race, all the cars had to stop at five o'clock every afternoon. Then for two hours, the cars sat in the hot sun to charge their batteries for the next day.

After setting up camp, Sunraycer's team planned the next day. They looked at maps to find out how

Road trains were another danger the cars faced.

248

many hills they would have to climb. And they checked the weather forecast. If the morning was to be clear but the afternoon cloudy, they'd go slower at first. That way they could save energy in the battery to use when the sun wasn't shining.

Days Ahead

On and on Sunraycer rolled. Kangaroos watched as it hissed across the desert. And each day the sleek solar car increased its lead.

Five and a half days from the start, Sunraycer crossed the finish line. The next solar car was two days and 600 miles (960 km) behind!

Sunraycer's sleek shape had worked just as the engineers had hoped. It had traveled an average of 42 miles (67 km) per hour for almost 2,000 miles (3,200 km). Not bad for a car getting its power from the sun!

Solar cars may never be as easy to use as gas-powered cars. They can do only what the sun and the weather allow them to do. But Sunraycer showed that sun-powered cars can make their way in this gas-guzzling world. ◆

René Just Haüy
(1743—1822)

When the French priest René Just Haüy accidentally dropped and broke a friend's calcite specimen, he was very embarrassed. But as Haüy picked up the mineral fragments, he noticed that all the pieces formed the same geometric shape.

René Just Haüy was born in Saint-Just, Picardy, France. Haüy became a Roman Catholic priest and studied at the colleges of Navarre and Lemoine where he became a botanist. The accident with the calcite redirected Haüy into mineralogy.

Haüy later tested more calcite and found it always took the shape of a rhombohedron, a slanted box. Haüy then hypothesized that each mineral is built up of layers or sections of a simple geometric shape. He proposed that this geometric shape is related to the internal crystalline form or molecular makeup of a mineral. For example, quartz crystals are always hexagonal (six-sided) and have a pyramidal top.

Haüy became a professor of mineralogy at the Museum of Natural History in Paris during the reign of Napoleon. He wrote the first important textbook on crystallography and established the science of crystallography. Hauy's work led to the development of many consumer products based on the characteristics of quartz. ◆

Marjory Stoneman
Douglas (1890—)

When Marjory Douglas's book, *The Everglades: River of Grass* was published, she became a proponent for preserving the Everglades. Today, Everglades National Park is the result of her efforts.

Marjory Stoneman Douglas was born in Minneapolis, Minnesota. In 1912 she graduated from Wellesley College in Wellesley, Massachusetts. As a reporter and editor for the *Miami Herald* in Miami, Florida, Douglas discovered the beauty and truth of the Everglades. She observed that the Everglades were not a forbidding swamp as previously thought, but a shallow river of grass that flowed southeast from Lake Okeechobee to the Atlantic Ocean.

Douglas later became an instructor at the University of Miami. At the university she wrote and published many short stories for the *Saturday Evening Post*. Many of her stories are about the Everglades.

Douglas's work has won her many wildlife awards. The National Park Service established a special conservation award named in her honor. The state of Florida even named its Department of Natural Resources building after her. ◆

Marisa Quiñones, GEOLOGIST

Marisa Quiñones is a geologist for an oil company. Her job is a lot like that of a detective; it requires skills in observation, logic, and attention to detail. She uses these skills and a variety of data to help her locate hidden oil reserves beneath the surface of the earth.

Seismic data is the single most important type of information that Quiñones relies on when looking for oil and natural gas. This type of data is generated by setting off dynamite in the ground or an air gun in the sea and recording the response of underground rocks to the generated sound waves. A seismograph records these sound waves as wavy lines that produce a cross-section of the earth's interior. Quiñones examines the seismic data using sophisticated computers and generates structure maps to locate oil "traps." Traps are areas where oil has accumulated.

After she examines the available data, Quiñones must determine the risks of finding oil and estimate how much may be found. If it seems there is a good chance of finding a large accumulation, she will recommend drilling a well.

At the well site, Quiñones examines rock chips ("cuttings") that come out of the borehole as the well is drilled. The cuttings allow her to unravel the subsurface geology, to predict any possible danger to the drilling operation, and to know when the trap has been reached. After the well is drilled, Quiñones analyzes well logs that are generated by running instruments down the hole. Well logs provide detailed information about the oil traps and the characteristics of the rocks in the subsurface. Favorable findings will determine whether the well will be completed and put into production.

Quiñones' work is just one aspect of the many careers that exist in geology. Geology is much more than just looking at rocks; it can involve finding water in dry countries that have limited resources, searching for precious rocks and minerals, or studying the earth's history. ◆

DISCOVER MORE
For more information about careers in geology, write to the

American Geological Institute
4220 King St.
Alexandria, VA 22302

COMPUTER CHIPS

Printed circuit

The flow of electric current in all computers relies on circuits. These have changed over the years. A computer chip in the late 1950s could hold only one circuit. Today a single chip can have more than 1 million circuits.

Chip Off the Wafer

A thin slice of silicon called a wafer is used to make computer chips. Hundreds of separate computer chips can be made from a single wafer.

Silicon is ideal for use in microcircuitry. Silicon is a semiconductor. That means sometimes it can conduct electricity and sometimes it cannot. Semiconductors are resistant to a wide range of temperatures. This natural resistance to heat makes silicon chips more efficient than other types of chips.

Controlling the Wafer

The electrons in the outer energy level of most atoms react readily with other atoms. This is not the case with silicon. The outer electrons of silicon are tightly bound to each other. By inserting materials in silicon that can "donate" or "steal" electrons, this situation can be changed. The chips' electrical conductivity is also changed by these insertions. This is called doping the chip. Doping allows the wafers' flow of electric current to be precisely controlled.

Great precision is required for computer chip production. Computer chips are "printed" in much the same way that photographic plates are made for printing magazines. This procedure is called optical lithography. Since the circuits are

252

microscopic, the chips are made under microscopes.

Smaller and Smaller

The size of computer chips continues to get smaller and smaller as the manufacturing technology improves. In the 1960s, the smallest circuits were about 25 micrometers (μm) across. That size was reduced to 2.5 μm during the 1980s. By the year 2000, some scientists predict that the size of a circuit could be reduced to as small as 0.25 μm!

These advances in miniaturization will be dependent upon new photographic processes, and the chemical and structural characteristics of the raw materials used to make the chips. ◆

Data processing center

A laptop computer

Physicist Luis Alvarez examined a thin layer of clay rock about 65 million years old. Alvarez knew that at about the same time the rock formed nearly all the dinosaurs had vanished in a mysterious mass extinction. He concluded that the clay had been deposited when a giant meteor struck the earth. Alvarez claimed that the dinosaurs died when a cloud of meteor dust covered the earth. What do you think? This unit will help you understand more about Alvarez's hypothesis and the history of the earth.

Science PARADE

The Rock Record

On his way into the Grand Canyon, science writer David Attenborough describes the pink, red, and blue layers of rock in the canyon wall. But at the bottom, he finds that he still has a scientific mystery on his hands. Perhaps you can help solve this mystery.

A mule will carry you in an easy day's ride from the rim to the very bottom of the Canyon. The first rocks you pass are already some 200 million years old. There are no remains of mammals or birds in them, but there are traces of reptiles. Close by the side of the trail, you can see a line of tracks crossing the face of a sandstone boulder. They were made by a small four-footed creature, almost certainly a lizard-like reptile, running across a beach. . . .

Halfway down the Canyon, you come to 400-million-year-old limestones. There are no reptiles to be found here, but there are the bones of strange armoured fish. An hour or so later—and a hundred million years earlier—the rocks contain no sign of backboned animals of any kind. There are a few shells and worms that have left behind a tracery of trails in what was the muddy sea floor. . . . By the late afternoon, you ride at last into the lower gorge where the Colorado River runs green between high rock walls. You are now a vertical mile below the rim and the rocks have been dated to the immense age of 2000 million years. Here you might hope to find evidence for the very beginnings of life. But there are no organic remains of any kind. The dark fine-grained rocks lie not in horizontal layers like all those above, but are twisted and buckled and riven with veins of pink granite.

Are signs of life absent because these rocks and the limestones directly above are so extremely ancient that all such traces have been crushed from them? Could it be that the first creatures to leave any sign of their existence were as complex as worms and molluscs?

from *Life on Earth: A Natural History*
by David Attenborough

For Your Journal

- Why do you think there are so many different layers of rock on the earth?
- Why do some rock layers contain the remains of only certain kinds of animals?
- Why are there missing layers in the rock record?

History in the Earth

Objectives

Determine *why uniformitarianism is important in science.*

Distinguish *younger rock layers from older rock layers based on position.*

Match *the characteristics of sedimentary rocks to the environment in which they were formed.*

If you stand at the edge of the Grand Canyon and look down, the Colorado River is almost 2 km below you. See how small the boat looks in the picture? There are several trails in Grand Canyon National Park that can take you from the rim of the canyon all the way down to the river. One of these trails, the Bright Angel Trail, winds along the canyon wall for nearly 13 km. Native Americans used this route hundreds of years ago. Later, prospectors and explorers used it to get to the river below. Imagine that it's your turn now! You can either ride a mule down the trail or go on foot. You have decided to hike.

Principles of Geology

Before you go, consider the river far below. Have you ever thought about what it would be like if water flowed downhill one day and uphill the next? Or if you never knew what color the sky would be when you looked outside each morning? What if you weren't sure whether gravity would be in effect tomorrow? If you think about it, there are a lot of things we assume will be the same from one day to the next. In fact, we could not live without this kind of sameness, or uniformity, in our lives. The following activity will help you find examples of uniformity in your own life.

Figure 8–1. You can learn a lot about the history of the earth by hiking to the bottom of the Grand Canyon.

Discover by Writing

Make a list of 20 things that are the same, or uniform, from one day to the next. Write the list in your journal. Then compare your list with your classmates' lists. Does everyone have the same things on their lists? Why are there so many differences on the lists? ✎

Uniformity is important for scientists, too. Scientists depend on the fact that some things will always be the same. In fact, a very important scientific principle is based on the sameness of certain processes. It is called the **principle of uniformitarianism,** and it states that processes that occurred in past times produced the same results as similar processes do today. It also tells us that processes occurring today will produce the same results in the future. For example, suppose we mix two chemicals together today and observe what happens. If we mix the same two chemicals together tomorrow or next year or a hundred years from now, we know that under the same conditions they will react the same way. Can you imagine what the world would be like without this uniformity?

Many scientific processes follow the principle of uniformitarianism. But what about the order in which things happen? Is this always the same too? The next activity may help to give you an idea.

Discover by Doing

Take a deck of cards, and let it fall to the floor. Some cards end up under others. Which cards landed first? Which ones landed last? ✎

Figure 8–2. Which of these rock layers is the youngest?

When you dropped the cards, the ones that were on the bottom of the deck settled to the floor first. Then they were covered by the cards that were on the top of the deck. The deposition of sediments on the earth's surface works the same way. Look at the layers of sandstone in the picture. The oldest layers are on the bottom. The layers that were deposited last are on top. Look again at the picture of the Grand Canyon at the beginning of the section. Which rock layers are the oldest?

The position of rock layers tells us something about the relative ages of the rocks—younger rock layers are found on top of older rock layers. Geologists call this the **principle of superposition.** Because things happen uniformly in our world, we assume that the principle of superposition has always been true.

▼ ASK YOURSELF

What would our world be like without the principle of uniformitarianism?

Puzzles in the Rocks

It's time to start your hike. You check your water supply and shoulder your pack. As you hike down the trail, you see many different layers of rocks within the canyon walls. You know that as you go down into the canyon, the rock layers get older. But what else can you learn by looking at the rocks?

Past Life Some of the rock layers you see have traces of sea animals in them, like those in the picture. Other layers have the remains of land animals. Some layers, like the ones in the picture below, look like cross sections of sand dunes. Still others are made up of very fine mud, like that found at the bottom of calm, deep water. What do all these different puzzle pieces tell you about the geological history of the Grand Canyon?

Figure 8–3. Animals similar to these live in the sea today. Uniformitarianism tells us that long ago these animals lived in the sea as well.

Figure 8–4. These rock layers were formed from sand blown by the wind.

As you stop to rest, you see a mule train approaching you. You back up against the canyon wall to let the mules pass, and you think about what you have seen so far. You have seen many geological puzzles during your hike. Much of geology is based on solving puzzles. What was the environment like when these particular sedimentary rocks were formed? You know the sediments here are older than those higher up the trail, but was the environment the same or different?

Figure 8–5. Time to move over!

If you were to see the remains of corals in limestone, you would assume, because of the principle of uniformitarianism, that a shallow ocean once covered the area. If you were to see layers of sandstone with deposits that look like those found in sand dunes, you would assume that the environment was very dry. If the coral limestone layer was under the sandstone layer, you might hypothesize that either sea level had fallen or the land had risen. In either case, the environment had changed from ocean to dry land.

Broken Records   As you hike down to the river, you see a geological record of what the past environments of the Grand Canyon were like. But you have to be careful. Some of the rocks may be missing. It's as if you let your deck of cards drop to the floor, and then someone—unseen by you—removes a few cards from the middle of the pile.

You can't be sure whether all the cards are there by looking quickly at the pile, and you can't be sure all the rock layers are there by looking quickly at the canyon walls. If some of the rock layers are missing, that means the geological record is broken. But how could layers of solid rock disappear?

Think about what happens to a beach when a storm strikes. Whole sections of beach are eroded, and the sand is redeposited somewhere else. This is just one way in which a break could occur in the geological record.

Figure 8–6. Locate the unconformity in these layers.

Notice the distinct break in the sedimentary rocks in the picture. Breaks are called *unconformities* because they do not conform to the geological record. Unconformities let you know that something is missing, but they make solving the puzzle of the geological history of an area more complicated.

 ASK YOURSELF

Why is much of historical geology like solving a puzzle?

SECTION 1 *REVIEW AND APPLICATION*

Reading Critically

1. Why is uniformitarianism so important in science?
2. Why do unconformities complicate the puzzles that geologists solve?

Thinking Critically

3. Suppose you find a sedimentary rock that contains corals. Where was this rock formed? How do you know?
4. Diagram a series of sedimentary rock layers that show that the level of the sea had dropped rapidly.

SKILL Interpreting Diagrams

▶ **MATERIALS**
  ● paper ● pencil

▼ **PROCEDURE**

1. Study the diagram. Which group of rock layers is older, group a or group b? How do you know, or how can you tell from the diagram?
2. Which numbered layer is the oldest of the horizontal layers in group a?
3. Which numbered layer is the oldest of the tilted layers in group b?
4. Were the tilted layers of group b level at one time in history? How do you know?
5. Did the movement along the fault in the group b layers take place before or after layer 9 was laid down? How can you tell?
6. Which layer was the first to form after the tilting of group b?
7. Which layers of rock must have been laid down before the magma, labeled "d," pushed its way in?

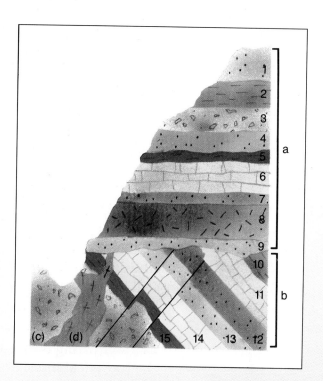

▶ **APPLICATION**

An unconformity might represent a time during which erosion was taking place at a particular location. The layer being eroded loses its flat appearance. As new layers form over this eroded layer, the bottom of the first new layer remains uneven. Which layers show periods of erosion?

✳ **Using What You Have Learned**

Look at the diagram again. Layers 6, 11, and 14 are deposits of limestone. Layer 3 contains the remains of ancient sea life. Layer 7 contains no such remains and is more fine grained and dense than layer 3. What can you say about the conditions that existed at the time layer 3 was laid down? What reasons can you give for layer 7 not having any remains?

Dating Rocks and Fossils

Objectives

Explain *the difference between relative age and absolute age.*

Describe *how fossils are made.*

Analyze *a geological diagram, and explain the sequence of events it shows.*

You can easily imagine the length of a week or a month. These are periods that most of us talk about all the time. How about a year? We know that some people are older in years than others. Middle school or junior high school students are usually older than elementary school students. Grandparents are older than parents, and great-grandparents are older still, maybe 80 or 90 years old. A few people even live to be more than 100 years old. Most of us would agree that 100 years is a very long time, but it's like the blink of an eye in geological time. In geology, we talk about millions of years. And some dinosaurs lived more than 100 million years ago! That's really hard to imagine. How are we able to assign ages to these very old geological events?

Radioactivity and Absolute Dating

Figure 8–7. How can you determine the absolute age of rocks?

You continue your hike into the Grand Canyon. You know that the deeper you go, the older the rock layers are. But all you can tell is how old one layer is in relation to another. This is *relative age*. What if you wanted to know how old a layer was in terms of years? You need to know the layer's *absolute age*.

Look at Table 8–1. It shows a few examples of radioactive elements used in absolute dating. Note that each element has a half-life. **Half-life** refers to the amount of time that it takes for half of the unstable element to become a more stable element. Knowing the half-life of an element allows you to determine the absolute age of rocks and minerals. In the next activity, you can learn more about radioactive elements and half-lives.

Table 8-1 Elements Commonly Used In Absolute Dating

Isotope	Symbol	Half-life (years)	Material
Tritium	H-3	12.3	Water, ice
Carbon-14	C-14	5730	Wood, bones, shells
Potassium-40	K-40	1.3×10^9	Rocks, minerals
Rubidium-87	Rb-87	4.9×10^{10}	Rocks, minerals
Thorium-232	Th-232	1.4×10^{10}	Rocks, minerals
Uranium-235	U-235	7.04×10^6	Rocks, minerals
Uranium-238	U-238	4.5×10^9	Rocks, minerals

ACTIVITY

How can half-lives be used to determine the age of a sample?

MATERIALS
100 pennies, shoe box, clock with second hand

PROCEDURE
1. Copy the table.

TABLE 1: DATA		
Shake	Time (seconds)	Pennies in box
0		100
1		
2		
3		
4		
5		
6		
7		
8		
9		
10		

2. Place 100 pennies in the shoe box so that all the pennies are heads up.
3. Record the time.
4. Cover the shoe box and shake it vigorously so that the pennies are well mixed.
5. Open the shoe box, and remove all the pennies that are tails up.
6. Record in your data table the time and the number of pennies remaining in the box.
7. Repeat steps 4–6 until there are no pennies left or until the table is complete. How many times did you shake the box before all the pennies were gone?

APPLICATION
1. Let's say your classmates stopped you, noted the time, and counted only 12 pennies remaining. How could they calculate the time that you had started the experiment?
2. Imagine that the shoe box is a fossilized bone that contains 25 units of carbon-14. When it was buried, it contained 100 units of carbon-14. How old is the bone?

One example of the absolute dating technique involves the use of carbon-14 (C-14). Some elements occur in several different forms, called *isotopes*. Carbon also occurs as carbon-12 (C-12). Some isotopes are very unstable and tend to change into more stable forms. This change is called *radioactive decay*. C-14 is unstable. It decays fairly quickly, forming nitrogen-14 (N-14), which is stable. All three, C-14, N-14, and C-12, are found naturally in organic, or once living, matter.

All living organisms have the same ratio of C-12 to C-14. But when an organism dies, the amount of C-14 starts to decrease. As time passes, the ratio of C-12 to C-14 changes. While the amount of C-14 is less every year, the amount of C-12 is the same. The half-life of C-14 is 5730 years. So 5730 years after an organism dies, it has only half of its original C-14. The diagram shows the amount of C-14 remaining after additional half-lives.

Carbon-14 remaining	Time
$1/2$	5730 years
$1/4$	11 460 years
$1/8$	17 190 years
$1/16$	22 920 years

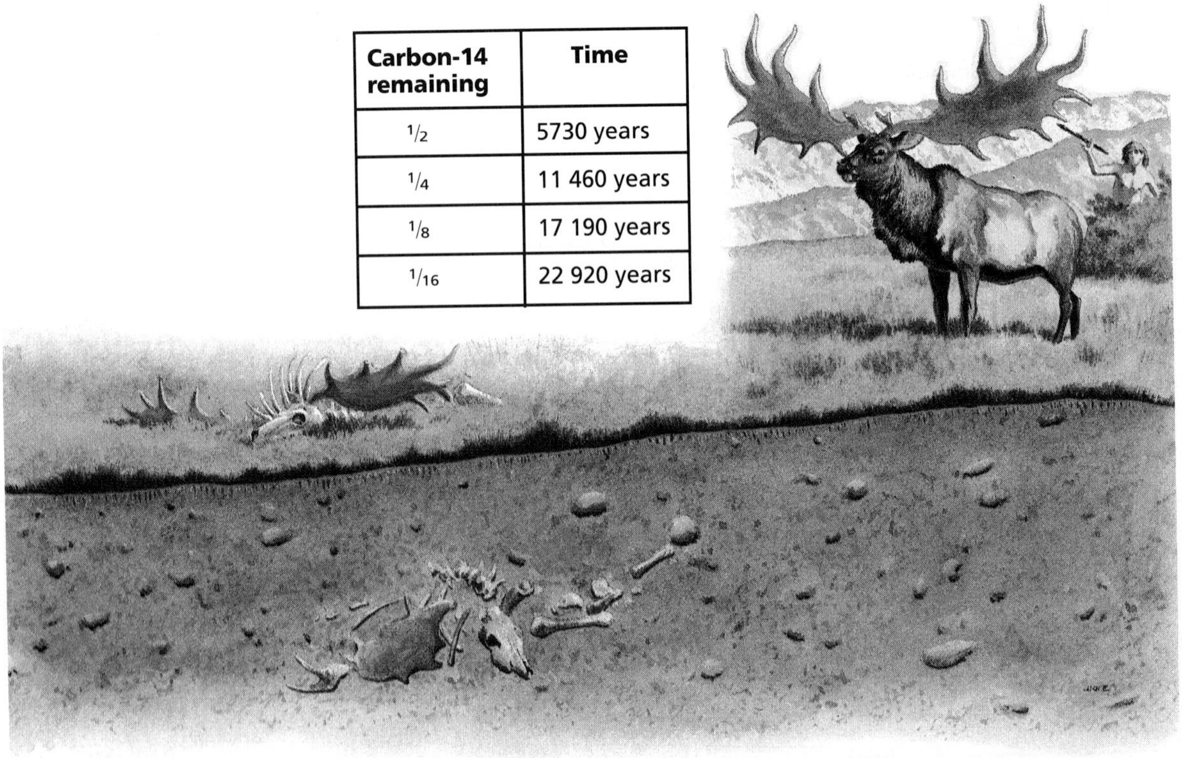

Figure 8–8. C-14 can be used to date any organic material.

You could take a bone from an animal that died within the last 50 000 years and send it to a lab that can measure carbon isotopes. The lab would tell you the ratio of C-12 to C-14. You could then figure out when the animal died.

 ASK YOURSELF

Why is C-14 useful for dating the remains of plants and animals?

A Rocky Date

Remember, to get the absolute age of a sample, you need to measure the amounts of both an unstable isotope and its stable decay product. C-14 is useful for dating relatively young, organic materials. Other isotopes, such as those of uranium, can be used to date older materials, such as rocks and minerals. Look again at Table 8-1, and explain why.

Dating igneous rocks is pretty straightforward—nothing has changed them since they were formed. You have to be careful, however, when dating sedimentary rocks. For example, would the absolute age of a grain of sand in a sedimentary rock be the same as the age of the sedimentary rock or the same as the age of the parent rock from which the sediment formed? The next activity will help you understand how you can determine the age of sedimentary rocks.

DISCOVER BY Problem Solving

Look at the diagram. It shows the result of four geological events:
1. Sedimentary layer A being deposited
2. Sedimentary layer B being deposited
3. Sedimentary layer C being deposited
4. An igneous intrusion

Which of the four events occurred first? Which occurred last? What is the relative age of each of the four events?

Now make a similar diagram as a puzzle for your classmates. Make it with more layers and more intrusions. Make it as complicated as you can, and see whether anyone can solve it. ✎

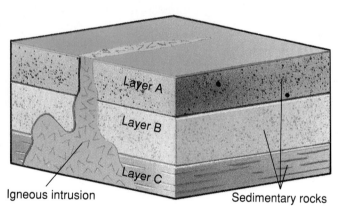

Igneous intrusion

Layer A

Layer B

Layer C

Sedimentary rocks

Geologists find natural puzzles that are harder to solve than almost any puzzle you could think of! And they look for every clue to solve their puzzles. The easiest way to date sedimentary rocks is to find out the age of certain plant or animal remains, called *index fossils,* that may be in the rocks. If you know when a plant or an animal died and was covered with sediment, you pretty much know when the rock formed.

Some organisms lived for only a short period of geological time. This doesn't mean they had short life spans, but that the species was not on the earth for very long. Once you know the absolute age of these index fossils, you know the age of the sedimentary rock layer as well.

 ASK YOURSELF

Explain how the age of a fossil can tell you the age of the sedimentary rock around it.

Fossils

Now you watch for certain fossils as you continue your hike. They can help you figure out the age of the rocks in the canyon walls. But what exactly are fossils? And how do they get embedded in rocks? Read the poem for some ideas.

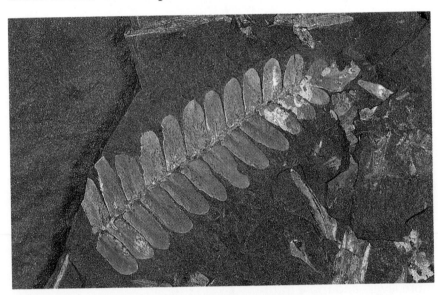

Figure 8–9. How did fossils get into these rocks?

Fossils are the remains or traces of plants and animals that lived in the past. Fossils can be as big as dinosaur bones or as small as bacteria. Bones and shells are obvious fossils. Tracks and burrows can also be fossils, since they are the traces of organisms.

Sometimes when original material such as a bone or shell survives, you can calculate the absolute age of the fossil. Bones and shells can remain in their original form for a long time. But bones and shells can also be replaced by minerals. Petrified wood is an example of this kind of fossil.

Several other processes can cause plants and animals to become fossils. For example, many animals from the Ice Age have been found frozen in glaciers, almost unchanged from the day they died. They are like the frozen meats and vegetables in your freezer. Other animals have fallen into naturally occurring tar pits and have been well preserved. The photo below shows part of the La Brea tar pits in California. Much of what we know about saber-toothed tigers comes from intact fossils found in these pits.

While no one has ever seen a live saber-toothed tiger, we have a good idea that they looked like this one. In fact, scientists can even tell what the big cats ate by looking at the contents of their stomachs. The remains of their prey are also preserved by the tar. In the next activity, you can try making your own "fossils."

Figure 8–10. This mammoth (above) was preserved in the frozen soil of Siberia for thousands of years.

Figure 8–11. The saber-toothed cat, whose fossil remains were found in the La Brea tar pits in California, probably looked like this (left).

DISCOVER BY Doing

Take a small piece of clay, and form it into a disk about 3 cm or 4 cm across. Carefully press your thumb into the clay. Then allow your "fossil" to harden overnight. Compare your print to those of your classmates. See whether you can match each print with the right person. ✎

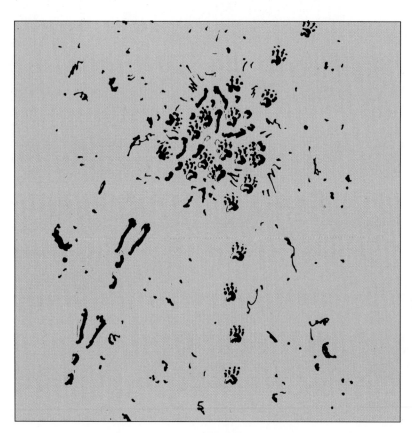

Figure 8–12. What do you infer happened here?

Although you cannot tell the absolute age of an animal without its physical remains, you can tell a lot of other things from the traces it leaves behind. The depth of dinosaur footprints, for example, can give you some idea of how heavy they must have been. Some prints include a center track. This tells you the animal had a tail that dragged behind, like some modern lizards. You can also learn a lot about fossil prints by studying modern tracks. Look at the tracks in the picture. What do you think happened here? Now see how creative you can be by doing the following activity.

DISCOVER BY Problem Solving

Your teacher will give you a drawing of some animal tracks that tells a story about the animal that made them. The tracks may show how the animal moved, or they may show what the animal was doing, such as hunting another animal. The tracks may be from more than one animal. They may even show what happened when the two animals came together. Discuss the tracks with your classmates. See how well you can solve this puzzle together. ✎

At last you have reached the bottom of the Grand Canyon. Now you can cool your feet in the Colorado River. It's a lot warmer down here than it was at the top, so the water feels refreshing.

You still have plenty of time to set up camp for the night, so you take some time to think about all you have seen. Your hike down the Bright Angel Trail has taken you back nearly half a billion years in geological time! What if the footprints you leave on the bank of the river don't get washed away before the sand turns to sandstone? Will future hikers be able to interpret your fossils correctly, or will your trip leave another puzzle for future geologists to solve?

Figure 8–13. You have just hiked through millions of years of Earth's history.

Figure 8–14. Trying to solve another geological puzzle

 ASK YOURSELF

Why are animals that have been frozen so useful to scientists?

SECTION 2 *REVIEW AND APPLICATION*

Reading Critically

1. How is absolute age different from relative age?

2. What can fossils tell you about the rocks in which they are found?

Thinking Critically

3. If a piece of wood contains only one-fourth of its original C-14, how old is that piece of wood?

4. How would you explain a formation in which the younger fossils are underneath the older ones?

INVESTIGATION

Making Fossil Molds

► MATERIALS
- a variety of shells • bones from a chicken or other animal • hacksaw
- petroleum jelly • plaster of Paris • water • tray • small hammer • needle-nosed pliers

▼ PROCEDURE

1. Wash and dry bones and shells to remove any soft tissue.

2. If necessary, cut any of the bones into pieces so that they will fit into the tray.

3. Coat the surface of each sample with a thin layer of petroleum jelly.

4. Prepare a mixture of plaster of Paris according to the directions on the package, and pour the mixture into the tray.

5. Submerge all of the shells and bones completely in the plaster.

6. Allow the plaster to dry for at least 24 hours.

7. Use the small hammer to crack the dried plaster and expose a shell or bone. Try to keep the plaster from breaking into more than a few pieces.

8. Use the pliers to gently remove the shell or bone from the plaster, leaving a mold of the item in the plaster.

9. Repeat steps 7-8 until all the shells and bones have been removed, leaving only the plaster molds.

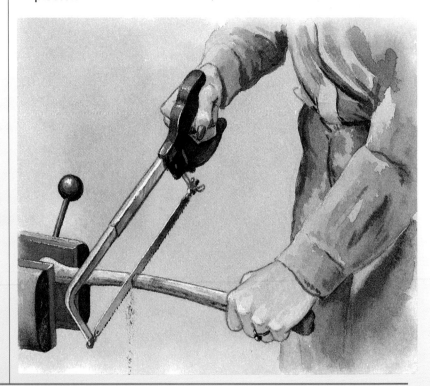

► ANALYSES AND CONCLUSIONS

1. Based on your experiences, what do you think are some of the problems geologists have in removing fossils from rocks?

2. Arrange your "fossils" in layers that tell the history of the rocks in which they were found. Have a classmate observe your layers and interpret the geological history from the "fossils."

► APPLICATION
How could these procedures be modified to make a cast of an animal's footprint?

✳ Discover More
Devise another way to make "fossils" of yourself or your classmates.

The Big Idea

Uniformitarianism is an important concept in science. In order to interpret the history of the earth, geologists must be able to know that there is a certain stability in the way scientific processes work. Because of this stability, geologists can make predictions and find the answer to puzzles, such as those found in the rock layers of the Grand Canyon. Much of geology involves solving puzzles by dating fossils, minerals, and rocks.

For Your Journal

Look at the answers you wrote in your journal at the beginning of this chapter. Now why do you think there are so many different layers of rock in the Grand Canyon? And why do some rock layers contain only one kind of fossil, while other layers seem to be missing altogether?

Connecting Ideas

Look at the fossils shown. In your journal, list them in the order in which they appeared on the earth. Then write a statement summarizing your ideas about fossils.

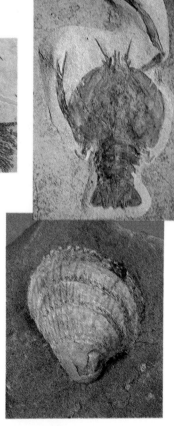

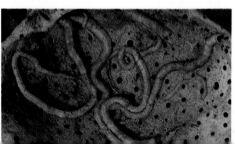

CHAPTER 8 REVIEW

Understanding Vocabulary

1. For each set of terms, explain the similarities and differences in their meanings.
   a) superposition (260), uniformitarianism (259)
   b) relative age (264), absolute age (264)

2. Explain how the terms radioactive decay (266), half-life (264), and fossil (268) are related.

Understanding Concepts

MULTIPLE CHOICE

3. Which of the following geological processes is not an example of the principle of uniformitarianism?
   a) evolution of life forms
   b) erosion of sedimentary rocks over millions of years
   c) deposition of sediments in an ocean
   d) mountain building

4. In a sequence of rocks that have not been disturbed by folding or other geological processes,
   a) older rocks will be on top of younger rocks.
   b) older rocks will be at the same level as younger rocks.
   c) only the youngest rocks will contain fossils.
   d) younger rocks will be on top of older rocks.

5. Rocks that contain dinosaur fossils most likely formed
   a) several years ago.
   b) in an ancient ocean.
   c) on land.
   d) due to wind erosion.

6. A limestone contains some fossilized corals. In what kind of environment was the limestone deposited?
   a) an ocean
   b) a desert
   c) a river bed
   d) a forest

7. Atoms that contain the same number of protons, but different numbers of neutrons, are unstable and are called
   a) fossils.   c) half-lives.
   b) radioactive.   d) isotopes.

8. A rock sample originally contained 100 grams of U-235. How much U-235 would remain after three half-lives?
   a) 100 grams   c) 25 grams
   b) 50 grams   d) 12.5 grams

9. A limestone contains many fossilized shells. What can these fossilized remains tell you?
   a) the absolute age of the rock
   b) the environment in which the rock formed
   c) the relative age of the shells
   d) the half-lives of the minerals in the rock

SHORT ANSWER

10. Which fossil might be more useful to a scientist studying mammoths, a mammoth preserved in ice or imprints of the same organism preserved in rock? Explain your answer.

11. Explain whether or not an absolute dating method can be used to determine the age of sandstone, gneiss, and granite.

Reviewing Themes

12. *Changes Over Time*
Why are radioactive isotopes often called "geological clocks?"

13. *Systems and Structures*
The Grand Canyon is much wider near the rim than at the bottom. How can you explain this?

Interpreting Graphics

14. Study the diagram below. Then write a brief geological history of what has happened, based on what you know about superposition.

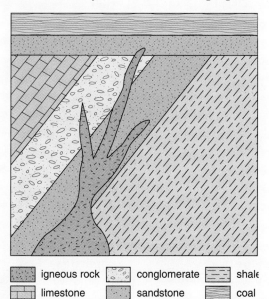

igneous rock	conglomerate	shale
limestone	sandstone	coal

15. Construct a diagram based on this description of the geological history of an area: A sandstone is the youngest rock layer in the area. Prior to the formation of the sandstone, a limestone was deposited. This limestone is younger than a shale. The shale is older than a chalk bed, but younger than a siltstone. All but the sandstone have been faulted. The rocks on the east side of the fault have moved down relative to the rocks on the west side of the fault.

Thinking Critically

16. What role do unconformities play in interpreting a formation of rocks?

17. A piece of ice from a glacier in Greenland has ⅛ of its original tritium. The half-life of tritium is 12.3 years. How old is the ice?

18. Would a worm likely be preserved as a fossil? Why or why not?

19. Why could C-14 not be useful in dating the oldest rocks in the Grand Canyon?

20. Why are fossils generally not found in igneous rocks?

21. This photograph shows an unconformity in a sedimentary rock formation. At many points in the rock record, there are places where fossils just seem to disappear. What do you think caused this gap in the rock record?

Discovery Through Reading

Jaroff, Leon. "Iceman", *Time* 140 (October 26, 1992): 62–69. "Iceman" discusses how the discovery of a frozen human provides clues about life on the earth over 5000 years ago.

CHAPTER 9

An Earth-History Calendar

It may be comforting to think that the dinosaurs did not deserve to die. But why, then, did they become extinct, while other species, such as humans, have survived? People like to think it's "good genes"— that "only the strong survive." But the record of the earth's history tells us that it may just be good luck.

The earth's past includes a graveyard of species. Though millions of kinds of plants and animals are alive on the earth today, living species represent only about 1 in 1000 of the species that ever lived. The other 999 of every 1000 have disappeared. They are dead, gone . . . extinct.

You'd think that scientists would have a good idea about why these species died, and they do—actually, several good ideas. One is the "falling bullet" theory, which states that species die out by chance. It says that life on earth is a lot like standing in a field of falling bullets. Some individuals get killed—by other organisms, natural disasters, or disease. Extinctions occur if too many individuals of a certain species die.

However, it's unlikely that many numerous species—such as trilobites—would die out by chance. So another theory states that it's not bad luck but bad genes that kill off a species. A species dies out because its members cannot compete with other species, or do not adapt to a changing environment. In other words, they deserve to die, because they are not fit to live.

Many scientists are fond of a third theory, which might be called extinction by accident. These scientists blame many extinctions on global catastrophes, such as meteors striking the earth. The fossil record hints that dinosaurs, for example, became extinct in the aftermath of a huge meteor shower.

by Philip Bishop

For Your Journal

- How has the earth changed during its long history?
- When and how did life originate on the earth?
- Why do some organisms become extinct, while others are able to evolve?

Secret Life

Calculate the relative lengths of the three divisions of Cryptozoic time.

Explain how Earth was formed and how living organisms may have originated and evolved.

Identify and **evaluate** the importance of the events of Cryptozoic time.

The earth has changed a lot in the 4.6 billion years since its origin. That's right, 4 600 000 000 years! It's impossible to imagine 4.6 billion years, isn't it? Actually, it's hard for people to imagine any time span much longer than their own lives.

The Earth-History "Year"

There are several problems in studying the earth's past. The first, as we've seen, is the concept of time. You can say that the earth is 4.6 billion years old, but a number that large is impossible to imagine. The second problem is that geologists can't find complete records of the earth's earliest history. Almost all the fossils that have been found so far are from the last 13 percent of the earth's past—87 percent of the earth's history left almost no fossil evidence.

Figure 9–1. Most of the earth's past left no fossil record.

Geologists call this early part of the earth's history *Crypto-zoic,* which means "secret life." The first part of the word *Crypto-zoic* is like the word *cryptic.* When something is cryptic, you have a hard time figuring out what it means. The last part of the word *Cryptozoic* is like the word *zoo,* which comes from a Greek word for "living creature." So Cryptozoic is a good word to describe the early history of the earth.

It's sometimes easier to understand long time spans if you use a familiar scale. A span of 4.6 billion years is too hard to imagine, but it's easy to imagine one year. So we'll put all of the earth's history on a calendar of one year. Let's call this an *earth-history calendar.* The earth-history calendar year starts with the formation of the earth at 12:00:01 AM New Year's Day. The end of the earth-history year is the present, or 11:59:59 PM New Year's Eve on the calendar. Everything that has taken place during the earth's past fits somewhere on the calendar.

Using an earth-history calendar, Cryptozoic time lasts from New Year's Day almost to Thanksgiving. That's most of the calendar year! Rather than trying to study all of this history in one segment, scientists break it down into smaller parts.

CRYPTOZOIC TIME		
Formation of the earth	**Origin of life**	**Evolution of complex organisms**

As shown in the figure, geologists divide Cryptozoic time into three parts. The first part includes the time of the earth's formation. The second part includes the origin of life. The third part includes the early evolution of organisms from simple cells to more complex plants and animals.

To make an earth-history calendar, you need to determine how many geological years fit into various time units, such as days and weeks, on a one-year calendar. For example, you already know that there are 4.6 billion geological years in one year of an earth-history calendar. That means that there are approximately 88 500 000 geological years in one week of the calendar and 150 geological years in one calendar second. In the next activity, you can calculate some other conversions that will be helpful in making your earth-history calendar.

If one calendar second equals 150 geological years, calculate the number of geological years that should be in each of the blank spaces in the chart. Copy the chart into your journal, and fill in the missing years.

4 600 000 000	geological years in a calendar year
	geological years in a calendar month
88 500 000	geological years in a calendar week
	geological years in a calendar day
540 000	geological years in a calendar hour
	geological years in a calendar minute
150	geological years in a calendar second

▼ ASK YOURSELF

What makes the study of Cryptozoic time so difficult?

Formation of the Earth

Geologists hypothesize that the planets in our solar system, including Earth, were formed from a cloud of dust circling the sun. Some of the particles of dust attached themselves to other particles. Slowly the dust particles came together into larger and larger masses. As the masses grew, each increased its own gravitational pull. Smaller masses began circling the larger masses. The nine largest masses became the planets, and the smaller masses circling them became satellites, or moons.

Figure 9–2. Earth's moon as it looks today.

As Earth formed, its surface may have been similar to the surface of its moon today. Craters, like those in the picture, and large plains of volcanic basalt may have marked its surface. But unlike its moon, Earth's surface was hot and there were large pools of bubbling lava.

Not only was Earth's surface hot, but its atmosphere was thick and heavy. Earth's early atmosphere was made up of water vapor, carbon dioxide, and hydrogen sulfide—a gas that smells like rotten eggs. As the earth cooled, water vapor condensed from the atmosphere, and it began to rain. There were severe storms with high winds and heavy downpours. Water filled the craters and all the low spots on the earth's surface. Although the rains cooled the earth's surface, its core remained hot—in fact, it is still hot today.

Figure 9–3. Storms of unimaginable intensity created Earth's oceans.

Millions of years of torrential rains created great oceans. There were giant land masses too, but they were shaped very differently from our modern continents. There may have been only one big continent and one huge ocean. The earth has definitely changed since then! In fact, the earth continues to be altered as continents move and the sizes and shapes of oceans change. But the change is so slow that we hardly notice it unless a volcano erupts or one of the continents shakes a little as it moves, causing an earthquake.

By the end of the first part of Cryptozoic time, late winter on the calendar, there were great oceans, continents with volcanoes, earthquakes, mountains and plains, and rivers and lakes. But there was no life. The seas were sterile, and not so much as a blade of grass grew on the land. It is not likely that life as we know it could have survived anyway, for the early earth provided none of the things that living organisms need.

 ASK YOURSELF

According to geologists, how did the planets of our solar system form?

Origin of Life

On an earth-history calendar, it is now sometime in April. There is still no life, but swirling in the waters of the oceans is a bubbling broth of complex chemicals. Some of them are carbohydrates, proteins, and nucleic acids—the chemicals of life. However, the progress from a complex chemical soup to a living organism is very slow.

Figure 9–4. The waters of the early oceans may have been a fermenting soup of complex chemicals.

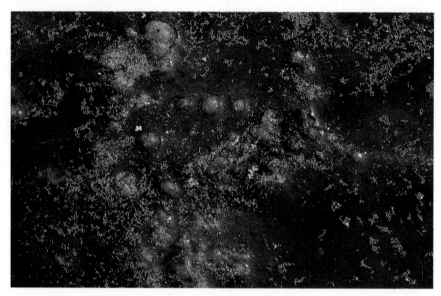

The earliest living organisms were probably simple cells that used the chemicals around them to obtain the energy they needed. Among these chemicals were *carbohydrates*—molecules made of carbon, hydrogen, and oxygen atoms. These early cells likely broke down carbohydrates by a process called *fermentation,* which requires no oxygen, to produce energy. Some bacteria and yeasts use this same process today.

Other early organic chemicals belonged to a group called *amino acids.* These chemicals are made of carbon, hydrogen, oxygen, and nitrogen atoms. Amino acids join together to make long, complex molecules called *proteins.* Much of the structure of modern cells is made up of proteins. In addition, many cellular processes require proteins.

One of the most important chemicals that evolved in the ancient seas was *chlorophyll.* Chlorophyll, whose molecules consist of a magnesium atom surrounded by carbon, hydrogen, oxygen, and nitrogen atoms, enabled early cells to make their own food. Organisms containing chlorophyll use energy from sunlight and carbon dioxide and water to make carbohydrates. Now that you've read about the chemicals of life, think about the following problem.

DISCOVER BY Problem Solving

When does a complex chemical become a living organism? Look in a life science book if you're not sure just what constitutes a living organism. Write your ideas in your journal.

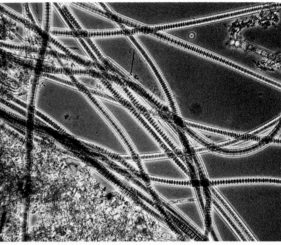

Figure 9–5. The fossils on the left look very much like the modern organisms on the right.

Of course there is no physical proof of when life originated on the earth, but the oldest fossils found so far are more than 3 billion years old. Fossils resembling modern bacteria have been found that are about 2 billion years old. They were among the first organisms to use chlorophyll to make their own food. The photos show some of these ancient fossils along with some similar, modern bacteria. In the next activity you will see some other simple organisms.

DISCOVER BY Doing

Put some fresh hay and water into a beaker or glass pan. Gently boil the mixture until the water has an amber color. Add some more hay and let the mixture sit for a day or two. Then look at a drop of the mixture under a microscope. Draw several of the organisms you see. Compare your drawings with those of your classmates. Where did these organisms come from? Write your ideas in your journal.

 ASK YOURSELF

Why was the development of chlorophyll important in the evolution of organisms?

Evolution of Complex Organisms

During the last part of Cryptozoic time, autumn on the calendar, simple organisms evolved into complex plants and animals. Though many different kinds of plants and animals evolved, their fossil remains are few. This may be because these organisms had almost no hard parts—shells, teeth, claws, and bones—that are readily fossilized. Although there is practically no fossil record of Cryptozoic life, new evidence is found each year. There is still much for geologists to discover and study.

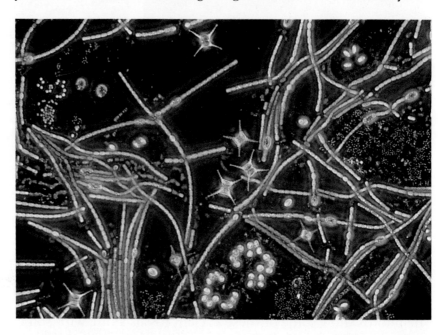

Figure 9–6. Single-celled organisms like these produced much of the atmosphere's oxygen.

Some early, bacterialike organisms evolved into cells similar to modern algae. Like the modern diatoms in the picture, these organisms used chlorophyll to make food by photosynthesis. In addition, these algae also produced oxygen, which began to accumulate in the atmosphere.

Later, some of these cells lost their ability to photosynthesize. They evolved a new process of getting energy from carbohydrates. Unlike fermentation, which produces energy without oxygen, the new process required oxygen. This process, called *respiration,* releases almost 10 times more energy than does fermentation. Almost all modern cells use respiration to obtain energy from food.

In about 150 million years—a relatively short period of geological time, less than two weeks on our earth-history calendar—simple, single-celled organisms evolved into multicellular organisms. These new organisms developed specialized cells for specific functions, such as reproduction.

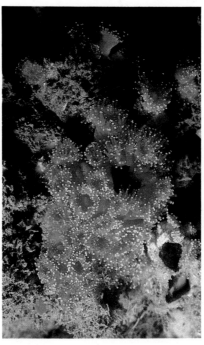

Figure 9–7. The earliest animals may have been similar to modern sponges (left) and corals (right).

Multicellular organisms evolved complex organ systems to carry out specific bodily functions. Your own body, for example, has a digestive system to break down food, an excretory system to get rid of waste, and several other systems for various functions.

Early multicellular organisms fed by eating algae or each other. Some evolved hard coverings for protection from enemies. The horseshoe crab in the picture is similar to some of these early shelled animals. The evolution of hard body parts marked the end of Cryptozoic time. The remains of animals could now be preserved. There was suddenly an abundance of fossil evidence. Life was no longer "secret."

 ASK YOURSELF

How is respiration different from fermentation?

Figure 9–8. A modern horseshoe crab

SECTION 1 *REVIEW AND APPLICATION*

Reading Critically
1. Describe one theory of how the earth was formed.
2. What important chemicals had to be present in order for life to evolve?

Thinking Critically
3. What were the most important changes that took place during Cryptozoic time?
4. Why was the process of respiration important to the evolution of multicellular organisms?

INVESTIGATION

Making a Geological Time Line

▶ MATERIALS
● construction paper ● scissors ● tape ● metric ruler ● markers

▼ PROCEDURE

1. Cut the construction paper along its length into strips 5 cm wide.
2. Tape the strips together end to end so that you have a ribbon about 5 m long.
3. Roll up the ribbon so that it is easy to handle. Unroll it as needed.

4. Using a scale of 1 mm equals 1 million years, measure, mark, and label each of the following: *Formation of the Earth* (4.6 billion years ago), to the *Origin of Life* (3.5 billion years ago), to the *Evolution of Complex Organisms*

(1.8 billion years ago), to the *End of Cryptozoic Time* (600 million years ago), to the *Present*.

▶ ANALYSES AND CONCLUSIONS
1. The first three parts of your time line are all part of Cryptozoic time. Why was this the time of "secret life"?
2. How many times longer is Cryptozoic time than the time after it?

▶ APPLICATION
On an earth-history calendar, what is the approximate date of each of the major events of Cryptozoic time?

✳ Discover More
Think of another device that could be used to study Cryptozoic time.

Visible Life

Since Cryptozoic time, thousands of different plants and animals have evolved. Most are now extinct, but a few of them, such as horseshoe crabs, are still around today. All organisms continue to change and new life forms are still evolving.

Paleozoic Life

As shown in the diagram, the time after Cryptozoic time is called *Phanerozoic time*. *Phanerozoic* means "visible life." Geologists divide Phanerozoic time into three parts, or eras: *Paleozoic*, *Mesozoic*, and *Cenozoic*.

PHANEROZOIC TIME		
Paleozoic (old life)	**Mesozoic** (middle life)	**Cenozoic** (recent life)

Paleozoic means "old life." The Paleozoic is the first era in which fossils are abundant in the geological record. The Paleozoic lasted from about 570 million years ago until about 225 million years ago.

One of the most common animals of the early Paleozoic was the trilobite. The name *trilobite* refers to the three lengthwise lobes of the animal. Trilobites lived from about 570 million years ago to about 250 million years ago—November and part of December on the calendar. They lived on the ocean bottom, burrowing through the mud for food. Look at the picture of the trilobite. What modern animal does it resemble?

Figure 9–9. Trilobites were common creatures of early Paleozoic seas.

Most trilobites were only a few centimeters long. Other animals, such as sea scorpions, were much larger—up to 2 m long. Sea scorpions lived during all but the early part of the Paleozoic. They were among the biggest animals in the oceans.

Figure 9–11. How is this late Paleozoic forest different from forests you've seen?

If you could have walked the land during the early Paleozoic, things would have looked really strange to you. There were no land plants or animals. Land plants didn't appear until the middle of the Paleozoic, and land animals appeared even later. By then, sharks, rays, and bony fishes had all evolved in the seas.

Near the end of the Paleozoic, huge forests began to appear on land. Look at the picture, and imagine walking through one of these forests. At first it seems like a tropical jungle, but you notice a big difference. There are no birds singing or monkeys screeching. Birds and mammals won't evolve until much later. You do hear a loud humming sound and observe a huge insect flying through the trees—a dragonfly with a 70-cm wingspan. That's nearly as long as your outstretched arm. Even the trees are different. Most of them are ferns, like the familiar houseplants, only they are 10 m tall.

Figure 9–12. Life at the end of the Paleozoic

By the end of the Paleozoic, amphibians were common and the land probably looked something like the picture above. There's one of those big dragonflies on the tree! Reptiles started to evolve but didn't spread very far until after the Paleozoic. The earth-history calendar is now into the middle of December. Before we leave the Paleozoic, try the following activity.

DISCOVER BY *Problem Solving*

Make a list in your journal of things you think would have to change in order for fish to evolve into land animals. Then compare your list with those of others in the class. ✐

▼ ASK YOURSELF

What animals evolved in the seas during the Paleozoic?

Mesozoic Life

Figure 9–13. Reptiles walked, ran, flew, and swam during the Mesozoic.

After the Paleozoic, reptiles became the ruling form of life on land. This marked the beginning of the *Mesozoic*, or "middle life," also known as the *age of reptiles*. The Mesozoic lasted from about 225 million years ago to about 65 million years ago. This was the time when dinosaurs roamed the earth.

Reptiles ruled the land. There were also great flying reptiles and reptiles that lived in the sea. Some of the biggest dinosaurs were gentle plant eaters that spent a lot of time grazing on the abundant vegetation. Other dinosaurs were like the ferocious, meat-eating dinosaurs of adventure films. They spent their time hunting prey. In the next activity, you will find out more about these two types of dinosaurs.

ACTIVITY
What was a day in the life of a dinosaur like?

MATERIALS
paper and pencil

PROCEDURE
1. Find a book or magazine that describes these giant reptiles. Your teacher may have books in your classroom.
2. Copy the chart into your journal.

3. Locate the information you need, and fill in the chart.

APPLICATION
1. Did both large reptiles live at the same time?
2. Did one of them eat the other?
3. Why do you suppose one spent a lot of time hunting prey?

	Tyrannosaurus	Apatosaurus
How big was it?		
How much did it weigh?		
What did it eat?		
What might have eaten it?		
When did it live?		
How did it move?		
How long did it live?		

Dinosaurs were not the only animals of the Mesozoic. In fact, they weren't even the only reptiles. In the excerpt that follows, you can get an idea of what life was like when dinosaurs ruled the earth.

Picture yourself going back through time—back some 75 million years to the days of dinosaurs. The place where you find yourself is now part of Montana. But it does not look like the Montana we know. For one thing, it is on the western shore of a warm, shallow sea. The sea stretches from the Gulf of Mexico to the Arctic, and it cuts North America in two. You are standing on the shore of that sea.

Earlier in the day, around dawn, mist rose gently from the sea, tinged pink by the rising sun. Now the mist has burned off and the day promises to be clear and hot. The air is salty, soft, and slightly damp. It smells like seashores everywhere.

Sometimes dinosaurs come down to the sea, to look around or graze on the plants that grow behind the beach. They may also feed on plants growing in the shallow waters between the shore and some nearby islands. But they never swim out to sea. Dinosaurs are not sea-going animals. Right now there are no dinosaurs in sight, but both sea and sky are full of life. The world of dinosaurs is shared by many other kinds of animals.

Overhead, patches of sky are dark with wings. Some are the wings of sea birds, flapping, soaring, gliding. Like today's birds, these have feathers. Unlike them, they have teeth, which they use in snatching fish from the sea.

Most of the wings belong to another kind of animal. They are long and narrow and made of skin that is covered with thin hair. These are the wings of flying reptiles, of pterosaurs. Pterosaurs appeared on earth at about the same time that the dinosaurs did and are related to them. Over the years there have been many sorts and sizes of pterosaurs. The smallest have the wingspan of sparrows. Before the pterosaurs die out with dinosaurs, some will be as big as small airplanes, with wingspans of 40 feet.

Figure 9–14. *Pteranodon* is the biggest pterosaur of the inland sea, with a wingspan of up to 23 feet.

The biggest pterosaur flying over the inland sea is Pteranodon. Like all reptiles, it grows throughout its life—older adults may have a wingspan of 23 feet. Pteranodon has a short body and hardly any tail. At the end of its neck is a small head with a huge crest and a big bill. Smaller pterosaurs have teeth, which they use to catch fish. Pteranodon does not. Perhaps it feeds like a pelican, scooping up fish and swallowing them whole. Perhaps it fishes in shallow waters, like today's long-legged herons, standing on one leg and snapping up fish and other creatures of the sea.

On land pterosaurs walk like birds, but they must run to build up speed for take-offs. Once in the air, the smaller pterosaurs flap their wings, glide, and soar. Pteranodon does little flapping because of its size. For this big creature, flapping flight takes huge amounts of energy. Pteranodon spends most of its time soaring on rising currents of air. It soars and glides, swoops down, then rises again to soar above sea and shore.

Figure 9–15. Smaller pterosaurs are hunting for fish in the same waters as *Hesperornis*, birds that cannot walk or fly.

Near one of the islands there is a sudden movement. A six-foot-long bird has popped to the surface. This is Hesperornis, a bird that cannot fly or walk. When a female must go ashore to lay her eggs, she lies on her belly and pushes herself along, as if she were sledding on the beach. It is a slow and clumsy way to move. But in the water, the birds are swift, skilled swimmers. At the surface, they sail along, kicking with their large webbed feet. To look for food, a bird flattens its tiny, stubby wings against its side and dives. It will chase any fast-moving fish it sees, grasping the prey with its sharp teeth.

Figure 9–16. A plesiosaur swims near a large shark, which is apparently not hungry.

Beneath the surface of the sea, many kinds of animals are going about their business, feeding or trying to escape becoming a meal for some other animal. Crabs scuttle across the sea floor. Snails graze on sea plants. Fishes flit everywhere. Some are sucking up tiny plants and animals from the sea floor. Some are nibbling on bigger plants. Others are looking for clams, shrimp, and small fish. Not far away, a shark swims lazily, looking around, perhaps not hungry.

The warm sea is also home to a number of reptiles. Only one looks familiar, a big sea turtle. The rest are animals no longer seen on earth.

Among the reptiles are the mosasaurs, relatives of today's monitor lizards. They look something like toothed whales. The smallest kind of mosasaur is 7 to 20 feet long, with a slim body and a fringed tail. It lives near the surface of the sea, paddling with its flipper-like legs and sometimes making shallow dives for food. It eats fish—and is itself sometimes eaten by other mosasaurs and sharks.

Another big reptile of the inland sea looks like a cross between a snake and a turtle, with a thick body and a long, slender neck. This is a plesiosaur. It swims by flapping its four huge flippers as if they were wings. When hunting, the plesiosaur may swim near the surface, raise its long neck, and look around. As soon as it spots a fish or squid, its long neck plunges into the sea and its sharp teeth close on the prey.

Some kinds of sea reptiles may give birth to live young in the sea. The young swim quickly to the surface to breathe. There they become prey for fish and other animals, unless they can hide in beds of seaweed.

Other reptiles of the sea lay eggs, and they must go ashore. The females haul themselves out of the water onto beaches. With their short legs, sea turtles are clumsy and slow on land. The other reptiles are even more clumsy, because they cannot walk. They must squirm and wriggle, hauling and pushing with their paddle-like legs to reach the places where they will dig their nests. The trip is a time of great danger, when the reptiles are easily preyed on by land animals. Land animals also eat reptile eggs, if they can find them, as well as newly hatched young making the long trip down the beach to the sea.

from *Living with Dinosaurs*
by Patricia Lauber and illustrated by Douglas Henderson

Figure 9–17. A mosasaur hauls herself up the beach to lay her eggs, while seabirds hunt for food, such as the sawfish that has been washed up from the sea.

Reptiles ruled the world during the Mesozoic. But during the late Mesozoic, birds and mammals were evolving. They were small at first and certainly not as dominant as the dinosaurs. But about 65 million years ago, all the giant reptiles became extinct. We may never know what caused them all to die—a huge meteorite impact, a worldwide climate change, too much oxygen in the atmosphere, or cosmic rays—but whatever the cause, by the end of the Mesozoic the dinosaurs were nearly gone and the mammals were ready to inherit the land. You can read more about the possible causes of the dinosaur's extinction in the Science Parade, on pages 302–303.

ASK YOURSELF

What was life like during the age of reptiles?

Figure 9–18. Although much smaller generally than the giant reptiles, mammals have become the dominant animals of the Cenozoic.

Cenozoic Life

If the Mesozoic is the age of reptiles, then the Cenozoic is the age of mammals. *Cenozoic* means "recent life," and during this time mammals filled the habitats left by the extinct dinosaurs and other reptiles.

You are now near the end of the earth-history calendar year. Although the Cenozoic began 65 million years ago, it represents only the last few days of the calendar. Toward the end of the Cenozoic, New Year's Eve on an earth-history calendar, an important event occured—the ice ages began. During many of these last hours of the calendar year, great glaciers have covered the northern continents.

The ice ages started about 3 million years ago. Each time the ice formed, it would take about 75 000 years to reach its farthest point. The ice would remain for less than 10 000 years and then retreat northward. Several times the glaciers formed, advanced, and retreated.

The ice ages occurred because the climate of the entire earth was much colder. As you might imagine, sea level was also different. When there was a lot of ice, sea level dropped. When the ice melted, sea level rose again.

Many scientists hypothesize that the earth is still in the ice ages—we are currently in a time between glaciers. If this is true, the earth may move into another glaciation within the next few thousand years.

Figure 9–19. During the ice ages, periods of glaciation alternated with periods of moderate climate.

One final note before leaving the Cenozoic. On your earth-history calendar, at a few minutes before midnight on New Year's Eve, human beings appeared. Together with every other living organism, humans probably evolved from simple cells that lived more than 4 billion years ago. What kinds of organisms do you think might evolve over the next 4 billion years?

ASK YOURSELF

Why would sea level change during the ice ages?

SECTION 2 *REVIEW AND APPLICATION*

Reading Critically

1. Why are there more fossils from the Phanerozoic than from the Cryptozoic?
2. How were forests of the Paleozoic different from forests today?

Thinking Critically

3. Why were there animals in the sea before there were animals on the land?
4. Why do organisms evolve? What happens if they don't?

SKILL *Ordering Data*

▶ **MATERIALS**
  ● tracing paper ● pencil

▼ **PROCEDURE**

1. Use tracing paper to make a copy of the evolutionary tree shown in the picture.
2. Using reference books, examine the physical features of the organisms labeled in this activity. Note their similarities and differences.
3. Write the name of each plant or animal group above the lettered circle on the tree.

A) Cyanobacteria
B) Red algae
C) Brown algae
D) Seed plants
E) Green algae
F) Fungi
G) Slime molds
H) Annelids
 I) Arthropods
J) Vertebrates
K) Primitive chordates
L) Echinoderms
M) Mollusks
N) Coelenterates
O) Sponges
P) Ciliates
Q) Amoeboids
R) Bacteria

4. Note the position of each group of organisms on the evolutionary tree. Decide which groups of organisms are most likely to have had common ancestors.

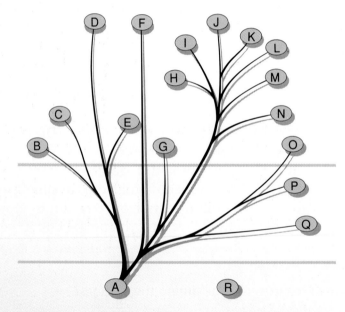

▶ **APPLICATION**
  1. Which group of organisms has evolved most independently of the others?
  2. Which group of organisms has the most "relatives" on its evolutionary branch?

✳ *Using What You Have Learned*
List the organisms from the tree that could be considered plants. Then list the organisms that could be considered animals. Are there organisms that could be on both lists? What are they? Why is there confusion as to which organisms could be plants and which could be animals?

The Big Idea

We know that the earth has been constantly changing since its formation 4.6 billion years ago. Volcanoes and earthquakes produce obvious changes in the earth's crust, but there are other, more subtle, changes as well. During the earth's long history, life originated from complex chemicals in ancient oceans, slowly evolved, and moved onto the land. Huge reptiles once ruled the earth. Now mammals, including humans, are the dominant animals.

For Your Journal

Think again about the questions you answered in your journal at the beginning of this chapter. How have your ideas changed about why some organisms became extinct while others evolved? Revise your journal entries.

Connecting Ideas

Make a record of the earth's history using an old calendar with large blocks. Place the items from the list under the proper dates on your calendar. Add as many drawings and descriptions as you want.

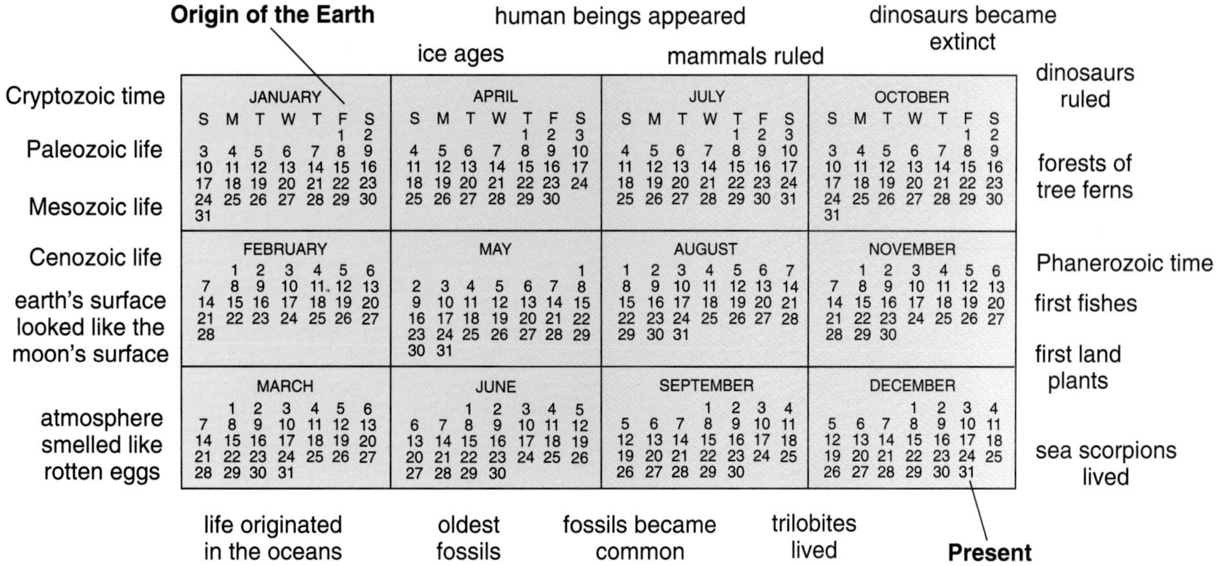

Origin of the Earth — human beings appeared — dinosaurs became extinct

ice ages — mammals ruled — dinosaurs ruled

Left labels:
- Cryptozoic time
- Paleozoic life
- Mesozoic life
- Cenozoic life
- earth's surface looked like the moon's surface
- atmosphere smelled like rotten eggs

Right labels:
- forests of tree ferns
- Phanerozoic time
- first fishes
- first land plants
- sea scorpions lived

JANUARY

S	M	T	W	T	F	S
					1	2
3	4	5	6	7	8	9
10	11	12	13	14	15	16
17	18	19	20	21	22	23
24	25	26	27	28	29	30
31						

FEBRUARY

S	M	T	W	T	F	S
	1	2	3	4	5	6
7	8	9	10	11	12	13
14	15	16	17	18	19	20
21	22	23	24	25	26	27
28						

MARCH

S	M	T	W	T	F	S
	1	2	3	4	5	6
7	8	9	10	11	12	13
14	15	16	17	18	19	20
21	22	23	24	25	26	27
28	29	30	31			

APRIL

S	M	T	W	T	F	S
				1	2	3
4	5	6	7	8	9	10
11	12	13	14	15	16	17
18	19	20	21	22	23	24
25	26	27	28	29	30	

MAY

S	M	T	W	T	F	S
						1
2	3	4	5	6	7	8
9	10	11	12	13	14	15
16	17	18	19	20	21	22
23	24	25	26	27	28	29
30	31					

JUNE

S	M	T	W	T	F	S
	1	2	3	4	5	
6	7	8	9	10	11	12
13	14	15	16	17	18	19
20	21	22	23	24	25	26
27	28	29	30			

JULY

S	M	T	W	T	F	S
				1	2	3
4	5	6	7	8	9	10
11	12	13	14	15	16	17
18	19	20	21	22	23	24
25	26	27	28	29	30	31

AUGUST

S	M	T	W	T	F	S
1	2	3	4	5	6	7
8	9	10	11	12	13	14
15	16	17	18	19	20	21
22	23	24	25	26	27	28
29	30	31				

SEPTEMBER

S	M	T	W	T	F	S
		1	2	3	4	
5	6	7	8	9	10	11
12	13	14	15	16	17	18
19	20	21	22	23	24	25
26	27	28	29	30		

OCTOBER

S	M	T	W	T	F	S
					1	2
3	4	5	6	7	8	9
10	11	12	13	14	15	16
17	18	19	20	21	22	23
24	25	26	27	28	29	30
31						

NOVEMBER

S	M	T	W	T	F	S
	1	2	3	4	5	6
7	8	9	10	11	12	13
14	15	16	17	18	19	20
21	22	23	24	25	26	27
28	29	30				

DECEMBER

S	M	T	W	T	F	S
		1	2	3	4	
5	6	7	8	9	10	11
12	13	14	15	16	17	18
19	20	21	22	23	24	25
26	27	28	29	30	31	

Bottom labels:
- life originated in the oceans
- oldest fossils
- fossils became common
- trilobites lived
- **Present**

Understanding Vocabulary

1. For each set of terms, explain the similarities and differences in their meanings.
   a) fermentation (282), respiration (284)
   b) carbohydrates (282), amino acids (282), chlorophyll (282)
   c) Paleozoic (287), Mesozoic (290), Cenozoic (296)

Understanding Concepts

MULTIPLE CHOICE

2. The division of Cryptozoic time that lasted the longest was the time of the
   a) formation of the earth.
   b) origin of life.
   c) evolution of complex organisms.
   d) dinosaurs.

3. Cryptozoic time ended about
   a) 4.6 billion years ago.
   b) 3.5 billion years ago.
   c) 1.8 billion years ago.
   d) 600 million years ago.

4. The earliest living organisms
   a) probably had hard body parts.
   b) probably produced energy by fermentation.
   c) were probably complex.
   d) probably obtained energy by respiration.

5. The oldest fossils found so far resemble modern
   a) bacteria.
   b) sponges.
   c) ferns.
   d) corals.

6. During the early Paleozoic, the most common kinds of animals were
   a) birds.   **c)** trilobites.
   b) dinosaurs.   **d)** fish.

SHORT ANSWER

7. Describe one characteristic of each division of Cryptozoic time.

8. Explain how and where glaciers fit on an earth-history calendar.

Interpreting Graphics

9. Copy the concept map below and complete it by identifying each era from the names of the dominant organisms listed.

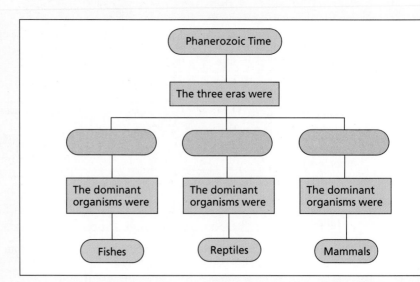

Reviewing Themes

10. *Environmental Interactions*
Explain why complex animals could not have evolved before organisms that made food by photosynthesis.

11. *Changes Over Time*
Describe the changes that occurred during the ice ages.

Thinking Critically

12. Why have scientists found very few fossils from Cryptozoic time?

13. Scenes from some movies show human beings battling with dinosaurs. Are scenes such as these scientifically accurate? Explain your answer.

14. The following statements summarize the geological history of the earth during Phanerozoic time. However, the events are not listed in the correct sequence. Rewrite the statements in the correct sequence.

a) Shallow seas cover the land.
b) Fish inhabit the seas and land plants appear.
c) Dinosaurs are dominant, and mammals begin to evolve.
d) Ice covers large parts of the continents and humans appear.
e) Great forests of tree ferns cover the land.
f) Trilobites inhabit the seas.
g) Amphibians are common and reptiles start to evolve.
h) Dinosaurs become extinct and mammals dominate.
i) Corals and sponges dominate the seas and simple land plants evolve.

15. This is a photo of a manatee, sometimes called a *sea cow*. Manatees and elephants are believed to have had a common ancestor. What kinds of environmental changes might cause the development of two such different-looking animals?

Discovery Through Reading

Erickson, Jon. *Ice Ages: Past and Future*. Tab Books, 1990. The author describes the great ice ages of the past and the factors that influenced them.

Science PARADE

HOW THE DINOSAURS DIED

For some unknown reason the dinosaurs disappeared 65 million years ago. The fossil evidence that we have provides few clues as to the cause of their extinction.

Struggling Theories

A current hypothesis has some scientists debating. The theory maintains that near the end of the Mesozoic era a giant meteorite or comet struck the earth. This could have caused global climate changes. The sun's rays would have been blocked as ash clogged the air, darkening the sky and lowering the temperature. Animals by the millions could have died, and many species could have become extinct as a result of the cold or lack of food. The descendants of those animals that survived would have become the present life forms on earth. Some scientists hypothesize that catastrophes like this occur about every 26 million years. They say that an unknown planet or star carrying along many comets passes near the solar system and causes these catastrophes. Whenever this alien body comes by the solar system, millions of creatures are killed by a rain of comets.

The hypothesis was developed in 1980 by Luis Alvarez after observing some sedimentary rock formations in central Italy. The sediments contain one of the best records of the boundry between

Ash-darkened sky

Seeking fossil clues

A dinosaur of the Jurassic Period

Large animals such as the dinosaurs would have had nothing to eat, and soon would have died too. The meteorite dust would have buried more than half of Earth's animal species as it eventually fell to Earth, like a black snow.

Searching for Evidence

Debates in the scientific community over these theories go on while the search for new evidence continues. Alvarez's research has inspired other scientists. Some have theorized that the same kind of global winter caused by the Cretaceous meteorite might occur after a nuclear war. Now evolutionary biologists, who once thought that the dinosaurs died out because of their small brains, believe that there may have been other reasons. Even astronomers want to know the cause of the catastrophe. If some unknown star or mysterious planet jolted a meteor into Earth's path 65 million years ago, will the same thing happen again?

The answer may affect the way humans view themselves. Humans are encouraged by evolutionary biology to believe that they survive because they are the most intelligent species on Earth. If a meteorite from space had not come crashing down, would the dinosaurs still be ruling the earth? The answers to questions like this lie waiting for you to find, somewhere in Earth's history. ◆

the Cretaceous period of the Mesozoic era and the Tertiary period of the Cenozoic era. Between the rocks of these two periods Alvarez discovered an unusual layer of clay. In this clay there was a high concentration of the element iridium, which is very rare in the earth's crust. Alvarez asked, "Why would the iridium level in this layer of clay be 25 times higher than normal?" He hypothesized that about 65 million years ago a meteorite slammed into the earth. He based his hypothesis on the fact that 1000 times more iridium is sometimes found in extraterrestrial bodies than in the earth's crust.

A Layer of Dust

Alvarez reasoned that the collision spread iridium-rich dust through the atmosphere. The layer of sediment that separates the rocks of the Cretaceous period from those of the Tertiary period was deposited by the dust.

Alvarez believed that this gigantic collision and the extinction of the dinosaurs happened at about the same time. He first calculated how much dust would have been thrown into the air by a meteorite collision. Then he determined how much sunlight the dust would have blocked. Alvarez claimed that as a result of the lack of light, plants and marine organisms would have died, causing a collapse of the food chain.

Alvarez's lab

MYSTERIOUS FOOTPRINTS IN STONE

The tracks went up sand dunes, but not down.

by Mary Elting from *Cricket*

"What animal would go uphill but not come down?" isn't just a riddle. It's a real question that scientists were asking about eighty years ago. They had discovered the footprints of a strange animal along a trail in the Grand Canyon of Arizona. These were not ordinary footprints, but fossil tracks, made when the animal walked on a sand dune that later hardened into stone.

Scientists examined hundreds of different sets of the fossil tracks, and all except three went *up* sand dunes—but not down! Why?

In 1934 a young geologist named Edwin McKee decided to solve the mystery. He had heard several explanations for the strange tracks. Some people guessed that the animals had all been migrating in the same direction—uphill. Others thought they were flying reptiles that took off into the air when they reached the top because they didn't know how to get down again.

None of these notions made enough sense to McKee. Instead of guessing, he decided to try an experiment. He knew the tracks were made by a very ancient reptile that lived about 250 million years ago. So he looked for a living reptile whose footprints looked like the fossil tracks. McKee discovered that a lizard called the chuckwalla made almost the same kind of track.

McKee's First Experiment

For his experiment, he built an open box and piled some sand into an artificial dune at one end. At the other end of the box he set down a chuckwalla. At first it did not move. He gave it a push, and off it dashed in such a hurry that its feet left only a blur of scuffed sand.

He tried a gentler tap to get the chuckwalla to walk, but halfway up the hill, the lizard stopped. Another little tap made it move again,

McKee built an artificial sand dune.

The chuckwalla left
no footprints at all!

One Question Leads to Another

Still, McKee had more work for his chuckwalla
to do. He wanted to find the answer to another
question: How did reptile tracks become fossils? Part of the answer was clear. After an animal's feet pressed down and left their tracks in
the sand, more sand covered the tracks without erasing them. But if the sand dunes were
dry, gusts of wind could ruffle them and blow
the tracks away. Was it possible that the
ancient animals had been walking in wet sand
that wouldn't blow away as easily?

McKee's Second Experiment

The chuckwalla traveled up and down McKee's
home-made dunes again and again—before
rain and after rain, morning and evening.
Finally, he sprayed a fine mist of water on some
tracks that the chuckwalla made in dry sand.
Then he left them to dry again. Wherever he
had sprayed, a very thin but firm crust had
formed over and around the prints—firm
enough to preserve them. At last McKee had
found an explanation.

McKee's Conclusion

Tiny dewdrops could have done what the
spray did. They must have made the sand firm
enough for the prints to hold their shape even
after a layer of new dry sand covered them.
Like a mold into which plaster has been poured,
the tracks and the new sand eventually hardened into rock. Millions of years later the rock
was broken open, and scientists saw the footprints, sharp and clear. Eddie McKee and his
chuckwalla had solved another puzzle. ◆

but with an angry swish of its tail that wiped
out all the prints.

What would make a lizard want to walk
steadily from one place to another? McKee
knew that the chuckwalla's movements often
depend on the temperature. Lizards are called
cold-blooded animals because their bodies
don't have a way of controlling their own heat.
Instead, their bodies are always about the same
temperature as the place where they happen to
be. When they get too cool in a shady spot,
they move out to warm themselves in the sun.

McKee decided to build a narrow trough six-
teen feet long, with a hill of white sand piled up
across the middle. He covered one end of the
trough, left the rest of it out in the sun, and put
the chuckwalla in the dark, cooler end. With no
urging, the lizard soon walked out toward the
sunshine and up the hill.

McKee's Discovery

At the top, the chuckwalla paused. But it did
not die or fly away. It braced its feet in the
sand, and using them as brakes, slid down the
other side—leaving no footprints at all! You
would never have known that an animal had
been there. The ancient reptiles must also have
walked up one side of the sand dunes and slid
down the other side. Eddie McKee had solved
the mystery of the one-way tracks.

Layers of dry sand covered
the firm tracks.

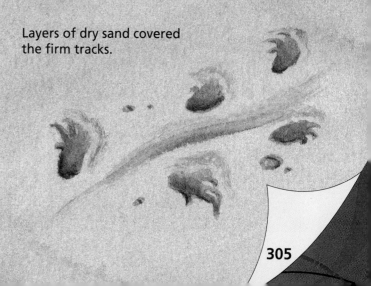

HAROLD UREY (1893—1981)

Harold Clayton Urey had an undying curiosity about all of the sciences. Instead of being interested in only one or two fields, like biology or chemistry, Urey tried to learn as much as he could about many different fields.

Urey received his bachelor's degree from Montana State University in 1917. He was interested at that time in zoology and chemistry. After earning his doctorate from the University of California at Berkeley, Urey traveled to Denmark, where he spent two years working with Niels Bohr, the famous physicist.

Urey's work in physics over the next decade led to his discovery of deuterium (doo TEER ee uhm), a heavy isotope of hydrogen. In 1934, Urey received a Nobel prize for this work.

In the early 1950s, Urey turned his attention to the study of geochemistry, astrophysics, and the origin of life. He wanted to know how the earth and the solar system came to be. His vast knowledge of biology, physics, and chemistry helped him conduct research and write many articles on geochemistry. Urey reviewed many hypotheses on how the sun and planets were formed.

He also studied the chemical reactions of gases that existed in Earth's primitive atmosphere. His work showed that amino acids could have formed spontaneously, thus supporting the theory that life could have started by itself on the primitive earth. ◆

WALTER ALVAREZ (1940—)

In the Yucatán peninsula of Mexico, Walter Alvarez may have found a massive crater. The crater dates back to the exact time he and his father believe a meteorite or comet slammed into the earth.

Alvarez and his father, Luis Alvarez, discovered a layer of iridium they think was left by the impact of a meteorite or comet 65 million years ago. They hypothesized that this impact might have led to the extinction of the dinosaurs.

Walter Alvarez is a geologist. He graduated with a bachelor's degree from Carleton College in 1962. Alvarez then earned a Ph.D. in geology from Princeton University. He has done research in Africa and Europe, but he is best

known for the work he and his father did on dinosaur extinction.

Their hypothesis that a meteorite or comet caused the extinction of the dinosaurs seemed reasonable, but was not accepted by many scientists because no one knew of any large craters dating back to that period. Alvarez and a team from the University of California at Berkeley were investigating impact debris in parts of the Caribbean. Interesting drill samples drew their attention to the northern Yucatán. A buried circular feature there may be a massive crater, measuring 180 kilometers in diameter. The crater would strengthen support for the Alvarez impact hypothesis. ◆

In Che Yang, HYDROLOGIST

In the hot Nevada sun, In Che Yang watches as a drilling crew brings a core sample to the surface. In Che Yang is project chief for a hydrochemistry study being conducted in Nevada. Yang wants to know how stable this rock environment might be. If the site is found to be suitable, the area may be used for the underground storage of hazardous wastes.

The core sample is marked and wrapped in plastic. Then the sample is sealed into special plastic containers to prevent the loss of any water or gas that may be trapped inside. Yang has the core placed in an ice chest, keeping it close to 8°C to help maintain its present chemical composition.

After the markings on the core are carefully recorded in the laboratory, the core is unwrapped and put into a compression cell—a six-inch-wide stainless steel cylinder which has a hole drilled in the middle to hold the core. The cell, with the core inside, is then placed in a mechanical press called a *load frame* which puts heavy pressure on the core. This forces out small amounts of gas and water hidden inside the rock's pores and cracks.

Yang has the gas analyzed in a *gas chromatograph* to see what it is made of. The water sample is also analyzed for its chemical components. Isotopes of hydrogen and oxygen, and carbon-14 content help determine its age.

In another bore hole, Yang has geologists lower a *packer,* which works like a series of balloons that divide the bore hole into several separate zones. A pump then pulls the "rock air" from each zone. The air flows into a glass vacuum trap that sits in a thermos filled with a very cold mixture of dry ice and alcohol. This "cold trap" condenses the water vapor, separating it from gases that do not condense. Condensed water is also analyzed for isotopes of hydrogen and oxygen to help identify its source and flow path.

Rainwater or melted snow may rapidly follow cracks in the rock (modern water), or slowly move through tiny pores in the rock itself (old water). The gases are saved for analysis, which includes carbon-14 dating to determine their age.

Each of the samples provides valuable data, which are recorded on graphs and charts. Yang uses his knowledge and training as an isotope hydrologist to interpret the graphs and charts. He can tell from the data what the climate was like in the past. Yang can also tell where precipitation came through the land surface and the amount of time it took for the water or gas to move through the rocks to where the sample was taken. Using this information, Yang hopes to protect our environment from the careless dumping of hazardous wastes. ◆

DISCOVER MORE

For more information about careers in geology and hydrology, write to the

Geological Society of America
P.O. Box 9140
3300 Penrose Place
Boulder, CO 80301

CARBON DATING

The atoms of carbon-14 (also called *radiocarbon*) are unstable and decay at a known rate. Their rate of decay can be used to determine the age of a fossil. Carbon dating is the most common way of dating fossils. The technique was developed in the 1940s, and can be used by scientists to find the age of organisms that lived up to 50 000 years ago.

Methods of Dating

There are two methods of carbon dating. One method involves burning a sample of the specimen to be dated to obtain carbon dioxide gas. The radiocarbon atoms in the gas release electrons as they decay into nitrogen-14. The number of electrons released are counted to determine the amount of radiocarbon present.

In the second method a small sample of the specimen is placed in a particle accelerator. The particle accelerator fires charged atoms from the specimen into a magnetic field. Carbon-14 atoms in the magnetic field are deflected and separated from other atoms of carbon. A carbon-14 detector then counts the individual carbon-14 atoms present. Some scientists think that this method is much better because it allows for the detection of very small amounts of radiocarbon.

Discovering a fossil

Dating Difficulties

Whether or not a fossil can be carbon dated depends on two things: how well preserved the fossil is and the age of the fossil. The detection of very small amounts of radiocarbon can be difficult because of background radiation. This naturally occurring radiation often interferes with radiocarbon detection. If a fossil is too old the instruments will not be able to detect radiocarbon in it.

There are other things that must be considered when deciding which method of carbon dating to use. For instance, when samples of a fossil are burned for carbon dating, a large sample must be used. This usually destroys the fossil. Sometimes fossils are not preserved well enough for carbon dating. They may be contaminated with recent carbon sources. This contamination would make any attempt at carbon dating inaccurate.

Greater Accuracy

To improve the accuracy of carbon dating, scientists compare measurements. A radiocarbon measurement is compared with a measurement from the rings of trees whose ages are already known. By doing this, the scientists are able to compensate for small differences in radiocarbon content that may have existed in the atmosphere in the past. In this way scientists are able to obtain a more precise dating of a fossil. ◆

Moving to the laboratory

Equipment in the laboratory

UNIT
4
METEOROLOGY

CHAPTERS

The rain blows horizontally, and the wind screams. Trees bend and snap under the force of winds in excess of 190 kilometers per hour. Their broken trunks and limbs fly dangerously through the air as the hurricane unleashes its power.

Just hours before, the sky was scattered with fluffy white clouds. The sun was shining and the trees gently swayed with the breeze. What brought on such a drastic change in the weather? How could you predict the formation of such a storm?

Science PARADE

CHAPTER 10

The Atmosphere

*W*iping sweat off her forehead, Pat stooped and fingered the sooty dirt of an almost blank planet. Her map said the third planet had once been home to intellegent life. But it didn't look promising. The heat was unbearable, and the odor was worse—like a gym locker full of skunks. Pat longed to run away from this filthy place. The methane gas was making her gasp. I must find some air! *she thought.*

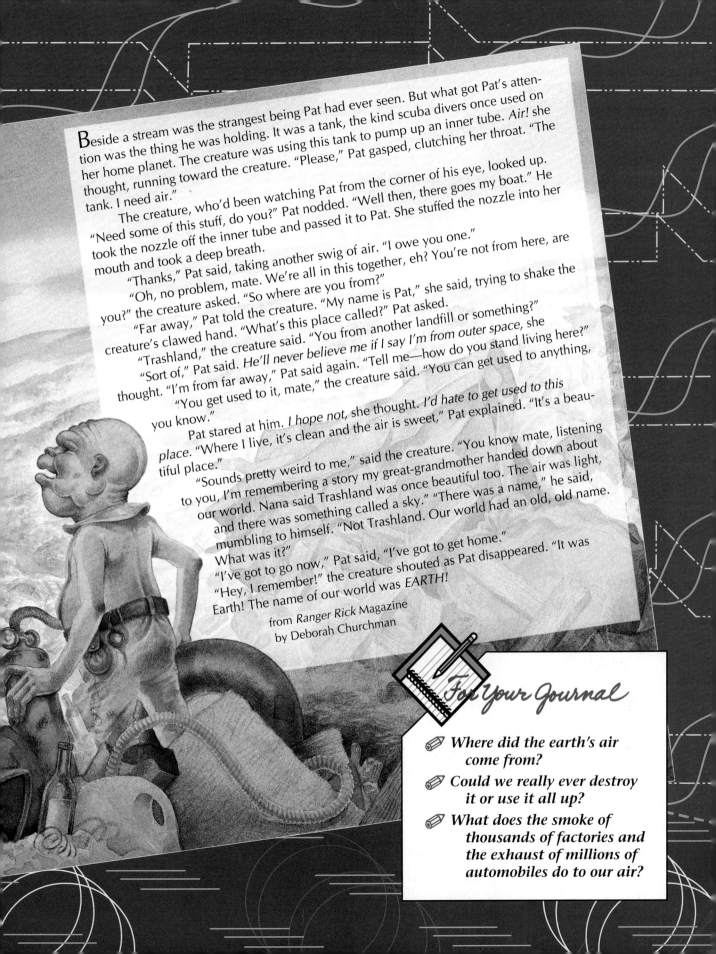

Beside a stream was the strangest being Pat had ever seen. But what got Pat's attention was the thing he was holding. It was a tank, the kind scuba divers once used on her home planet. The creature was using this tank to pump up an inner tube. *Air!* she thought, running toward the creature. "Please," Pat gasped, clutching her throat. "The tank. I need air."

The creature, who'd been watching Pat from the corner of his eye, looked up. "Need some of this stuff, do you?" Pat nodded. "Well then, there goes my boat." He took the nozzle off the inner tube and passed it to Pat. She stuffed the nozzle into her mouth and took a deep breath.

"Thanks," Pat said, taking another swig of air. "I owe you one."

"Oh, no problem, mate. We're all in this together, eh? You're not from here, are you?" the creature asked. "So where are you from?"

"Far away," Pat told the creature. "My name is Pat," she said, trying to shake the creature's clawed hand. "What's this place called?" Pat asked.

"Trashland," the creature said. "You from another landfill or something?"

"Sort of," Pat said. *He'll never believe me if I say I'm from outer space,* she thought. "I'm from far away," Pat said again. "Tell me—how do you stand living here?"

"You get used to it, mate," the creature said. "You can get used to anything, you know."

Pat stared at him. *I hope not,* she thought. *I'd hate to get used to this place.* "Where I live, it's clean and the air is sweet," Pat explained. "It's a beautiful place."

"Sounds pretty weird to me," said the creature. "You know mate, listening to you, I'm remembering a story my great-grandmother handed down about our world. Nana said Trashland was once beautiful too. The air was light, and there was something called a sky." "There was a name," he said, mumbling to himself. "Not Trashland. Our world had an old, old name. What was it?"

"I've got to go now," Pat said, "I've got to get home."

"Hey, I remember!" the creature shouted as Pat disappeared. "It was Earth! The name of our world was EARTH!

from *Ranger Rick* Magazine
by Deborah Churchman

For Your Journal

- *Where did the earth's air come from?*
- *Could we really ever destroy it or use it all up?*
- *What does the smoke of thousands of factories and the exhaust of millions of automobiles do to our air?*

Composition and Structure of the Atmosphere

Objectives

Analyze *the characteristics of the atmosphere that make it stable.*

Name *and* ***describe*** *the layers of the atmosphere.*

Explain *the relationship between atmospheric pressure and altitude.*

Time is running out. Actually, it would be more precise to say that our air is running out. You have been selected as a member of a research team. The mission of this team is to determine the criteria for selecting another planet suitable for human habitation. The first step in this process is to establish what characteristics make Earth habitable. Your specific task is to assemble and organize all available data about Earth's atmosphere. You already know we need air to breathe. But what else is there about air that makes life on Earth possible?

The Atmosphere, Then and Now

Air near the earth's surface is part of the **atmosphere**—the ocean of gases that surrounds the planet. As you begin your research, you realize that you need to know how Earth's atmosphere developed in order to assess another planet's present or future ability to sustain life.

Figure 10–1. Earth's atmosphere gently envelopes the planet.

A look at the history of Earth's atmosphere reveals that it has evolved over time. The atmosphere is much different today than it was when the earth first formed. Scientists theorize that a cloud of dust and gases condensed to form the sun and its planets about 4.6 billion years ago. Whatever atmosphere Earth had then was dissipated and carried off into space by high-energy radiation from the sun, called the *solar wind*. Solar wind is composed of particles whose speed has been accelerated by the sun's magnetic fields. These magnetic fields are indicated by sunspots, and vary in strength as the number of sunspots varies. If there are many sunspots, the solar wind is very strong.

After the solar wind scattered Earth's original atmosphere, a new atmosphere began to form. Volcanic eruptions released gases that had been trapped inside the earth during its formation. Most of the lighter gases that were released during these eruptions, such as hydrogen, escaped into space. Energy from the sun triggered chemical reactions with the remaining gases, which in turn produced a second atmosphere.

Figure 10–2. Earth's second atmosphere probably consisted of gases released during volcanic eruptions.

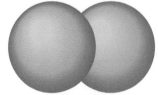

Oxygen gas
(O_2)

Ozone gas
(O_3)

Figure 10–3. Ozone contains three atoms of oxygen rather than two atoms as in an oxygen molecule.

As you study the composition of this primitive atmosphere, it surprises you to learn that water vapor played an important role in its development. Water vapor was very abundant and produced torrential rains. The rains began a chain reaction that continued the evolution of the atmosphere.The rain collected in low spots on the earth's surface and produced a primitive ocean.

Sunlight caused some of the water vapor to break up into hydrogen and oxygen. The hydrogen escaped into space, as it had before. However, the oxygen atoms remained and reacted to produce oxygen molecules and **ozone,** a form of oxygen. Molecules of oxygen consist of two oxygen atoms, while an ozone molecule consists of three atoms of oxygen.

Ozone collected in a layer about 30 km above the earth's surface. This ozone layer absorbed most of the ultraviolet radiation from the sun and prevented this radiation from reaching the earth's surface. This protective ozone shield made it possible for living things to evolve and thrive on Earth. Ultraviolet radiation is harmful to living things. The research team unanimously agrees to include an ozone layer as an essential criterion for planet selection.

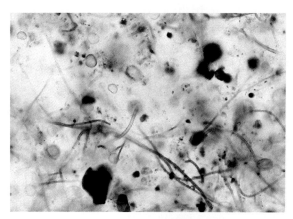

Figure 10–4. Bacteria like these were the likely source of gases in the atmosphere.

Life Appears Life on Earth most likely began in the waters of the primitive ocean. The earliest forms of life were probably bacteria that used ammonia gas as an energy source for their life functions. Over time, bacteria evolved that could produce their own food. But your data reveals that these bacteria also released gases into the atmosphere. What were these gases?

DISCOVER BY *Doing*

Light a small candle and cover it with a glass jar. **CAUTION: Be careful with the flame.** How long does the candle burn? Why does the flame go out? Now place a healthy green plant under the jar in strong sunlight or under a lamp. After about 20 minutes, light the candle again. Then, quickly move the jar from the plant to the candle, keeping the jar upside down. Again time how long the candle burns. Explain any difference in burning time. ✏

As the activity demonstrated, the candle burned somewhat longer when the plant was present. Since a fuel requires oxygen to burn, the plant must have been releasing oxygen. This is what probably happened in the early atmosphere. The new bacteria released oxygen gas as a by-product. As the number of plants and other organisms that produced oxygen increased over time, some of this oxygen dissolved in the ocean waters, and some became part of the evolving atmosphere.

As the variety of living things increased, each new life form interacted with its environment and changed it. Our atmosphere today is much different than the one that surrounded Earth millions of years ago. Table 10-1 summarizes the data the research team has gathered about Earth's atmospheres.

Table 10-1	**Comparison of the Primitive Atmosphere and the Present Atmosphere**	

Gas	Percent of Molecules	
	Primitive Atmosphere	**Present Atmosphere**
Carbon dioxide (CO_2)	92.2	0.03
Nitrogen (N_2)	5.1	78.1
Sulfur dioxide (SO_2)	2.3	*
Hydrogen sulfide (H_2S)	0.2	*
Methane (CH_4)	0.1	*
Ammonia (NH_3)	0.1	*
Oxygen (O_2)	0.0	20.9
Argon (Ar)	0.0	0.9

*Trace amounts

In addition to the gases listed in the table, today's atmosphere also contains small amounts of other gases, such as water vapor. Although the amount of water vapor in the air varies from time to time and from place to place, the composition of today's atmosphere remains fairly constant. However, certain human activities could change this.

Chemical Cycles As your research team discovers, the release of oxygen into the atmosphere was just the beginning. There are many chemical cycles that interact with the atmosphere. Try the next activity to find out why these cycles are important and how human activities can affect them.

DISCOVER BY Researching

Obtain several reference books from your school or public library, and read about the water cycle, the carbon dioxide-oxygen cycle, and the nitrogen cycle. Write short descriptions of these cycles in your journal. Explain how these chemical cycles involve the atmosphere, and how human activities can affect these cycles. ✎

The water cycle, carbon dioxide-oxygen cycle, and nitrogen cycle help keep the earth's atmosphere in working order. During the process of photosynthesis, green plants remove carbon dioxide from the atmosphere and add oxygen to it. During respiration, animals remove oxygen from the atmosphere and release carbon dioxide. Some kinds of bacteria help remove nitrogen from the atmosphere, while other kinds help to replace nitrogen. All these living things interact with the environment, and these interactions keep the atmosphere in balance. Human activities can upset this balance if they pollute the atmosphere.

So far, only one observation really troubles the research team, and that is the condition of the ozone layer. There is a hole in the ozone layer over Antarctica and a thinning of this layer over the Arctic region. As you learned earlier, the ozone layer provides a protective shield against ultraviolet radiation from the sun. Certain human activities appear to be causing the deterioration. Chemicals added to the atmosphere by industries and by gases used in refrigerators, air conditioners, and some aerosol sprays may react with the ozone and destroy it.

In recent years, many nations have passed laws prohibiting or limiting practices that are harmful to the atmosphere. The research team agrees that potential new planets must be investigated for the presence of atmosphere-damaging chemicals.

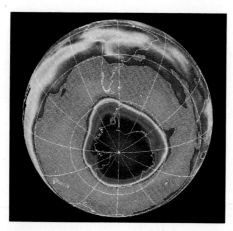

Figure 10–5. Scientists have discovered a hole in the ozone layer above Antarctica.

 ASK YOURSELF

What caused the earth's primitive atmosphere to change?

Structure of the Atmosphere

Your research team delves deeper into its mission, carefully examining the structure of Earth's atmosphere. You expect this information to allow you to screen out other candidates for colonization. Try the following activity to reproduce the team's first model of Earth's atmosphere.

DISCOVER BY Doing

Make layers of air by placing a small dish of ice and a lighted candle under a glass jar. **CAUTION: Be careful with the flame.** When the candle goes out, notice where the smoke collects inside the jar. Draw what you see. Why do you think the smoke collects in one layer?

As with the model, Earth's atmosphere is made up of distinct layers, each with its own characteristics. In the next activity, you can develop a model that reveals more about these characteristics.

DISCOVER BY Doing

Pour small amounts of colored water, motor oil, cooking oil, and alcohol into a container, and tightly close it. Shake the container vigorously for 10 seconds. Observe what happens to the mixture as it settles.

Like the concoction you made, the layers of the atmosphere have different characteristics. How do you think the atmosphere is different from the model in the activity? The diagram and descriptions on the next pages summarize the characteristics of the layers of Earth's atmosphere.

Layers of the Atmosphere

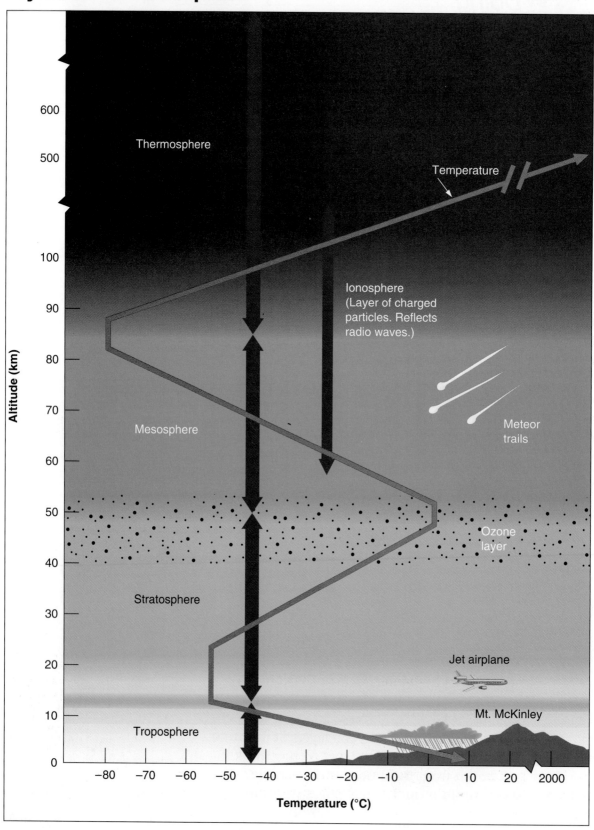

Troposphere The bottom layer of the atmosphere, the one closest to the earth, is called the *troposphere*. Because the earth's surface is warmer than the air above it, the bottom portion of the troposphere is warmer than the rest of this layer. This uneven heating makes the air in this layer circulate. As you may remember, *convection* is the transfer of heat by the circulation or movement of solids, liquids, or gases. When air is heated, it expands and becomes less dense than the air around it. The warm air is pushed up by the cooler, more dense air. As the cooler air is heated, it, too, begins to be replaced by cold air. This circulation creates weather systems. Almost all of the earth's weather takes place in the troposphere. Temperature generally decreases with altitude in the troposphere. However, some 12 km above the surface the temperature levels off to about −60°C.

Stratosphere The next layer of the atmosphere is called the *stratosphere*. The temperature in this layer does not change much for the first 10 km or so. Then it begins to rise steadily. This increase in temperature is due to the presence of ozone. Most of the ozone in the atmosphere is located in this layer, with its main concentration at an altitude of about 42 km. As ozone absorbs ultraviolet radiation, heat is produced. Temperature rises to a maximum of about 8°C near the upper boundary of the stratosphere.

Mesosphere In the third layer of the atmosphere, the *mesosphere,* temperatures again decrease with altitude. This is due to the increasing distance from the earth's warm surface. At the top of this layer, temperatures are about −80°C.

Thermosphere The outermost layer of the atmosphere is called the *thermosphere*. In this layer, gas molecules are so far apart that the solar wind easily passes through. When high-energy particles of the solar wind collide with gas molecules, the molecules speed up, causing the temperature of this layer to increase steadily. Such collisions also cause charged particles, called *ions*, to form. The formation of ions occurs most frequently in the lower part of the thermosphere, which is known as the *ionosphere*. It is in this region that colorful displays called *auroras* occur.

Figure 10–6. Auroras occur in the ionosphere.

The region of the thermosphere beyond the ionosphere is called the *exosphere*. In this region gas molecules are so far apart that few collisions of any kind occur.

 ASK YOURSELF

Compare the different layers of the atmosphere.

Gravity, Weight, and Air Pressure

At first, the research team wonders why the gases of the atmosphere don't just escape into space. What holds them all together? Try the following activity to see whether you can determine why the atmosphere does not simply float away.

ACTIVITY

What keeps the atmosphere from escaping?

MATERIALS

safety goggles, laboratory apron, gloves, graduate, empty metal can, bowl, water, Bunsen burner or alcohol lamp, matches or burner lighter, wire gauze, tripod

PROCEDURE
1. **CAUTION: Put on safety goggles, a laboratory apron, and gloves before you begin this activity.**
2. Measure about 20 mL of water in the graduate, and pour the water into the can.
3. Fill the bowl nearly to the brim with cold water.
4. Light the burner or lamp, and adjust it to produce a low flame.
5. Set the can on the wire gauze atop the tripod. Place the entire assembly over the flame.
6. Watch for steam to begin escaping from the can. Continue heating the

can for 30 seconds after you first observe the steam.
7. Extinguish the flame.
8. Quickly place the can upside down in the bowl of water. Record your observations.

APPLICATION
1. What happened to the can when it was placed in the cold water? Explain.
2. How does what happened to the can relate to the earth's atmosphere not escaping into space?

The research team realizes that the earth's atmosphere is held around the planet by the same force that causes an apple to fall from a tree. That force is *gravity.* Gravity pulls the gas molecules toward the earth. The influence of the earth's gravity is strongest near the surface. More than 75 percent of the entire atmosphere is found in the troposphere, within 15 km of the earth's surface.

The pull of gravity on gas molecules gives them weight. This weight causes the air to push against the earth's surface. The force of this push is called air pressure, or **atmospheric pressure.** It was atmospheric pressure that caused the empty can to be crushed. Why was the air pressure greater on the out-side of the can than on the inside?

The research team lists a barometer as a necessary tool for evaluating possible planets. A *barometer* (buh RAHM uh tuhr) is an instrument used to measure atmospheric pressure. Any planet being considered for colonization would have to have the right atmospheric pressure.

As this diagram of a simple mercury barometer shows, air pushing down on the surface of the mercury in the dish will force some mercury up into the tube and hold it there. At sea level, air at 0°C will support a column of mercury 760 mm high. This is known as normal atmospheric pressure. It is also referred to as one atmosphere of pressure, or 1003 millibars.

The research team notes that the barometer indicates that atmospheric pressure decreases with altitude. This decrease oc-curs because the higher you go, the less air there is above you pushing down. At the top of Mount Everest, the highest place on Earth, atmospheric pressure is less than one-third of what it is at sea level.

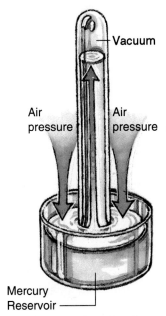

Figure 10–7. A simple mercury barometer

 ASK YOURSELF

Why does atmospheric pressure decrease with altitude?

SECTION 1 *REVIEW AND APPLICATION*

Reading Critically

1. How did Earth's primitive atmosphere differ from today's atmosphere?
2. Why is the ozone layer of the atmosphere so important?

Thinking Critically

3. Why do the amounts of oxygen and carbon dioxide in the atmosphere remain constant?
4. If a barometer reads 760 mm of mercury at sea level and 660 mm at an altitude of 1000 m above sea level, what might it read in Death Valley, which is nearly 100 m below sea level?

SKILL Communicating Data Using a Graph

▶ **MATERIALS**
- graph paper • pencil

▼ **PROCEDURE**

1. In the troposphere, air temperature decreases at an average rate of 6.5°C per km of altitude. Using a sheet of graph paper, make a graph like the one below that will show this trend.
2. Title your graph. Write the title at the top of the grid.
3. Label the axes. Label the *X*, or horizontal, axis "Temperature". Label the *Y*, or vertical, axis "Altitude".
4. Select a scale for the X axis. Start with 20°C. You may wish to have each square represent 5°C, or you may select some other scale.

5. Select a scale for the Y axis. The bottom of the scale should be 0 km. The top of the scale should be 12 km. Select an appropriate value for the boxes so that the scale is divided into equal intervals.
6. Plot your data. The data should show a 6.5° decrease in temperature for every km increase in altitude.
7. Once you have plotted several points on your graph, draw a line to connect the points. You now have a visual representation of how temperature changes with altitude in the troposphere.

▶ **APPLICATION**

Look at your graph. What pattern do you observe? Are some patterns easier to understand when the data is shown on a line graph? Why or why not?

✳ *Using What You Have Learned*

Use your graph to predict the temperature at an altitude of 16 km if the trend continues.

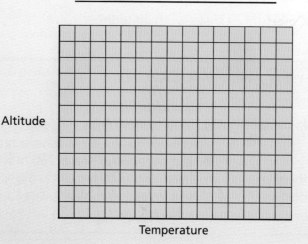

Altitude

Temperature

Radiant Energy and Heat

The research team now knows the mixture of gases in the atmosphere that make it possible for humans to live on Earth. Using this data, you will be able to eliminate many planets without further investigation. Ozone, oxygen, and certain cycles that maintain the atmosphere are on the "must have" list. You are now ready to move on to the next step in the selection process: temperature. How hot is too hot, and how cold is too cold? The team prepares to collect more data.

Objectives

Evaluate *the reasons why the earth's surface heats unevenly.*

Describe *what happens to radiant energy that reaches Earth.*

Explain *how radiant energy is transformed into heat.*

Heating the Earth's Surface

If you have ever placed your hand on the hood of a car that has been parked in the sun or walked barefoot on a sidewalk on a sunny day, you know that sunlight can be changed into heat energy. In a similar way, but on a much larger scale, sunlight heats the earth's surface. However, the earth's surface is not heated uniformly. There are two main reasons for this uneven heating: the angle at which the sun's rays strike the surface of the earth and the types of materials the rays strike. The research team needs to determine how this information affects their criteria for finding a suitable planet.

Figure 10–8. Because the earth's surface is curved, the sun's rays strike places near the equator straight on and places near the poles at a slant.

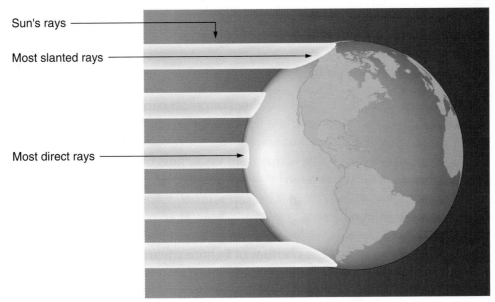

Sun's rays

Most slanted rays

Most direct rays

Although the sun's rays travel in straight lines, the angle at which the rays strike the earth's surface is affected by the earth's shape. The curved shape of the earth's surface causes the sun's rays to strike different parts of the surface at different angles. As you can see in the preceeding diagram, the sun's rays strike the surface near the equator straight on. As you move away from the equator, the curve of the earth's surface causes the sun's rays to arrive at more of a slant. The polar regions receive the most slanted rays from the sun.

The tilt of the earth's axis also affects the angle at which the sun's rays strike the surface. As Earth revolves around the sun, the tilt of the earth's axis causes the angle at which the sun's rays strike different parts of the surface to change throughout the year. You can see this change in the diagram below. In July the sun's rays strike the Northern Hemisphere at less of a slant than they do in January. In September and March, the sun is directly overhead at the equator.

Figure 10–9. The tilt of Earth's axis affects how sunlight strikes the surface.

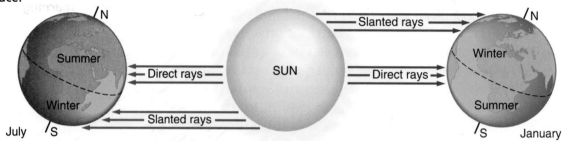

The research team at first thinks that this information about slant and tilt is not important. They are puzzled because they know that the amount of energy in any given ray is the same. Therefore, how can the earth's tilt or shape affect that? Try the next activity to find out.

DISCOVER BY *Doing*

Obtain a sheet of graph paper and a flashlight. Hold the light about 20 cm from the paper, and shine it directly onto the graph paper. Check the size, shape, and brightness of the lighted area of the paper. Count the number of lighted squares, and enter this information in your journal. Without moving the light, tilt the paper away from the light at an angle of about 45°. Again, observe the size, shape, and brightness of the lighted area. Count and record the number of lighted squares. Now assume that the light can be changed to heat. In your journal, explain why direct rays will heat the surface more than slanted rays. ✏

As your experiment demonstrated, direct rays—those that strike at little or no slant—light the earth's surface better than slanted rays do. Although all of the sun's rays carry the same amount of energy, the energy of the slanted rays is spread out over a larger area than the energy of the direct rays. The area lighted by the sun's rays increases as the rays become more slanted. To determine a planet's suitability for life, its shape and tilt in relation to its energy source must be examined.

What other information about energy sources must be considered? Radiant energy from the sun strikes the earth. This energy travels in the form of waves. Some of these are high-energy waves, such as visible light and invisible ultraviolet waves. Low-energy waves, such as infrared and radio waves, are also radiated by the sun. For radiant energy to produce heat, it must be absorbed by something. The illustration shows what happens to the radiant energy received by the earth. Notice that 30 percent of the energy, including some of the radiant energy that does reach the Earth's surface, is not absorbed. It is reflected away, mostly by clouds and surfaces covered with snow or ice.

Figure 10–10. The distribution of radiant energy reaching the earth

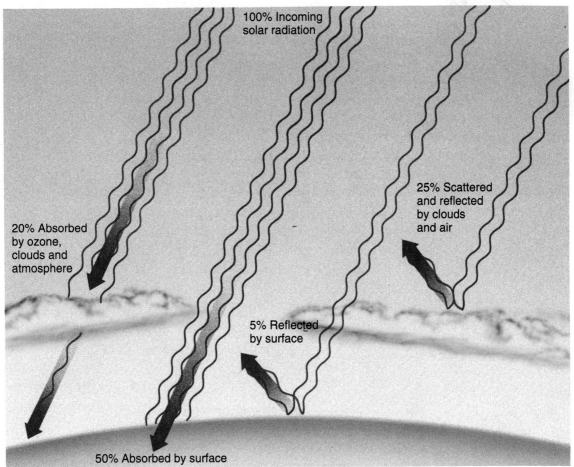

100% Incoming solar radiation

25% Scattered and reflected by clouds and air

20% Absorbed by ozone, clouds and atmosphere

5% Reflected by surface

50% Absorbed by surface

When radiant energy *is* absorbed by the earth's atmosphere or its surface, it is changed into heat energy. However, not everything on the earth absorbs radiant energy equally. For example, dark-colored materials absorb better, and thus heat up faster, than light-colored materials. Solids heat up and cool down faster than liquids. Land—soil and rock—heats up and cools down faster than water. Since land and water are the two components of the earth's surface, the surface is unevenly heated.

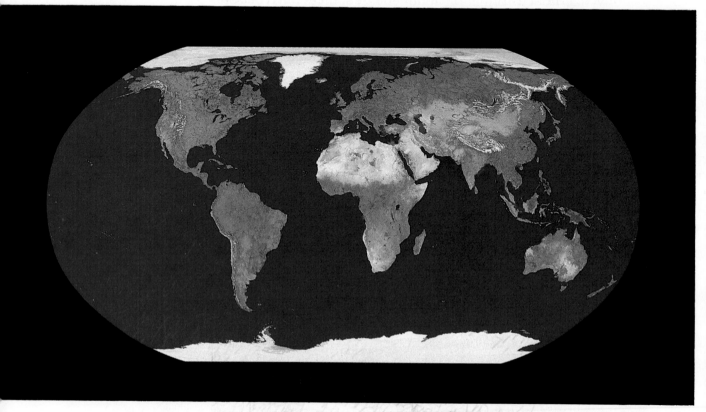

Figure 10–11. Because the earth's surface is made up of different materials, it is heated unevenly by radiant energy.

For Earth to remain livable, the amount of radiant energy received must be balanced by the amount of energy returned to space. About 30 percent of the incoming radiant energy is reflected into space. Most of the 70 percent that is absorbed by the earth and its atmosphere is reradiated back into space in the form of infrared waves. This balance between incoming radiant energy and outgoing heat energy is known as the **radiation balance.**

 **ASK YOURSELF**

Why does the earth's surface heat up unevenly?

Heating the Atmosphere

Radiant energy heats the earth's surface. Some of this heat energy, in turn, heats the atmosphere. Heat energy from the earth's surface is transferred to the atmosphere in various ways. For example, air in contact with the earth's surface is heated by conduction. **Conduction** is the transfer of heat from one material to another by direct physical contact. Thus, air near the earth's surface is usually warmer than the air a few meters above the surface. But most of the heat energy in the atmosphere is transferred by *convection.*

You can see convection at work in the diagram shown here. Although the stove is located in one corner, it can heat the entire room. Air touching the stove is heated by conduction. Air near the stove is heated by radiation. Convection currents carry the heated air to all parts of the room.

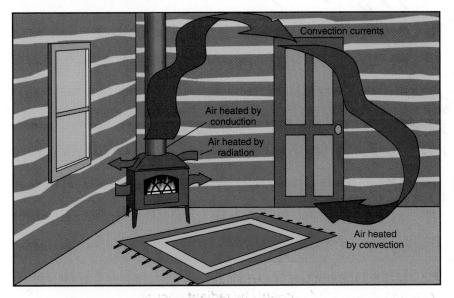

Figure 10–12. How does this stove heat the entire room?

Radiation is the transfer of heat by waves. If you hold your hand near a hot object, you can feel the heat coming from the object. This heat is in the form of infrared waves. These low-energy waves are easily absorbed by many other materials. Some components of air are excellent absorbers of infrared waves. Carbon dioxide gas and water vapor, for example, trap infrared radiation, which would normally escape into space, and hold it near the earth's surface. This same effect can be seen in a greenhouse or in a car with the windows closed. Like carbon dioxide, glass lets high-energy waves through but traps the returning infrared radiation. As a result, the temperature inside the greenhouse or car becomes much warmer than the temperature outside.

Figure 10–13. The glass in a greenhouse lets light energy in, but keeps heat energy from going out. Carbon dioxide in the atmosphere, as well as ozone and water vapor, acts like the glass in a greenhouse, trapping heat near the earth's surface.

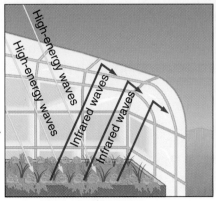 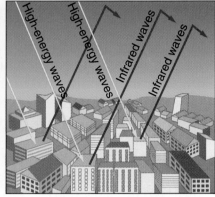

Earth's radiation balance is delicate and can be upset by seemingly minor changes. For example, a decrease of a few degrees in worldwide temperatures could lead to an increase in the amount of ice and snow on the surface. Snow and ice reflect radiant energy, remember. This could lead to still lower temperatures and, perhaps, another ice age. The radiation balance has shifted many times throughout Earth's history.

In the past, upsets of the radiation balance have been brought about by natural events, such as extensive volcanic activity. Today, however, human activities threaten that balance. For example, the burning of fossil fuels increases the carbon dioxide content of the atmosphere. Increased levels of carbon dioxide trap more and more infrared radiation. The increased retention of infrared radiation heats up the atmosphere. The carbon dioxide keeps the heat in as well, resulting in a gradual warming of the atmosphere. Over time, this warming may increase the **greenhouse effect,** the warming of the earth like the inside of a greenhouse. This warming could cause the ice that now covers Antarctica and Greenland to melt. The resulting rise in ocean levels would flood coastal regions around the world.

 ASK YOURSELF

How is heat transferred from the earth's surface to the atmosphere?

SECTION 2   *REVIEW AND APPLICATION*

Reading Critically

1. What must happen to radiant energy for it to be changed to heat energy?

2. What is a greenhouse effect?

Thinking Critically

3. Why is most of the radiant energy received by the earth returned to space as infrared radiation?

4. Describe how your life would be changed if worldwide temperatures were to increase or decrease by several degrees.

INVESTIGATION

Comparing Radiant Energy and Color

▶ MATERIALS
- shoe box with lid ● scissors ● black construction paper ● white construction paper
- tape ● thermometers, 2 ● lamp ● stopwatch

▼ PROCEDURE

1. Cut the lid of the shoe box as shown, and use it to divide the box into two equal sections.
2. Line one side of the box with black construction paper and the other side with white construction paper. Use tape to secure the paper in place.

3. Cut small holes in the box, and insert thermometers as shown.
4. Center the lamp directly above the box, about 3 cm above the rim.
5. Record the temperatures shown on both thermometers at 0 minutes. Then turn on the lamp.

6. Observe and record temperatures at 1-minute intervals.
7. After 10 minutes, turn off the lamp. Continue recording temperatures at 1-minute intervals until each thermometer has returned to the temperature recorded at 0 minutes.

TABLE 1: RADIATION AND TEMPERATURE		
Time	°C (dark)	°C (light)
0 min.		
1 min.		
2 min.		
3 min.		
4 min.		
5 min.		
6 min.		
7 min.		
8 min.		
9 min.		
10 min.		

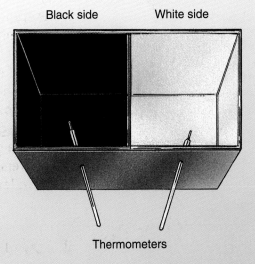

Black side   White side

Thermometers

▶ ANALYSES AND CONCLUSIONS
1. Which side of the box heated up faster?
2. Which side of the box was hotter at the end of 10 minutes?
3. Which side of the box returned to the starting temperature first?
4. Compare the effects of the radiant energy from the lamp on light and dark surfaces.

▶ APPLICATION
If you lived in the tropics, what color car do you think would be the best to buy? Explain.

✳ Discover More
Put a piece of plastic wrap over the top of the box and repeat the Investigation. How do the results differ? Why? Does this change your car-color decision?

Air in Motion

Objectives

Differentiate between local winds and planetary winds.

Explain the relationship between air pressure and wind direction.

Name and **describe** the planetary winds.

The last atmospheric criterion the research team evaluates is air movement. Air that is moving horizontally—parallel to the earth's surface—is called **wind.** The research team is amazed at the variability of wind activity on Earth. Its data indicates that winds change from day to day and from place to place. Winds can blow in any direction and can range in strength from a gentle breeze to a raging force strong enough to uproot trees and destroy buildings. However, all winds that move along the earth's surface have two things in common. They are set in motion by uneven heating of the earth's surface, and they tend to blow from regions of higher to lower air pressure. The research team categorizes wind into two types—local winds and planetary winds.

Figure 10–14. Wind is the horizontal movement of air.

Local Winds

Local winds are winds that affect a limited area. If you have ever been to the shore of an ocean or a large lake, you know that a breeze is usually blowing. During the day, land heats up faster than water, and the land heats the air above it. As the warm air rises, the pressure it exerts on the earth's surface decreases, thereby creating an area of low pressure over the land. Air over the water is cooler and denser than air over the land, creating an area of high pressure over the water. This cool, dense air moves toward the land, creating a sea breeze.

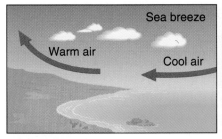
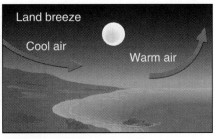

Figure 10–15. Sea breezes and land breezes are local winds.

At night, the reverse occurs. Land cools faster than water. Thus the pattern of circulation is reversed. Air moves from the land toward the sea, creating a land breeze. Notice in the illustration above that winds are named for the direction they are blowing from and that air always moves from regions of high to low air pressure. Sea breezes and land breezes are examples of local winds. Local winds are set in motion by surface features of a particular area, such as landforms, water, and ice.

Figure 10–16. The cold, dry winds that stream from the Antarctic ice cap and the puffy clouds that form over tropical islands are both the result of local winds.

▼ **ASK YOURSELF**

During the day, why is air pressure over land lower than it is over water?

Coriolis Force

Have you ever tried to walk across the platform of a turning carousel or down the aisle as a bus or train takes a curve? If so, you know that the motion of the surface can make it difficult to move in a straight line. Now think about the earth's surface. The surface and its atmosphere are moving because of the earth's rotation. What effect, if any, does the earth's rotation have on objects traveling on or near its surface? If you try the next activity, you can get an idea.

Using an old turntable covered with paper, try drawing a straight line from the center to the edge of the turntable as it moves. What happens? ✏

Now imagine a hypersonic plane traveling due south from the North Pole toward the shuttle landing strip at Cape Canaveral, Florida. The plane is expected to reach the cape in one hour. If the earth did not rotate, the plane would be over Cape Canaveral at 1 P.M. sharp. However, the earth does rotate, and in one hour it will rotate through 15° of longitude. Thus, while the plane is in flight, Cape Canaveral will move about 1500 km to the east and part of Mexico will move into the target area. The plane will now be somewhere over Mexico! Because of the earth's rotation, the plane will appear to have followed a curved path. Try the following activity to help you understand why this happens.

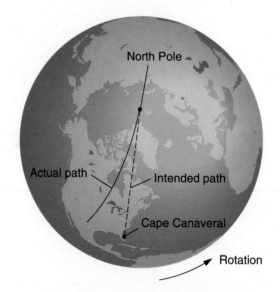

Figure 10–17. Because of the earth's rotation, objects traveling across the earth's surface in the Northern Hemisphere seem to curve toward the right.

DISCOVER BY Problem Solving _____

Earth makes one complete rotation in 24 hours. A person living on the equator travels farther in that time than does a person living at a latitude of 45°N. This means that the person on the equator must travel faster than the other person. Explain why this is so. (HINT: Draw two circles, one inside the other, to help you figure this problem out.) ✏

The apparent deflection of objects moving across the earth's surface due to the earth's rotation is called the **Coriolis** (coh ree OH lihs) **force.** Like the plane, winds also seem to follow a curved path as they move across the earth's surface. Because of the Coriolis force, winds in the Northern Hemisphere seem to curve to the right; winds in the Southern Hemisphere seem to curve to the left. The farther a wind travels, the more it is influenced by the Coriolis force. Local winds are affected very little. Winds that travel long distances are strongly influenced by the Coriolis force.

 ASK YOURSELF

How does the Coriolis force influence winds?

Planetary Winds and Pressure Belts

Planetary winds are part of a global pattern of air circulation. Remember that the earth's surface is heated unevenly. Air rises in warm equatorial regions and sinks over cold polar regions. Thus, you might expect air to circulate in one huge convection pattern. In such a system, the planetary winds would blow from the poles toward the equator. However, this does not happen. In the next activity you can find out what does.

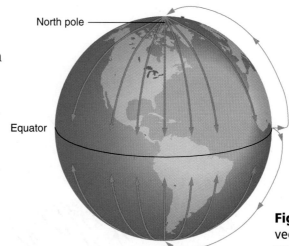

North pole

Equator

South pole

Figure 10–18. Convection heating alone would cause winds to blow from the poles directly to the equator.

DISCOVER BY *Doing*

Place two electric fans about 1 m apart, blowing toward each other. Light a candle or a match and then carefully blow it out so that there is a lot of smoke. Direct the smoke into the area midway between the two fans. In your journal, describe what happens to the smoke in the space where the winds from the two fans meet. What might happen in the atmosphere where two wind systems meet?

As warm air rises over the equator, it cools down. Eventually, it stops rising and spreads out toward the poles. By the time this air reaches the latitudes of 30° north and south, most of it is cool enough to sink to the earth's surface. The rest travels all the way to the poles before it sinks. This rising and sinking air creates pressure belts across the earth's surface. The planetary winds move between these pressure belts. The illustration shows the locations of the pressure belts and planetary winds.

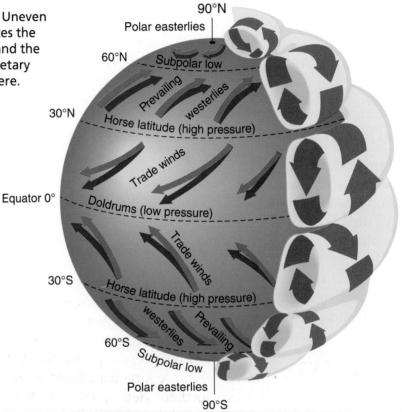

Figure 10–19. Uneven heating produces the pressure belts and the pattern of planetary winds shown here.

Doldrums and Horse Latitudes

Near the equator, warm rising air creates a belt of low pressure called the *doldrums*. This is a region of warm temperatures and very little wind.

At the latitudes of 30° N and 30° S, sinking air produces regions of high pressure, fair weather, and calm wind conditions. The lack of wind is also responsible for the name *horse latitudes* given to these regions. At times, sailing ships carrying horses to the Americas became stranded in these regions for days and even weeks at a time. Lack of feed and water forced crew members to throw the horses overboard, thus leading to the name "horse latitudes." To be becalmed in these areas could be fatal, as it was in "The Rime of the Ancient Mariner."

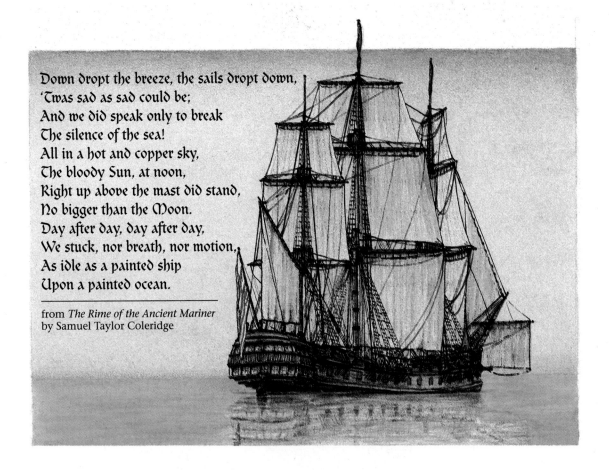

Down dropt the breeze, the sails dropt down,
'Twas sad as sad could be;
And we did speak only to break
The silence of the sea!
All in a hot and copper sky,
The bloody Sun, at noon,
Right up above the mast did stand,
No bigger than the Moon.
Day after day, day after day,
We stuck, nor breath, nor motion,
As idle as a painted ship
Upon a painted ocean.

from *The Rime of the Ancient Mariner*
by Samuel Taylor Coleridge

Trade Winds and Prevailing Westerlies   The air that sinks at the horse latitudes spreads out north and south along the earth's surface as planetary winds. Winds moving from the north or from the south toward the equator are called the *trade winds*. These winds were so named because early traders sailed these winds from the Old World to the Americas. Trade winds are really northeasterly in the Northern Hemisphere, due to the Coriolis force. Winds moving toward the poles, called the *prevailing westerlies*, also curve due to the Coriolis force. These winds helped early traders return to Europe.

Polar Highs and Subpolar Lows   The air that sinks near the poles is very cold and dense, creating high-pressure systems at the poles. This air, moving toward the equator, is deflected by the Coriolis force to form the *polar easterlies.*

In the middle latitudes, the polar easterlies meet the prevailing westerlies. The warmer air of the westerlies is forced up over the colder, denser air of the polar easterlies, creating low pressure belts called the *subpolar lows*. The mixing of cold and warm air along the subpolar low is responsible for much of the weather that occurs in mid-latitude regions of the Northern Hemisphere.

Jet Streams   Above the earth's surface near the top of the troposphere, narrow bands of high-speed winds called **jet streams** circle the globe. These "rivers of wind" form above the subpolar lows. The speed of jet streams varies from as little as 60 km/h in the summer to more than 300 km/h in the winter. The jet streams do not follow regular paths around the globe. They form broad loops that may extend almost to the tropics.

Figure 10–20. A jet stream follows a wavy course around the earth.

The research team concludes that planetary wind systems are essential in planet selection because of their interaction with weather phenomena. Will the team members ever locate another planet with similar conditions and circumstances? The complex, interrelated chain of events that has occurred on Earth may make it a nearly impossible task. Earth may just be one of a kind.

▶ **ASK YOURSELF**

What are the names and locations of the planetary winds?

SECTION 3   *REVIEW AND APPLICATION*

Reading Critically
1. Why do winds turn to the right in the Northern Hemisphere and to the left in the Southern Hemisphere?
2. What causes the planetary winds to blow?

Thinking Critically
3. If the earth's axis were not tilted, how would the heating of the earth's surface be different?
4. What would the planetary wind system be like if the earth did not rotate?

The Big Idea

The atmosphere is the ocean of air that surrounds the earth. A complex chain of events has occurred over time that results in an atmosphere that offers protection, sustenance, and opportunity for growth. The atmosphere has a layered structure, each layer dependent on the others to create an environment that provides wind and temperature stability. Radiant energy, the earth's shape and the tilt of its axis, and the presence of land and water all interact to create an atmosphere that supports life.

For Your Journal

Look again at the questions you answered in your journal at the beginning of this chapter. Have any of your ideas changed? Revise your journal to show how your ideas have changed.

Connecting Ideas

Copy this concept map into your journal and complete it.

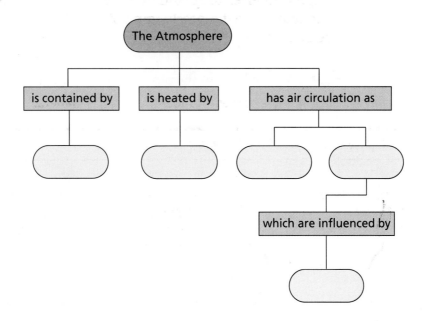

Understanding Vocabulary

1. Explain how the terms in each set are related.
   a) conduction (327), radiation (327), radiation balance (326)
   b) atmosphere (314), atmospheric pressure (321), barometer (321)
   c) jet stream (336), Coriolis force (333)
   d) greenhouse effect (328), ozone (313)

Understanding Concepts

MULTIPLE CHOICE

2. In the Northern Hemisphere, the trade winds blow from
   a) north to south.
   b) east to west.
   c) northeast to southwest.
   d) south to north.

3. Where do the sun's rays strike the earth at the greatest slant?
   a) polar regions
   b) Northern Hemisphere
   c) Southern Hemisphere
   d) equator

4. Most of the ozone in the atmosphere is found in the
   a) troposphere.
   b) stratosphere.
   c) mesosphere.
   d) exosphere.

5. Atmospheric pressure decreases with altitude because
   a) the temperature decreases.
   b) there are fewer gas molecules in the atmosphere.
   c) the force of gravity increases with altitude.
   d) the gas molecules are different.

6. Animals help keep the gases in the atmosphere in balance by
   a) removing carbon dioxide and adding oxygen.
   b) removing ozone.
   c) removing oxygen and adding carbon dioxide.
   d) removing nitrogen.

SHORT ANSWER

7. Describe how the sun's radiation heats the earth's atmosphere.

8. Why is it important not to disturb the earth's radiation balance?

Interpreting Graphics

9. Look at the picture. Is the air pressure greater over the land or over the water? Why?

10. In which direction will the wind blow? Why?

Cool Air

Warm Air

Reviewing Themes

11. *Systems and Structures*
Compare the effects of local winds and planetary winds on the weather.

12. *Environmental Interactions*
What is the effect of carbon dioxide on the earth's radiation balance?

Thinking Critically

13. Describe the Coriolis force on Venus, which rotates very slowly and in a direction opposite to that of Earth.

14. Why should skiers protect themselves from sunburn even though the sun's rays are the least direct in winter?

15. Why are the coldest winter nights usually cloudless?

16. How would jet streams affect the flight times of an airplane flying from Los Angeles to New York City and back to Los Angeles again?

17. Athletes at the 1968 Olympics in Mexico City, which is 2240 m above sea level, were extremely tired because of a lack of oxygen. Yet, many world records were set that year. Explain this apparent contradiction.

Discovery Through Reading

Berreby, David. "Acid-Flecked Candy Colored Sunscreen." *Discover* 13 (January 1992): 44–46. This article discusses the effects of the eruption of Mount Pinatubo on global climates and the ozone layer.

WATER IN THE AIR

*S*omething about fog appeals to a writer's imagination. Because this misty water in the air takes so many shapes, poets and storytellers can see in it what they want. Poet Carl Sandburg had a gentle view of fog. Perhaps he stood outside his home in the evening and imagined the mists as a great cat, with its tail curled around his front door.

FOG

The fog comes
on little cat feet.

It sits looking
over harbor and city
on silent haunches
and then, moves on.

Carl Sandburg

*I*n his story "The Strange Case of Dr. Jekyll and Mr. Hyde," Robert Louis Stevenson gives the sooty London fog an air of murderous danger. A lawyer pursuing the evil Mr. Hyde sees fog in "a marvellous number of degrees and hues of twilight; for here it would be dark like the back-end of evening; and there would be a glow of a rich, lurid brown, like the light of some strange conflagration (fire)." It seems as if Mr. Hyde could pounce on you at any moment!

For Your Journal

- What causes fog, mists, and clouds?
- How does water get into the air?
- Why do some clouds form precipitation, while others do not?

Evaporation and Condensation

Objectives

Compare and contrast *the processes of evaporation and condensation.*

Distinguish *between absolute humidity and relative humidity.*

Describe *the relationship between air temperature and relative humidity.*

Ms. Marina enjoys putting on shows of illusion. Nothing beats pulling a rabbit out of a hat. People always love that trick. But being an illusionist is only a hobby for Ms. Marina. During the day, she is a science teacher at Newton Junior High School. Ms. Marina's students are always asking her to do illusions, but she keeps telling them that science beats illusions any day of the week, especially when it comes to making things disappear and reappear.

Water's "Disappearing Act"

"Think about water," Ms. Marina tells her class one day. "What are some of the ways you use water? You drink it, you bathe in it, you brush your teeth with it, you wash your clothes in it, and you cook your food with it. You probably take it for granted because it's always there. But if you take the time to study water, it shows itself as a fascinating substance. It is the only substance commonly found in nature in three different phases—solid, liquid, and gas. And water has a way of 'disappearing' into thin air at times and 'reappearing' out of the same air at other times. So you see, water really has some unusual properties."

Figure 11–1. Unlike a rabbit out of hat, water can be found in nature in three different phases—solid, liquid, and gas. But the gas cannot be seen.

Ms. Marina points out the window. It has just rained, and puddles of water can be seen everywhere. "In a few hours," she tells the class, "most of the puddles will be gone. Where does the water go? Think about wet laundry hanging on a clothesline. In time, the laundry will dry. Where does the water go? Let's try an activity to find some answers."

Figure 11–2. Where does the water go when wet clothes dry?

 *Doing*

Obtain two small, transparent, plastic drinking glasses. Add 100 mL of warm water to each glass. Place one glass under a heat lamp and the other in a cool, shaded area. After the glasses have stood for 24 hours, measure the amount of water remaining in each glass. Record the measurements in your journal, and explain the results of this activity.

As you may have inferred from the activity, the water in the glasses goes into the air. This occurs by evaporation. **Evaporation** is the process by which a liquid changes to a gas. When water evaporates, it changes to an invisible, odorless gas called *water vapor*. Evaporation takes place at the surface of a liquid. To picture what happens during evaporation, imagine that you have a shoe box half full of small plastic beads. As you shake the box gently, the energy passes to the beads, and they vibrate back and forth. When you supply more energy to the system, by shaking the box more vigorously, the beads bounce all around in the box. Some of the beads near the top may gain enough energy to escape from the box.

In the shoe-box model, the beads represent water molecules and the shaking represents heat energy. In nature, as heat is added, water molecules speed up, causing the temperature of the water to increase. When molecules near the surface of the water gain enough energy, they escape into the air as water vapor. The more the water is heated, the faster evaporation takes place. In the previous activity, water evaporated faster in the glass under the heat lamp. Something else happens during evaporation. In the next activity, you can learn what it is.

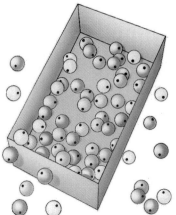

Figure 11–3. Shaking the box vigorously gives some of the beads enough energy to escape from the box.

ACTIVITY

How can you show that evaporation is a cooling process?

MATERIALS
thermometers in notched corks (3), rubber bands, cotton or gauze, ring stand, clamps (3), 100-mL beakers (2), rubbing alcohol, water, wax pencil

PROCEDURE
1. Use rubber bands to fasten a piece of cotton or gauze carefully to the bulb of each thermometer.
2. Attach the clamps to the ring stand at a height of 10-20 cm above the table. Insert the corks into the clamps and tighten.
3. Half fill one beaker with rubbing alcohol and the other beaker with water. Label each beaker with a wax pencil. Read and record the temperature on each thermometer.
4. Lift the beakers so that the bulb of one thermometer is in the alcohol and the bulb of the second thermometer is in the water. The third thermometer remains dry. Be sure the cotton or gauze on each thermometer bulb is saturated. Return beakers to the table.
5. At 30-second intervals, record the temperature of each thermometer.

APPLICATION
1. What happened to the temperature readings of the three thermometers?
2. If any temperature changes were observed, which thermometer showed a faster change in temperature? Explain.

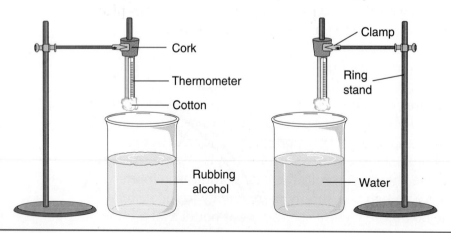

What is unusual about evaporation is that it seems to do contradictory things. Heat makes water evaporate faster, but evaporation is a cooling process. Think about how you feel when you come out of the water after swimming. As water evaporates from your skin, your skin feels cool. The reason for this cooling effect is that during evaporation molecules of liquid take in heat from their surroundings—in this case your skin—and change it into the energy of motion. In the same way, water takes heat from the surrounding air as it evaporates, thus cooling the air.

Have you ever used a sponge to soak up water? A sponge consists of fibrous material that contains many tiny holes, or pores. When a sponge is placed in water, water enters the pores. Eventually, the pores become full, and the sponge cannot hold any more water. The sponge is said to be *saturated* with water.

In some ways, air near the earth's surface acts like a sponge. Recall that air is a mixture of gases. All gases are made up of molecules separated by empty spaces. When water evaporates, the molecules of water vapor enter the air and fill in some of the empty spaces. Sometimes all the spaces between air molecules become filled with water molecules, just as the pores of a saturated sponge are filled with water. When this happens, the air is saturated with water vapor.

Saturated air contains all the water vapor it can hold. For example, if the air over a pond is saturated, the number of water molecules that leave the surface of the pond and enter the air as a gas is offset by an equal number of water molecules that leave the air and return to the surface of the pond as a liquid. But air seldom becomes saturated. Before that happens, winds usually bring in drier air and carry the moist air away.

In nature, enormous amounts of water evaporate from the oceans. Evaporation from lakes, rivers, animals, and plants also adds water vapor to the air. Most of the evaporation takes place in the tropics, where temperatures are always warm.

Figure 11–4. The spaces, or pores, in a sponge allow it to soak up water.

Figure 11–5. Much of the water vapor in the air evaporates from warm, tropical oceans.

 **ASK YOURSELF**

Why does most evaporation take place in the tropics?

It's All Relative

Ms. Marina isn't finished describing the amazing properties of water. "You know that water can disappear and that it can make you shiver as it evaporates. But did you know that water can also make you feel sticky and damp without its raining a drop?" How is that possible?

The amount of water vapor, or moisture, in the air varies from day to day and from place to place. You know this from experience. On some days the air feels crisp and dry. On other days it feels sticky and damp. On which type of day do you think air contains the most water vapor?

Warm air can hold more water vapor than cold air. To understand why, look at the diagrams. They show a cylinder with a movable piston at the top. The temperature of the air in the cylinder at the left is 10°C. When the air is heated, the molecules speed up and move farther apart. The air expands and takes up more space. The same thing happens to air near the earth's surface. When the air is heated by the sun, it expands. There is more space between molecules. Thus, warm air can hold more water vapor than cold air.

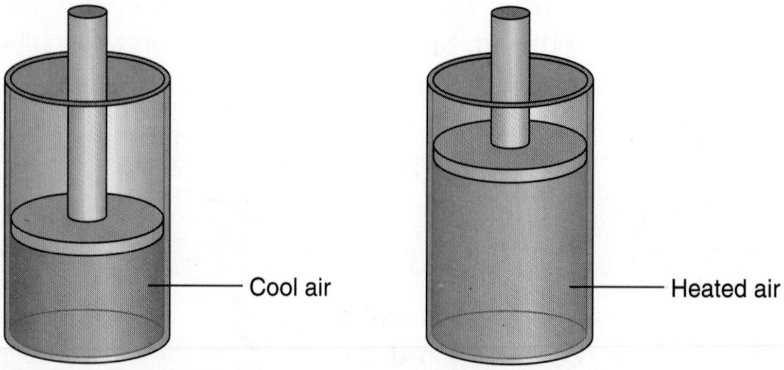

Figure 11–6. The diagrams show a cylinder containing air before being heated (left) and after being heated (right). What happens when the air is heated?

Cool air — Heated air

Water vapor in the air is called **humidity.** The amount of water vapor the air contains at any given time and place is called *absolute humidity.* Since most of the time air is not saturated—there is still some room for more water vapor—the amount of moisture in the air is usually expressed as relative humidity. **Relative humidity** is the amount of water vapor actually in the air compared with the maximum amount the air could hold at that particular temperature.

Relative humidity is expressed as a percentage. For example, at 30°C, 1 m³ of air can hold 30.4 g of water vapor. If the air actually contains this much moisture, it is saturated and the relative humidity is 100 percent. However, if the air contains only

13 g of water vapor, it contains 13/26, or 1/2, as much moisture as it could hold, and the relative humidity is 50 percent. How much water vapor is in the air if the relative humidity at 30°C is 25 percent?

Meteorologists use an instrument called a *psychrometer* to measure relative humidity. This instrument consists of two thermometers. The bulb of one thermometer is covered with a damp cloth. This is the wet-bulb thermometer. The other thermometer is the dry-bulb thermometer. As air passes over the wet-bulb thermometer, evaporation of water from the damp cloth cools the thermometer bulb. At the same time, the dry-bulb thermometer measures air temperature.

When air is saturated, no net evaporation occurs. Therefore, unless the air is saturated, the temperature reading of the wet-bulb thermometer will always be lower than that of the dry-bulb thermometer. The difference between the two thermometer readings is called the *wet-bulb depression*. The drier, or less humid, the air is, the greater the wet-bulb depression will be.

The wet-bulb depression and the air temperature are used to find relative humidity in a table like the one shown. For example, suppose the wet-bulb reading is 14°C and the dry-bulb reading is 20°C. The wet-bulb depression is 6°C. In Table 11-1, find 6°C in the row across the top and 20°C (the air temperature) in the column at the left. The number where the 6°C-column intersects the 20°C-row is the relative humidity, which is 51 percent. In the following activity, you can determine some relative humidity readings on your own.

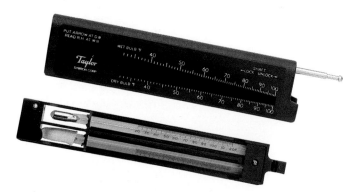

Figure 11–7. A psychrometer

Table 11-1 Relative Humidity

Dry-Bulb Temperature (°C)	Wet-Bulb Depression (°C) (Dry-bulb temperature minus wet-bulb temperature)																			
	1	2	3	4	5	6	7	8	9	10	11	12	13	14	15	16	17	18	19	20
-10	67	35																		
-8	71	43	15																	
-6	74	49	25																	
-4	77	55	33	12																
-2	79	60	40	22																
0	81	64	46	29	13															
2	84	68	52	37	22	7														
4	85	71	57	43	29	16														
6	86	73	60	48	35	24	11													
8	87	75	63	51	40	29	19	8												
10	88	77	66	55	44	34	24	15	6											
12	89	78	68	58	48	39	29	21	12											
14	90	79	70	60	51	42	34	26	18	10										
16	90	81	71	63	54	46	38	30	23	15	8									
18	91	82	73	65	57	49	41	34	27	20	14	7								
20	91	83	74	66	59	51	44	37	31	24	18	12	6							
22	92	83	76	68	61	54	47	40	34	28	22	17	11	6						
24	92	84	77	69	62	56	49	43	37	31	26	20	15	10	5					
26	92	85	78	71	64	58	51	46	40	34	29	24	19	14	10	5				
28	93	85	78	72	65	59	53	48	42	37	32	27	22	18	13	9	5			
30	93	86	79	73	67	61	55	50	44	39	35	30	25	21	17	13	9	5		
32	93	86	80	74	68	62	57	51	46	41	37	32	28	24	20	16	12	9	5	
34	93	87	81	75	69	63	58	53	48	43	39	35	30	26	23	19	15	12	8	5
36	94	87	81	75	70	64	59	54	50	45	41	37	33	29	25	21	18	15	11	8
38	94	88	82	76	71	66	61	56	51	47	43	39	35	31	27	24	20	17	14	11
40	94	88	82	77	72	67	62	57	53	48	44	40	36	33	29	26	23	20	16	14

DISCOVER BY *Calculating*

If the dry-bulb reading is 34°C and the wet-bulb reading is 30°C, what is the wet-bulb depression? What is the relative humidity? If the wet-bulb depression is 10°C and the air temperature is 20°C, what is the relative humidity? ✎

On a hot, humid day, the air feels heavy and your clothes stick to your skin when you sweat. In addition to making you feel uncomfortable, the combination of high temperature and high humidity can be harmful to your health. Because the air is nearly saturated with water, sweat does not evaporate from your skin. When sweat cannot evaporate, the body's natural cooling system does not work very well. Strenuous exercise can cause your body to become overheated. A body temperature rise of just two or three degrees can cause heat exhaustion. A further increase in body temperature can lead to a total collapse and may even be fatal. In hot, humid weather, it is important to drink plenty of fluids and to avoid strenuous exercise.

Figure 11–8. The runner is suffering from heat exhaustion brought on by strenuous exercise on a hot, humid day.

 **ASK YOURSELF**

What is the relative humidity of saturated air?

Water from "Thin Air"

"What other tricks can water do?" Ms. Marina's class asks her.

"You know that it can disappear," she answers, "but it can also reappear. Imagine it is a hot, sticky day. You fill a tall, dry glass with ice-cold water and set it on the kitchen counter while you look around for a snack. When you pick up your drink, the outside of the glass is covered with water. Where did this water come from? It must have come from somewhere. Let's do another activity to find the answer."

Figure 11–9. Water drops form when warm, moist air is cooled as it touches the sides of the cold glass.

DISCOVER BY Observing

For several days in a row, observe the following weather conditions: temperature, humidity, wind, and cloudiness. Make observations around noon, sunset, and before you go to bed. Record the times and conditions in your journal, using such terms as dry or humid, and cloudy or clear. Each morning, see whether water has formed on the grass, lawn furniture, or cars. In your journal, describe the relationship between the weather conditions you observe and the presence of water. ✎

You may have inferred that the water on the glass came out of the air. This process, called **condensation,** occurs when a gas changes to a liquid. In the case of the cold glass, water vapor in the air changed to liquid water on the outside of the glass. Condensation is the opposite of evaporation. If air at 40°C cools to 10°C, the molecules slow down and move closer together, and the air contracts.

Cooling air is something like squeezing a wet sponge. When you squeeze a sponge, its pores become smaller and water is forced out. When air cools and contracts, there is less space between molecules to hold water vapor. If moist air contracts enough, some of the water vapor will be forced to leave the air as liquid water. The air near the cold glass was cooled by the temperature of the ice-cold water, forcing some of the water vapor in the air to condense and appear as liquid water on the outside of the glass.

The temperature at which condensation occurs is called the *dew point.* This name is derived from a common example of condensation—the formation of *dew.* At night, the temperature usually drops and the ground cools rapidly. Air near the ground also cools, and sometimes it cools enough for water vapor to condense, forming dew. If the dew point is below 0°C, water vapor condenses out of the air as tiny ice crystals, called *frost.*

The dew point changes along with the relative humidity of the air. Look at Table 11-2. It is similar to Table 11-1, except that the numbers in the boxes give dew-point temperatures rather than relative humidity. As Table 11-2 shows, the drier the air, the greater the difference between air temperature and dew point. Humid air has to cool only a few degrees before condensation begins, since there is only a small difference between the air temperature and the dew point.

Figure 11–10. Water vapor often condenses on smooth surfaces, such as windowpanes. Because the dew point is below 0°C, the water vapor forms frost.

Table 11-2

Dew Points

Wet-Bulb Depression (°C)																				
	1	2	3	4	5	6	7	8	9	10	11	12	13	14	15	16	17	18	19	20
−10	−15	−22																		
−8	−12	−18	−30																	
−6	−9	−14	−23																	
−4	−7	−11	−17	−30																
−2	−5	−8	−13	−20																
0	−2.5	−6	−10	−15	−25															
2	−0.5	−3	−7	−11	−18	−30														
4	2	−1	−4	−7.5	−12	−17														
6	4	1.5	−1	−4	−8	−14	−22													
8	6	4	1	−1.7	−4.5	−9	−15	−20												
10	8	6	4	1	−1.5	−5	−9.5	−15	−28											
12	10	9	6	4	1	−2	−5.5	−10	−16	−30										
14	12	11	8	6	4	1	−2	−6	−10	−17.5										
16	14	12.5	10.7	8.5	6	4	1	−2	−6	−10	−18									
18	16	14.5	13	11	9	6.5	4	1	−2	−4.5	−10	−18								
20	18	16.7	15	13	10.5	9.5	7	4.5	2	−1	−5	−10	−18							
22	20	18.7	17	16	13.5	11.5	10	7.5	5	2	−1.5	−5	−10	−18						
24	22	20.7	19	17.5	16	14	12	10	8	5	2.5	−1	−5	−10	−18					
26	24	22.7	21	19.5	18	16.5	15	13	10.5	8	6	3	−1	−5	−10	−18				
28	26	24.7	23	22	20	19	17	15	13	11	9	6	3	−1	−5	−10	−18			
30	28	26.7	25	24	22	21.5	20	18	16	14	12	10	6	3	−1	−5	−10	−18		
32	30	28.7	27	26	24	23	22	20	18	17	15	13	10	6	3	−1	−5	−10	−18	
34	32	30.7	29	28	26	25	24	22	20	19	17	15	13	10	6	3	−1	−5	−10	−18
36	34	32.7	31	30	28	27	26	24	22	21	19	17	15	13	10	6	3	−1	−5	−10
38	36	34.7	33	32.5	30	29	28	26	24	23	21	19	17	15	13	10	6	3	−1	−5
40	38	36.9	35	34	32	31	30	28	26	25	23	21	19	17	15	13	10	6	3	1

Dry-Bulb Temperature (°C)

ASK YOURSELF

Why would dew be less likely to occur in a dry, warm place than in a cool, humid place?

SECTION 1 REVIEW AND APPLICATION

Reading Critically

1. Why do you feel chilly after swimming?

2. What is the relationship between dew point and relative humidity?

Thinking Critically

3. What usually happens to the relative humidity after the sun goes down? Explain.

4. How is it possible for dew to form on an automobile and yet not form on a nearby sidewalk on the same night?

SKILL Reading a Chart

► **MATERIALS**
- windchill chart ● paper ● pencil

There is an important relationship between apparent air temperature and wind speed. Moving air tends to speed up evaporation and remove heat from a location. Thus, a strong wind on a cool day makes the air feel colder than it really is. This combination of temperature and wind is known as *windchill*.

▼ **PROCEDURE**

1. Part of a windchill table is presented below. The windchill is found by locating the column with the current air temperature and the row with the wind speed. The box where the two intersect gives the windchill.

2. If the temperature is 20°F and the wind speed is 10 knots, the temperature feels like it is 5°F. What does the temperature feel like if the actual temperature is –20°F and the wind speed is 20 knots?

► **APPLICATION**
Use the data in the chart to find the windchill for each of the following conditions:

1. 15°F —10 mph
2. 10°F —20 mph
3. 5°F —30 mph
4. 0°F —10 mph
5. –5°F —20 mph
6. 10°F —30 mph
7. –15°F —10 mph
8. –20°F —20 mph
9. –25°F —30 mph
10. –30°F —10 mph

✳ **Using What You Have Learned**
Choose a specific windchill, such as –25°F, and see how many combinations of wind speed and temperature will produce the same apparent temperature.

Wind Speed		Cooling Power of Wind Expressed as "Equivalent Chill Temperature"																
Knots	MPH	Temperature (°F)																
Calm	Calm	40	35	30	25	20	15	10	5	0	–5	–10	–15	–20	–25	–30	–35	–40
		Equivalent Chill Temperature																
3–6	5	35	30	25	20	15	10	5	0	–5	–10	–15	–20	–25	–30	–35	–40	–45
7–10	10	30	20	15	10	5	0	–10	–15	–20	–25	–35	–40	–45	–50	–60	–65	–70
11–15	15	25	15	10	0	–5	–10	–20	–25	–30	–40	–45	–50	–60	–65	–70	–80	–85
16–19	20	20	10	5	0	–10	–15	–25	–30	–35	–45	–50	–60	–65	–75	–80	–85	–90
20–23	25	15	10	0	–5	–15	–20	–30	–35	–45	–50	–60	–65	–75	–80	–90	–95	–100
24–28	30	10	5	0	–10	–20	–25	–30	–40	–50	–55	–65	–70	–80	–85	–95	–100	–110

TABLE 1: WINDCHILL CHART (APPARENT TEMPERATURE)

INVESTIGATION

Determining Dew Point

▶ MATERIALS
● small soup can ● water ● thermometer ● ice cubes ● flashlight ● pencil

▼ PROCEDURE

1. Fill the can about half full of room-temperature water.
2. Read and record the temperature of the air and the temperature of the water in the can.
3. Add several ice cubes to the water, making sure not to splash any water over the rim of the can. Stir constantly with a pencil.

4. While one lab partner watches the temperature of the ice and water mixture, another partner should shine the flashlight on the can to assist in watching for the first sign of condensation to appear on the outside of the can.

5. As soon as condensation starts, read and record the temperature of the water. This temperature is the dew point.
6. If time permits, repeat the experiment outside the school building or at home. Be sure to start with water at room temperature.

▶ ANALYSES AND CONCLUSIONS
1. Why did the water have to start out at room temperature?
2. Why did the ice and water mixture have to be constantly stirred?
3. Why was it important to detect the exact instant that condensation first appeared on the outside of the can?
4. Why was it necessary to read the temperature at the instant that condensation started?

▶ APPLICATION
If the dew point were close to the air temperature, what would this indicate about the relative humidity?

※ Discover More
Devise a way for people to survive in the desert without water, using the processes of evaporation and condensation.

Clouds and Precipitation

Ms. Marina's class is impressed by the interesting properties of water, but they still like her illusions, especially the one where a person seems to float in thin air. Ms. Marina says, "Water can do that too—float in thin air—and without illusion! Every cloud you see is full of water, and it just floats along. Think about it. What causes clouds to float? And why do they sometimes release their water as rain and snow?"

Up, Up, and Away

Hot-air balloons create a colorful sight as they drift gracefully across the sky. But what causes hot-air balloons to rise?

Figure 11–11. Although clouds and balloons may appear to levitate, it is not an illusion.

The answer is that the air in the balloons is heated. Heated air expands and becomes less dense than the cooler air surrounding the balloon. This cooler air pushes up on the balloon, causing it to rise and float.

The same process works in nature. Recall that air is warmed by the surface beneath it. The warmed air rises and is replaced by cooler, denser air. As warm air rises, it begins to cool, and its relative humidity increases.

If the rising air becomes cool enough, condensation will take place. Tiny water droplets form on microscopic solids in the air, such as dust, smoke, and salt crystals. These solids are called *condensation nuclei*. If enough condensation occurs, a cloud forms. A **cloud** is a collection of water droplets or ice crystals in the atmosphere.

Figure 11–12. Fog often forms where cold water and warm air meet.

Most clouds form some distance from the earth's surface. However, under certain conditions clouds form at or near the surface. These clouds are called *fog*. There are several different types of fog, but they all form when warm, moist air is cooled. In the next activity, you can learn more about the conditions necessary for the formation of fog.

ᴅɪsᴄᴏᴠᴇʀ ʙʏ *Researching*

Several areas in the world are noted for their foggy conditions, including San Francisco, California; the Grand Banks in the North Atlantic; and the British Isles. Obtain materials from your school or local library that describe the conditions responsible for the formation of fog in these areas. Prepare a brief report in your journal, and present it to the class. ✐

▶ ASK YOURSELF

Why do clouds form?

That One Looks Like an Elephant

"Not only can water form clouds," continues Ms. Marina, "but it can form a wide variety of cloud types—one for every occasion. There are the white, puffy clouds that you often see on summer afternoons; high, whispy clouds that look like feathers or horses' tails; and low, gray clouds that cover the entire sky. Some clouds mean rain is coming, while others mean fair weather will continue.

Conditions in the atmosphere determine the types of clouds that form. These conditions include air temperature, air pressure, humidity, and wind speed and direction. Let's make some observations of clouds and associated conditions of the atmosphere."

DISCOVER BY *Observing*

Every day for a week, describe in your journal the different kinds of clouds you see. Along with each cloud description, record the temperature, air pressure, humidity, wind speed and direction, and the time. At the end of the week, try relating the different cloud types with the weather conditions. ✐

Clouds can be classified into three main groups, based on their development. Two of these groups are cumulus and stratus. **Cumulus** (KYOO myoo luhs) **clouds** are white, puffy, fair-weather clouds. These clouds form when hot, humid air rises. On summer days, convection currents sometimes rise to the top of the troposphere, where the air is very thin. The result is the formation of towering clouds with spreading tops. Such clouds are called *cumulonimbus clouds* (*nimbus* refers to precipitation), or thunderheads. **Stratus clouds** are extended cloud layers that seem to cover the entire sky. If the layer is thick enough and produces rain or snow, the clouds are called *nimbostratus*.

Sometimes clouds form in air much colder than 0°C. These clouds are made up of tiny ice crystals instead of water drops. They usually form in the upper troposphere and are called **cirrus clouds.** The various clouds of the troposphere are summarized in Table 11-3.

ASK YOURSELF

Why do different types of clouds form?

Table 11-3 **Cloud Types**

Cirrus	Elevation: 7 km to 13 km Thin, wispy clouds made of ice crystals. Associated with fair weather but may precede storms.	Elevation: 0.5 km to 2 km Gray, rolled clouds in a continuous layer. May produce light rain or snow.	**Stratocumulus**
Cirrocumulus	Elevation: 7 km to 13 km Thin clouds, which look like ripples or waves. Associated with fair weather but may precede storms.	Elevation: 0.5 km to 2 km Low, gray clouds that cover the entire sky. Usually produce light snow or drizzle.	**Stratus**
Cirrostratus	Elevation: 7 km to 13 km Thin clouds that cause halos around the sun or moon. Associated with fair weather but may precede storms.	Elevation: 0.5 km to 2 km Thick, dark clouds that block out the sun. Usually produce steady, heavy rain or snow.	**Nimbostratus**
Altocumulus	Elevation: 2 km to 7 km Light gray clouds in rolls or patches. Associated with fair weather but may precede storms.	Elevation: 0.5 km to 10 km White, puffy clouds usually associated with fair weather. May produce rain or snow showers.	**Cumulus**
Altostratus	Elevation: 2 km to 7 km Gray clouds that usually produce light rain or snow. May precede warmer weather.	Elevation: 0.5 km to 20 km Towering clouds, sometimes with flattened tops. Produce thunderstorms, high winds, heavy rain, and hail.	**Cumulonimbus**

What Goes Up Must Come Down

The phrase "what goes up must come down" is usually applied to objects that leave the earth's surface and return in the same physical state—a batted ball or a space shuttle, for example. However, water, which leaves the earth as a gas (water vapor), often returns as a liquid (rain) or a solid (snow).

Rain and snow, along with sleet and hail, are forms of precipitation. **Precipitation** is water that falls to the earth's surface as a liquid or a solid. Precipitation is part of the *water cycle*. The water cycle begins with evaporation—liquid water becomes a gas. The next part of the water cycle is condensation—water vapor changes back to a liquid or a solid. The cycle is completed as precipitation returns water to the earth's surface.

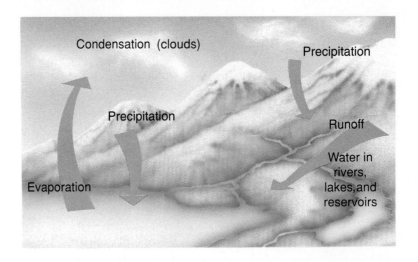

Figure 11–13. Precipitation is part of the water cycle.

The amount of precipitation varies widely around the world. For example, in some areas of the Atacama Desert in Chile, precipitation has seldom been recorded. At the other extreme, on Mount Waialeale in Hawaii, the average yearly rainfall is more than 11 m. That is more than enough water to cover a three-story building!

Figure 11–14. Annual rainfall amounts vary around the world, ranging from very dry, as in deserts (right) to very wet, as in tropical rain forests (left).

It's Pouring   Once a cloud forms, water droplets tend to collide with one another and join together to form larger droplets. In calm air the droplets remain fairly small, and the precipitation is usually referred to as drizzle. In turbulent air, some droplets, carried by air currents, travel up and down through the cloud, growing larger.

When a droplet becomes too heavy for air currents to support, it falls to the earth as a raindrop. The diagram shows the relative sizes of cloud droplets and raindrops.

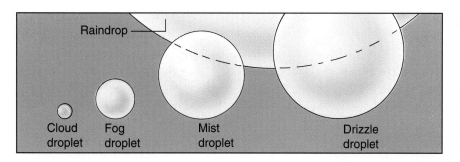

| Raindrop | | | |
| Cloud droplet | Fog droplet | Mist droplet | Drizzle droplet |

Figure 11–15. This diagram shows the relative sizes of droplets in a cloud.

No Two Flakes Are Alike

At temperatures below 0°C, condensation produces tiny ice crystals. In a cloud, these crystals join together to form snowflakes. You may have heard that no two snowflakes are alike. That may never be proven, but snowflakes do come in a wide variety of shapes and sizes.

A snowflake is much less dense than a raindrop. On the average, 1 m of snow, when melted, will produce only 10 to 15 cm of water. A heavy rainfall drops much more water than does a heavy snowfall.

The Mail Must Go Through

Sometimes air temperature near the ground is below freezing, while the temperature in the clouds is above freezing. Under such conditions, precipitation that leaves the clouds as rain freezes as it passes through the cold air near the ground. Frozen rain is called *sleet.*

Sleet is not the same as freezing rain. Freezing rain occurs when the ground, and objects on the ground, are colder than 0°C. On striking the cold surface, the rain freezes, coating everything with a layer of ice.

Figure 11–16. Snowflakes form in a variety of shapes and sizes.

Figure 11–17. Although it can make walking or driving dangerous, freezing rain can produce beautiful landscapes.

The Gang's All Here   Hail and sleet are not the same thing. Hail is usually associated with severe thunderstorms. Temperatures near the tops of cumulonimbus clouds, or thunderheads, are often below freezing. Ice crystals formed in the cloud top become coated with a thin layer of water as they fall through the cloud. Strong updrafts carry the raindrops back up to the freezing areas of the clouds, where the water freezes. Several round trips through the clouds can produce hail the size of marbles. Even baseball-sized or larger hail is observed occasionally.

Figure 11–18. Compare the sizes of the hailstones to those of the golf ball and baseball on the left. Small hail sometimes accumulates like snow (right).

 **ASK YOURSELF**

Why don't all clouds produce precipitation?

SECTION 2   *REVIEW AND APPLICATION*

Reading Critically

1. What kinds of clouds usually produce precipitation?
2. Describe the process by which hail is formed.

Thinking Critically

3. How would the water cycle be affected if water that evaporates from the oceans and returns to the land as precipitation were not returned to the oceans?
4. When water evaporates, heat is removed from its surroundings. Condensation is the opposite of evaporation. How is the temperature affected when water condenses?

---

*H*IGHLIGHTS

The Big Idea

The water cycle is a system of energy exchange. Humidity, clouds, and precipitation are all parts of the water cycle. When water evaporates, heat is removed from the surroundings. When water vapor condenses, heat is added to the surroundings. Each interaction involves an exchange of heat and a change of phase. But each part of the water cycle has its limits. The air can only hold so much water.

Look back at what you wrote in your journal as you began this chapter. Have any of your original ideas about fog and precipitation changed? If so, revise your journal to show what you have learned.

Connecting Ideas

Copy this concept map into your journal and fill in the empty spaces. Extend the map to show how evaporation and condensation are important processes in the water cycle.

WATER CYCLE

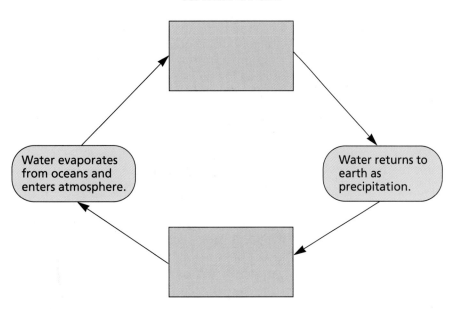

Understanding Vocabulary

1. Explain how the terms in each set are related.
   a) evaporation (345), condensation (351)
   b) humidity (348), relative humidity (348), absolute humidity (348)
   c) precipitation (359), dew (352)
   d) cloud (357), cumulus clouds (358), stratus clouds (358), cirrus clouds (358)

Understanding Concepts

MULTIPLE CHOICE

2. Which of the following does *not* describe evaporation in the atmosphere?
   a) Surrounding air is cooled.
   b) Heat increases the rate of evaporation.
   c) Water molecules escape to the atmosphere.
   d) Energy is changed to heat.

3. Relative humidity is highest when
   a) the temperature is near the dew point.
   b) the temperature is rising.
   c) the wind blows.
   d) the temperature is falling.

4. As warm air rises, it cools, and water vapor in the air will
   a) evaporate.   c) turn to rain.
   b) condense.   d) precipitate.

5. Water that falls to the earth is called
   a) snow.   c) precipitation.
   b) dew.   d) condensation.

6. Which of the following is a type of frozen rain?
   a) fog
   b) sleet
   c) snow
   d) drizzle

SHORT ANSWER

7. Explain the process of cloud formation.

8. How do snow and sleet differ?

Interpreting Graphics

Answer the following questions using the chart below.

9. The dry-bulb temperature is 28°C. The relative humidity is 59 percent. What is the wet-bulb depression?

10. Suppose you are exercising. Will sweating cool you more if the dry-bulb temperature is 24°C and the wet-bulb depression is 4°C, or if the dry-bulb temperature is 32°C and the wet-bulb depression is 11°C?

Wet-Bulb Depression

Dry-Bulb Temperature	1	2	3	4	5	6	7	8	9	10	11	12	13	14	15	16	17	18	19	20
16	90	81	71	63	54	46	38	30	23	15	8									
18	91	82	73	65	57	49	41	34	27	20	14	7								
20	91	83	74	66	59	51	44	37	31	24	18	12	6							
22	92	83	76	68	61	54	47	40	34	28	22	17	11	6						
24	92	84	77	69	62	56	49	43	37	31	26	20	15	10	5					
26	92	85	78	71	64	58	51	46	40	34	29	24	19	14	10	5				
28	93	85	78	72	65	59	53	48	42	37	32	27	22	18	13	9	5			
30	93	86	79	73	67	61	55	50	44	39	35	30	25	21	17	13	9	5		
32	93	86	80	74	68	62	57	51	46	41	37	32	28	24	20	16	12	9	5	

Reviewing Themes

11. Energy
Explain how water changes from its liquid form in a lake or ocean into water vapor in the atmosphere. Then describe the process that changes water vapor into precipitation that falls back to the earth.

12. Environmental Interactions
How can rain be falling when the relative humidity at ground level is less than 100 percent?

Thinking Critically

13. What happens to the energy released during condensation?

14. Consider the following situation: Near the ground, the temperature is –10°C. At 1000 m, the temperature is 10°C, and at 5000 m, where condensation is taking place, the temperature is –25°C. Describe the precipitation that would form and any changes to it that would occur as it fell.

15. Since hail and sleet are both made of ice, why is there no sleet in the summer and no hail in the winter?

16. Fog often develops in the early morning and then "burns off" after the sun comes out. Explain the process that causes the fog to disappear.

17. The cloud shown below has a flattened top, sometimes called an anvil because of its shape. What do you think causes this anvil shape?

18. You know that hail forms as raindrops make round trips between warm and cold layers inside a cloud. Look at the number of hailstones in this picture. What must be happening inside the clouds to allow hail of this number to form?

Discovery Through Reading

McMillan, Bruce. *The Weather Sky*. Farrar Straus Giroux, 1991. The author describes clouds during different seasons, explains how they form, and how clouds can be used to predict the weather.

CHAPTER 12

WEATHER SYSTEMS

Humans have always thought about the weather—through myths, poetry, and, of course, a science of weather. Understanding weather can give us a feeling of control over the uncontrollable forces of wind, rain, and snow. To the poet Emily Dickinson, an approaching thunderstorm was like a drama building toward a climax—quivering leaves, scurrying wagons, the pale claw of lightning. Young Emily must have had quite a thrill when that lightning bolt reached past her house and split a nearby tree!

A Thunder-Storm

The wind began to rock the grass
With threatening tunes and low—
He flung a menace at the earth,
A menace at the sky.

The leaves unhooked themselves from trees
And started all abroad;
The dust did scoop itself like hands
And throw away the road.

The wagons quickened on the streets,
The thunder hurried slow;
The lightning showed a yellow beak,
And then a livid claw.

The birds put up the bars to nests,
The cattle fled to barns;
There came one drop of giant rain,
And then, as if the hands

That held the dam had parted hold,
The waters wrecked the sky,
But overlooked my father's house
Just quartering a tree.

Emily Dickinson

For Your Journal

- How do air masses affect weather?
- What kind of weather produces thunderstorms?
- How do we predict what the weather will be?

Frontal Weather

Everybody talks about the weather, but nobody does anything about it.

Charles Dudley Warner

Part of this quotation was made in fun, because there is really nothing anyone can do about the weather. But we can plan for it. Today more than ever we at least can understand weather systems. This understanding can help us make better choices. For example, if Dorothy from *The Wizard of Oz* had understood the weather in Kansas, would she have run away that afternoon? To some extent, you plan your daily activities around the weather.

Weather Factors

If you have ever watched a weather report on television or seen a newspaper weather map, you know that on any given day, different parts of the country experience different weather conditions. You can see this in the satellite photograph shown here. At the time the photograph was made, large areas of the United States were covered by clouds, while other areas were clear.

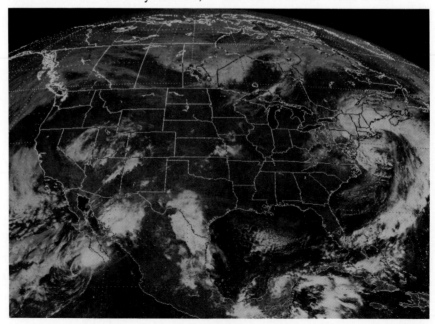

Figure 12–1. Satellites provide a view of weather systems as they move across the earth's surface.

What exactly is weather? **Weather** can be defined as the state of the atmosphere at a given time and place. Changes in weather occur because air in the troposphere is always moving. The driving force behind this motion is heat energy from the sun. Recall that the sun heats the earth's surface unevenly. This uneven heating creates differences in air temperature, air pressure, humidity, and density. These differences in turn cause convection currents to form and winds to blow. All of these factors—temperature, pressure, moisture, and winds—combine to produce weather conditions. As air in the troposphere circulates, air with one set of weather conditions moves out of an area and air with another set of conditions replaces it.

Figure 12–2. These photographs clearly show the movement of weather in the troposhere.

ASK YOURSELF

What factors combine to produce weather?

Air Masses and Fronts

As air at the bottom of the troposphere moves slowly across the earth's surface, that air takes on the characteristics of the surface beneath it. For example, air over a tropical ocean becomes warm and humid, while air over land in a polar region becomes cold and dry. Bodies of air that form in this way are called air masses. An **air mass** is a large body of air that has nearly the same temperature and humidity throughout. Most weather systems involve air masses.

Air masses are classified according to temperature and humidity. *Tropical* air masses originate over low-latitude regions, while *polar* air masses originate over high-latitude regions. Air masses that form over oceans are called *maritime air masses;* and air masses that form over land are called *continental air masses.* The area where an air mass takes on its characteristics is called a *source region.*

The map below shows the source regions for air masses that influence weather in North America. Capital letters *T* and *P* are often used to show tropical and polar air masses, and small letters *m* and *c* are often used to show maritime and continental air masses. For example, an air mass forming over warm ocean waters is designated as *mT* for maritime Tropical. What are the temperature and moisture characteristics of such an air mass?

When a large air mass covers a region, the weather there will remain much the same for some period. Eventually, however, the air mass will move on, and another air mass will move

Figure 12–3. An air mass is named for the source region over which it forms.

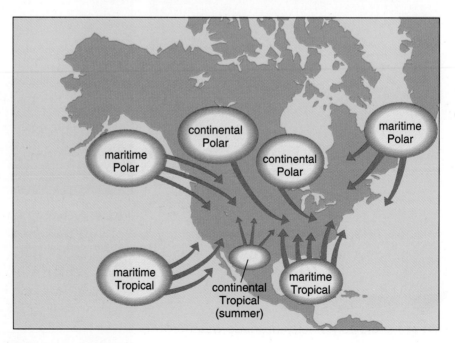

in to take its place. The weather will then be influenced by the characteristics of the new air mass. At some point, the boundary between the two air masses will pass through the region. The boundary, or area, between two different air masses is called a **front.** The weather along a front is usually unsettled or stormy. In the next activity you can learn more about weather conditions along fronts.

ACTIVITY
How can you make weather observations?

MATERIALS
thermometer, aneroid barometer, magnetic compass, anemometer

PROCEDURE

1. Copy Table 1 into your journal. For five consecutive days, make weather observations at the same time each day, and record them in the table.

TABLE 1: WEATHER OBSERVATIONS					
	DAY 1	DAY 2	DAY 3	DAY 4	DAY 5
Temperature					
Pressure					
Sky					
Clouds					
Precipitation					
Wind					

2. Use the thermometer to measure the temperature in a shady location, and record it in the table.

3. Use the barometer to measure the air pressure, and record it in the table.

4. Observe the sky, and record your observations in the table, using the terms *clear, partly cloudy, partly sunny,* and *cloudy.*

5. If clouds are present, determine the types from the photographs on page 359, and record the cloud names in the table.

6. Indicate any precipitation that is falling or has fallen since your previous observation.

7. Using the compass and the anemometer, measure the wind conditions and record them in the table.

APPLICATION

1. Describe any relationship you may have found between wind direction and air temperature.

2. Describe any relationship you may have found between air pressure and cloud cover or precipitation.

At the earth's surface, a front is about a kilometer wide, but it may be hundreds of kilometers long. Temperature difference between two air masses may be more than 10°C. While there may be some mixing of air, one thing always occurs along a front—warm air is forced up over cooler air, which is denser. The rising, warm air produces clouds, which often bring precipitation. There are four types of weather fronts: cold fronts, warm fronts, occluded fronts, and stationary fronts.

Cold Fronts   A *cold front* forms when a cold air mass moves into and replaces a mass of warmer air. Along a cold front, the cooler, denser air wedges itself under the warmer air, forcing it to rise rather rapidly, as shown. As the warm air rises, it cools, clouds form, and precipitation occurs within a few kilometers of the front. The precipitation along a cold front is usually heavy but brief, lasting only a few hours. Thunderstorms are often associated with cold fronts. Fair, cooler weather usually follows the passage of a cold front.

Figure 12–4. Cold and warm fronts are shown below. When a front passes through, precipitation often occurs.

Cold Front

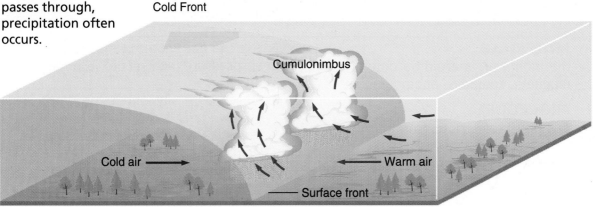

Warm Front

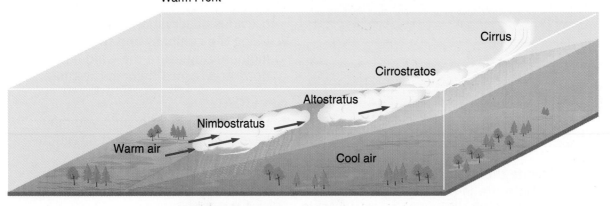

Warm Fronts   When warmer air overtakes a colder air mass, a *warm front* forms. As the front continues to move, the warm air rises over the denser cold air. Water vapor condenses and precipitation often occurs. The slope of a warm front is much gentler than that of a cold front. The pattern of cloud formation is also different along a warm front than it is along a cold front, and precipitation falls over a larger area. The precipitation is usually steady but not as heavy as that associated with a cold front. Warmer, humid weather often follows the passage of a warm front.

Occluded Fronts

Sometimes a cold front catches up with a warm front along the ground. The cold air forces the warm air up, where it becomes trapped above the two cooler air masses, forming an *occluded front*. Steady precipitation often occurs until the two cool air masses have combined.

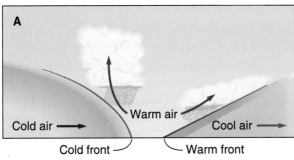

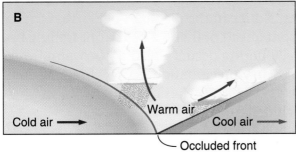

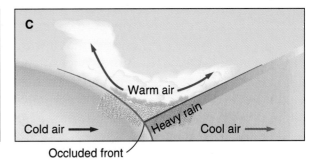

Figure 12–5. An occluded front forms when a warm front is overtaken by a cold front and warm air is trapped above the cooler air masses.

Stationary Fronts

Occasionally a front—either cold or warm—stops moving. When this happens, a *stationary front* forms. Fronts usually remain stationary for only a day or so, but a stationary front can last for several days. Areas located along a stationary front may experience extended periods of precipitation, until the front starts moving again. The next activity will help you to understand what happens when different air masses meet.

DISCOVER BY *Problem Solving*

An air mass that formed over central Canada moves over the United States and into a region occupied by an air mass that formed over the tropical Atlantic Ocean. Name and describe the characteristics of each air mass. Tell what kind of front will form and the weather conditions that will be found along that front. ✐

➤ ASK YOURSELF

Why does unsettled weather usually occur along a front?

Highs and Lows

In describing weather, the terms *high* and *low* refer to atmospheric pressure, or air pressure. Earlier you learned that warm, rising air produces low pressure at the surface. Cooler, denser air often moves in to take its place. Remember that the Coriolis force causes air to be deflected, or to curve. In the Northern Hemisphere, winds circulate in a counterclockwise direction toward the center of low pressure, as shown below. Weather systems that circulate around low-pressure centers are called *cyclones,* or *lows*.

Lows often form along fronts. Pressure or temperature disturbances along a front cause a "wave" to form, which eventually develops into a low-pressure center. As air moves in toward this center, a counterclockwise circulation develops and a low-pressure system forms. Notice in the diagrams below that the system has a warm front ahead of it and a cold front trailing behind. Widespread precipitation occurs ahead of the warm front, and a narrow band of heavy precipitation occurs along the cold front. Eventually, the cold front catches up with the warm front, forming an occluded front.

Figure 12–6. These diagrams show the stages in the formation of a low-pressure system.

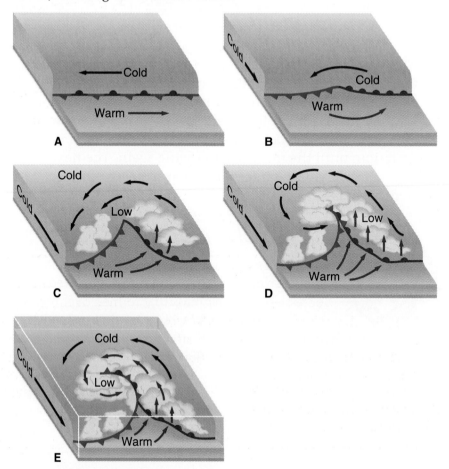

Areas of low pressure are balanced in the troposphere by areas of high pressure. In these areas, cool, dry air sinks and spreads out across the earth's surface. Again, the Coriolis force causes these winds to be deflected, so that winds circulate outward from the center of high pressure in a clockwise direction, as shown. Weather systems that circulate around high-pressure centers are called *anticyclones,* or *highs.* As the air in a high-pressure system sinks, it becomes warmer and the relative humidity decreases. For this reason, the weather associated with highs is usually fair and dry.

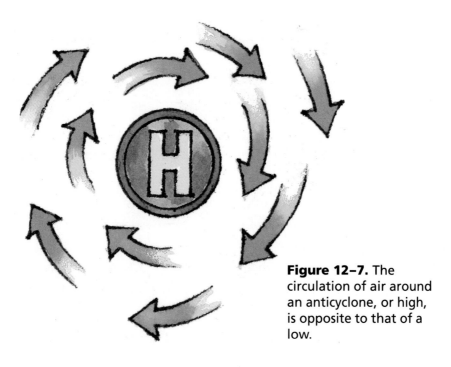

Figure 12–7. The circulation of air around an anticyclone, or high, is opposite to that of a low.

 ASK YOURSELF

Why are cyclones called lows and anticyclones called highs?

SECTION 1   *REVIEW AND APPLICATION*

Reading Critically

1. Describe the characteristics of each of these air masses: (a) continental Tropical; (b) maritime Polar; (c) maritime Tropical; (d) continental Polar.

2. Describe the stages in the formation of a cyclone.

Thinking Critically

3. Why might the slope of a cold front be steeper than that of a warm front?

4. What kind of front forms where a maritime Polar air mass collides with a continental Tropical air mass? Describe the weather you would expect to find along such a front.

INVESTIGATION

Modeling Air Masses and Cloud Formation

▶ **MATERIALS**

- scissors ● tape ● fish tank (small) ● cardboard (larger than fish tank)
- pencil ● thermometers (2) ● ring stands (2) ● test-tube clamps (2)
- paper towels ● 100-mL beaker ● ice ● Petri dish ● safety goggles
- laboratory apron ● candle ● matches ● crucible ● wire gauze
- 1000-mL beaker

▼ PROCEDURE

1. Using the scissors and tape, trim the piece of cardboard to make a lid for the top of the fish tank.

2. With a pencil, make two holes, one on each side of the lid, so that the thermometers can be inserted into the tank as shown. Cut two larger holes, each about 5 to 10 cm in diameter, on either side near the front edge of the cardboard.

3. Cut paper towels into small strips, and pack them loosely into the 100-mL beaker.

4. Place ice in the Petri dish, and put it toward the front on one side of the tank.

5. **CAUTION: Be sure to wear safety goggles and a laboratory apron during the rest of this Investigation.**

6. Shorten the candle (if necessary) to no more than 5 cm long. Melt the bottom of the candle just enough so that it can be stuck to the inside of the crucible.

7. Light the candle, and put it toward the front on the other side of the tank.

8. Position the lid so that one of the larger holes is over the ice and the other is over the candle.

9. Place the wire gauze on the hole over the ice.

10. Secure the thermometers to the ring stands and insert one on each side of the tank.

11. Ignite the paper towels in the small beaker and allow them to burn for several seconds. Place the smaller beaker on top of the wire gauze.

12. Invert the larger beaker, and use it to cover the smaller beaker.

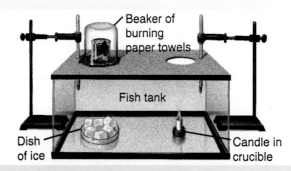

Beaker of burning paper towels

Fish tank

Dish of ice

Candle in crucible

▶ ANALYSES AND CONCLUSIONS

1. What happened to the smoke that accumulated in the large beaker?
2. Describe how the smoke moved inside the tank.
3. What temperatures were recorded by the thermometers?
4. Explain why the clouds of smoke moved as they did.

▷ APPLICATION

Explain how this Investigation relates to the formation of clouds that form along a front.

✳ *Discover More*

Do research to find out how air pollution affects the temperature of air masses.

Severe Weather

Objectives

Explain how lightning is produced.

Describe the formation of thunderstorms, tornadoes, and hurricanes.

Compare and contrast the physical characteristics of thunderstorms, tornadoes, and hurricanes.

The voice that beautifies the land!

The voice above,
The voice of the thunder
Within the dark cloud
Again and again it sounds
The voice that beautifies the land

From "Twelfth Song of the Thunder"
    A Navajo song

Does severe weather always mean bad weather? Not necessarily, as this Navajo song explains. Severe weather is clearly serious weather and can be dangerous weather, but it isn't always bad.

Thunderstorms

Under certain conditions, rainstorms are accompanied by lightning and thunder, strong winds, and maybe even hail. Such storms, of course, are thunderstorms. A **thunderstorm** is a severe weather system with cumulonimbus clouds, strong winds, heavy rain, lightning, and thunder. Like all storms, thunderstorms involve great amounts of energy. This energy is produced by condensation. As air rises and cools, water vapor condenses, producing clouds and releasing heat energy. The heat causes the air to expand, increasing the speed of the upward-moving air currents. The stages in the development of a thunderstorm are shown below.

Figure 12–8. The illustration (right) shows the development of a thunderstorm. These cumulonimbus clouds (left) show a characteristic flattened top.

Lightning is one of the most frightening and dangerous characteristics of a thunderstorm. **Lightning** is a large electrical discharge, or spark, that travels between two oppositely charged surfaces. Have you ever received a mild shock when you reached for a doorknob on a cool, dry day? If so, you have been "struck" by a very weak form of lightning. Friction between your shoes and the floor causes an electrical charge to build up on your body. When your hand comes close to the doorknob, the charge releases, or discharges, in the form of a spark.

In a thundercloud, friction occurs between raindrops and ice crystals as vertical currents carry them up and down through the cloud. The upper portion of the cloud becomes positively charged, and the lower portion becomes negatively charged. As the charge at the bottom of the cloud increases, negative charges in the ground beneath the cloud are repelled, or pushed away, and the ground takes on a positive charge. When the difference between oppositely charged regions becomes great enough, lightning travels between those regions. Lightning may travel between the cloud and the ground, between different clouds, or between different areas of the same cloud.

Lightning will always follow the shortest path available between charged surfaces. Therefore, lightning often strikes the highest object in an area, because that object is closest to the bottom of the cloud. For this reason, you should never stand under a tree during a thunderstorm. If you are caught outdoors during a thunderstorm, seek shelter in the lowest area possible, but do not go in or near water. If you are in a building, stay away from windows and exterior walls and do not touch or use any electrical appliances.

ASK YOURSELF

Where do thunderstorms get their energy?

Figure 12–9. Lightning can be spectacular and dangerous. The Empire State Building in New York City is protected by a huge lightning rod that safely channels the lightning into the ground.

Tornadoes

In *The Wizard of Oz,* Dorothy, her dog Toto, and her entire house are carried away from their home in Kansas by a violent storm. Although called a cyclone in the movie, the storm was really a tornado. A **tornado** is an intense storm with very high winds that circle a small center of extremely low pressure.

Tornadoes form within severe thunderstorms along cold fronts that separate two very different air masses. One air mass must be warm and humid, and the other must be cool and dry. The advancing cool air wedges under the warm, humid air, forcing it to rise rapidly. As moisture condenses from the rising air, a line of thunderstorms called a *squall line* forms along the front. The energy released during condensation produces very strong convection currents along the squall line, and many small, very intense low-pressure cells develop. If the contrast between the two colliding air masses is great enough, some of the cells may intensify, forming tornadoes.

A tornado starts as a small low-pressure center, or *vortex,* within a thunderhead. The tornado develops down from the bottom of the thunderhead as air below the cloud is drawn into the vortex. As the rising air cools, clouds form along the wall of the vortex, making the funnel shape of the storm visible. As is true of all cyclonic systems in the Northern Hemisphere, the air moving into the vortex spins in a counterclockwise direction.

Figure 12–10. This series of photographs shows the development of a tornado.

The average tornado has a diameter of between 150 m and 600 m. Tornadoes generally move parallel to the squall line at speeds of from 40 km/h to 60 km/h. The length of a tornado's path along the ground varies from a few meters to more than 20 km. However, tornadoes do not always remain in contact with the ground. A tornado may touch down briefly and destroy a single house in a neighborhood, leaving nearby houses undisturbed. A tornado lasts, on the average, only five to ten minutes.

A combination of high winds and low pressure makes tornadoes extremely destructive. Horizontal winds within a tornado may exceed 500 km/h, while vertical currents may move at speeds of more than 300 km/h. These wind speeds are estimates based on damage caused by tornadoes. Precise measurements of wind speeds within a tornado are not possible, because measuring instruments are usually destroyed by the high winds.

Few structures can withstand the fury of a tornado. When a tornado occurs, you should huddle under a stairway or under a strong table or workbench in a cellar or basement. If no cellar is available, stay in a corner of an interior room or in an interior doorway, away from any windows.

Figure 12–11. Tornadoes that form over water are called waterspouts.

Tornadoes can form anywhere, even over water, as shown. Such tornadoes, called *waterspouts,* are less powerful than those that form over land, because the temperature difference between air masses over water is usually not very large. Tornadoes occur most frequently in a region of the United States known as Tornado Alley, which receives as many as 300 tornadoes a year. Tornado Alley extends from the Gulf Coast to the Great Plains states of Iowa and Minnesota.

 **ASK YOURSELF**

Under what conditions do tornadoes develop?

Hurricanes

Imagine two fast-moving vehicles—a race car traveling at a speed of 250 km/h and an 18-wheel tractor-trailer traveling at 125 km/h. Which vehicle do you think would cause more destruction if it went out of control? The same sort of comparison can be made between different kinds of storms. Tornadoes are considered to be the most violent storms, because they have the highest winds and the lowest central pressure. However, when it comes to sheer destructive power, tornadoes rank a distant second behind hurricanes.

A **hurricane** is a large, powerful cyclonic storm that develops over warm ocean waters. Like all cyclones in the Northern Hemisphere, hurricane winds circulate in a counterclockwise direction around a central region of low pressure. In the western Pacific Ocean these storms are called *typhoons,* and in the Indian Ocean they are called *cyclones.*

Hurricanes usually form in the regions of the trade winds, between 20° N and 20° S latitudes. You may find it useful to look back at Figure 10–19 on page 336 to refresh your memory about planetary winds.

The "hurricane season" for the United States is usually late summer and early fall, when water temperatures in the Northern Hemisphere are warmest. During this period, the southeast trade winds are located north of the equator, where they collide with the northeast trade winds. The result of these colliding wind belts is the formation of low-pressure areas known as *tropical depressions.* All hurricanes begin as tropical depressions. An average of only six Atlantic depressions develop into full-blown hurricanes each year.

Figure 12–12. This photograph clearly shows the near-circular shape of two hurricanes, with their well-defined eyes.

Hurricanes, like thunderstorms and tornadoes, get their energy from the condensation of water vapor. As a tropical depression develops, central pressure decreases and more warm, moist air rises. The air cools, and condensation releases heat, which further reduces the air pressure. As this cycle repeats itself over and over, the storm grows in size and intensity.

A fully developed hurricane is almost circular and may be 600 km or more in diameter. Air pressure in the center of the vortex is about 10 percent lower than that outside the vortex. The diagram shows a cross section of a hurricane. At the center of the storm is the *eye*—a core of warm, relatively calm air about 10 km to 20 km across. The highest winds circle the eye in a band 20 km to 50 km wide. This band of high winds, known as the *eye wall*, extends upward from sea level to the top of the troposphere.

Beyond the eye wall, spiral bands of clouds circle the center of the storm. The strength of the winds decreases as you move farther from the eye. The violent circulation of air within a hurricane can also cause tornadoes to form, making a hurricane even more dangerous.

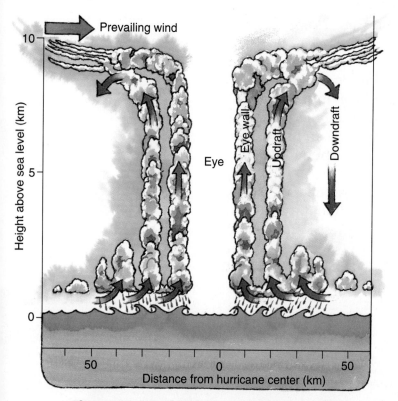

Figure 12–13. This diagram shows a cross section of a fully developed hurricane.

The destructive potential of a hurricane is tremendous. The speed of the sustained, or steady, winds of most hurricanes ranges from 120 km/h to 150 km/h, but they may reach as high as 300 km/h. However, high winds are not the only source of damage and destruction. Much hurricane damage is caused by water. Over the ocean, the winds cause a dome of water called a *storm surge* to build up. A storm surge may be 1 m to 5 m high and up to 80 km wide. Strong winds produce waves several meters high on top of this dome of water. Flooding causes tremendous damage when the waves and storm surge move onto shore. Flooding may be increased by the heavy rains that usually accompany a hurricane.

One of the most destructive hurricanes in U.S. history struck the east coast of south Florida on August 24, 1992. This storm, with sustained winds of more than 220 km/h, wiped out entire communities and caused more than $20 billion in damage to buildings and crops. Hurricane Andrew crossed Florida, entered the Gulf of Mexico, and moved on to cause extensive damage in Louisiana two days later. Try the following activity to find out how America prepares for hurricanes.

Researching

The National Hurricane Center is located in Coral Gables, Florida. Find references to this center in your school or local library, and read about its history and the important work done at the center. Present a brief report to your class.

Hurricanes, moved by the prevailing winds, travel at speeds of 5 km/h to 25 km/h. Fortunately, the movement of a hurricane can be tracked and its time and place of arrival predicted quite accurately. Thus, people are warned to evacuate low-lying coastal areas well in advance of a hurricane's arrival, keeping loss of human life at a minimum.

Figure 12–14. These photographs show damage caused in Florida and Louisiana by Hurricane Andrew.

ASK YOURSELF

Why do most hurricanes develop during late summer and early fall?

SECTION 2   *REVIEW AND APPLICATION*

Reading Critically
1. What is the source of energy for all storms?
2. What causes lightning?

Thinking Critically
3. Why do hurricanes that form over the Pacific Ocean seldom affect the west coast of the United States?
4. How are hurricanes different from other cyclones?

Predicting Weather

Explain *how changes in air pressure and wind direction can help in the prediction of upcoming weather.*

Recognize *various symbols used on weather maps.*

Utilize *a weather map to make weather predictions.*

People are interested in the future. They would like to know what's going to happen tomorrow, or next week, or next year. Most people are interested in knowing about future weather conditions as well. Meteorologists can make short-term weather predictions based on certain facts, but, in the long-run, people just have to wait and see what happens.

Local Weather Conditions

The steps in predicting the weather are the same as those used in carrying out any investigation. First, a meteorologist collects information about atmospheric conditions from a number of different sources. Next, the information is carefully studied and analyzed. Finally, the meteorologist reaches a conclusion based on the results of this study and makes a weather forecast. A **forecast** is a prediction of future weather conditions.

Figure 12–15. Knowing what the weather will be helps you decide what to wear.

At this point you know enough about atmospheric conditions to do a little forecasting of your own. You know about air masses, fronts, and the cloud types associated with those fronts. You know about pressure systems and wind directions around those systems. You know that high pressure is associated with fair weather and low pressure with unsettled or stormy weather. And, finally, you know that weather systems generally move from west to east across the continental United States because of the prevailing westerlies.

Local weather conditions change as weather systems move in and out of your area. Some of these changes begin to show up quite early, and there are certain clues, or signs, you can watch for that will help you to predict how the weather is going to change.

Air Pressure   A change in air pressure is a helpful clue in predicting weather changes. Recall that air pressure is measured with a barometer. Because air rises along a front, air pressure almost always decreases as a front approaches. So a falling barometer indicates that air pressure is decreasing, a sign that unsettled or stormy weather is coming. A rising barometer means that air pressure is increasing, a sign that the weather will probably improve.

Rows and floes of angel hair
And ice cream castles in the air
And feather canyons ev'rywhere
I've looked at clouds that way

But now they only block the sun
They rain and snow on ev'ryone
So many things I would have done
But clouds got in my way

From "Both Sides Now"
by Joni Mitchell

Clouds   On a hot summer day, a buildup of cumulonimbus clouds is a sure sign that thunderstorms will strike somewhere in the local area. A buildup of such clouds across the entire horizon might indicate the approach of a cold front. Behind the front, the weather will probably be fair and dry.

On a clear, pleasant day, the appearance of thickening cirrus clouds high in the sky is often the first sign of an approaching warm front. As the front moves closer, air pressure will decrease and other cloud types and precipitation associated with the front will appear. After the front passes, the weather will probably be warmer and more humid. The next activity will help you to predict what will happen as a front passes.

DISCOVER BY *Problem Solving*

Using your knowledge of atmospheric conditions, fronts, and high- and low-pressure systems, predict what will happen to the air pressure and cloud cover at the four locations (A, B, C, and D) shown as the fronts approach and then move past those locations. ✎

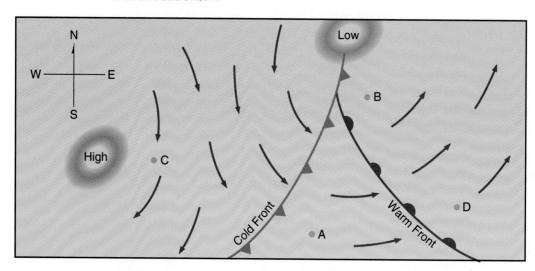

Winds Changes in wind direction and wind speed can also indicate upcoming weather changes. The diagram in the Discover above shows two fronts and two weather systems moving in an easterly direction. At location A, as the cold front ahead of the high-pressure system passes, the winds will shift and become northwesterly. At location B, the winds will come in from the south and then switch to northerly as the front passes. At location C, which has had northwesterly winds, calm conditions will prevail. Meanwhile, at location D, winds will shift from southwesterly to westerly as the warm front passes.

 **ASK YOURSELF**

How can a knowledge of air pressure be used to make weather predictions?

Weather Maps and Forecasts

As stated earlier, meteorologists base their forecasts on information from many sources. These sources include personal observations of local weather conditions, information from weather stations around the world, and satellite photographs. Computers are used to organize and analyze the information. In the United States, the National Weather Service (NWS) is responsible for gathering and analyzing weather data and making weather forecasts. The data collected by the NWS are made available for local meteorologists to use in making local forecasts. The data are also used to produce weather maps. A weather map shows the weather conditions over a large area of the earth. Some weather maps show the weather over the entire earth.

Weather maps that show conditions at the earth's surface are called *surface maps*. The NWS produces surface maps based on data collected from more than 800 stations in the United States. On these maps, each station is identified by a station model. As shown, symbols are used to represent many different weather conditions at each station, such as cloud cover, air pressure, winds, and type of precipitation. NWS weather maps also include lines called *isobars,* which connect points of equal air pressure. Isobars are similar to contour lines on a topographical map. In the following activity, you can learn more about constructing weather maps.

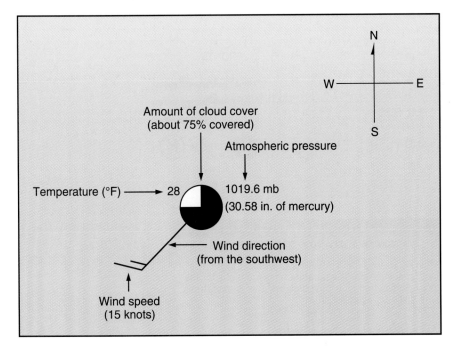

Figure 12–16. Symbols on a station model of a surface map show weather conditions at that station.

The only weather maps most people see are those printed in newspapers or shown on television. These maps are usually simplified versions of NWS surface maps. As shown below, some newspaper maps use color and such everyday terms as *showers* and *flurries* to make their maps more attractive and easier to interpret.

Figure 12–17. This map shows the symbols used on simplified weather maps found in many newspapers.

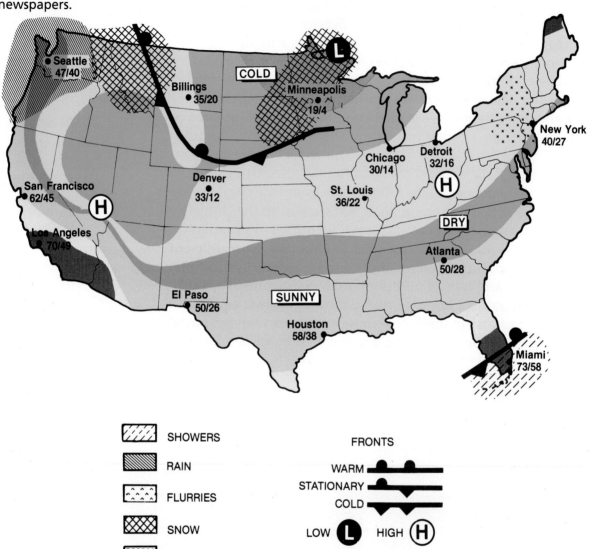

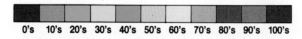

SHOWERS

RAIN

FLURRIES

SNOW

ICE

FRONTS

WARM

STATIONARY

COLD

LOW **L** HIGH **H**

0's 10's 20's 30's 40's 50's 60's 70's 80's 90's 100's

Shown are noontime positions of weather systems and precipitation. Temperature lines are highs for the day. Forecasted individual high and low temperatures are given for selected cities.

Surface maps can be useful tools for making fairly accurate weather forecasts. If the systems shown on this map continue to move at the same speeds and directions, the low will be over Detroit one day later. Where will the low be located two days later? The next activity will help you to use a weather map to make predictions of future weather.

DISCOVER BY *Doing*

Use tracing paper to trace two copies of the weather map on page 388. Include all state boundaries and selected cities. Do not trace any weather symbols. Use these blank maps to show where you think the weather systems and fronts will be on the two days following the one shown. Use colored pencils to indicate the temperature zones on your map. Indicate predicted high and low temperatures for each city. ✏

Naturally, forecasting weather is not simply a matter of moving weather systems a few hundred kilometers to the east each day. If it were, then weather forecasts would always be accurate. New air masses are always forming somewhere, and storms are developing. Unexpected changes are also taking place that make weather forecasting both challenging and a little uncertain. Even though we can plan for many weather conditions, Dorothy probably couldn't have avoided her tornado after all. And if she had, think of what she would have missed!

▼ ASK YOURSELF

Where do meteorologists obtain the information they use to make their forecasts?

SECTION 3 *REVIEW AND APPLICATION*

Reading Critically

1. What weather change might a falling barometer indicate?
2. What kinds of information do weather maps provide?
3. How are the following shown on weather maps? (a) highs and lows (b) rain, snow, sleet (c) temperatures, pressures, wind speeds

Thinking Critically

4. Imagine that your weather has been clear and dry for a couple of days. On the third day, you notice that the barometer is falling slowly and cirrus clouds have moved in from the west. What weather changes might you predict for the next day?
5. If you were located south and west of the center of a low-pressure system, what would you expect the wind direction to be?

SKILL *Analyzing Weather Reports*

▶ **MATERIALS**
- paper • pencil

▼ **PROCEDURE**

1. Meteorologists use symbols to describe weather patterns and conditions on the earth's surface. The symbols are used to represent high- and low-pressure systems and different kinds of fronts. Symbols are also used to indicate different atmospheric conditions at selected locations on the map.

2. Study the weather report and map below.

3. Try to determine the weather conditions at each city shown on the map.

Weather Report

A cold front moving to the southeast extends from the Pacific Ocean through California, Nevada, and Idaho to a low-pressure center over Montana. A high-pressure center is located to the north and west of the front. Extending from the low over Montana is a warm front moving to the northeast. This front extends through Wyoming and Colorado into New Mexico. A large high-pressure center is located over Kentucky, Indiana, and Illinois. It is snowing lightly over the New England states. Here are the sky conditions and temperatures for the following cities:

Atlanta—clear, 54/35
Boise—snow, 28/10
Boston—light snow, 32/22
Chicago—clear, 48/32
Denver—rain, 44/36
El Paso—cloudy, 52/44
Knoxville—partly cloudy, 45/30
Los Angeles—cloudy, 56/48
New York—flurries, 38/28
Pittsburgh—clear, 47/39
St. Louis—clear, 60/48
Tampa—cloudy, 78/68

▶ **APPLICATION**

On a map of the United States like the one shown above, use symbols to indicate the weather conditions described in the accompanying weather report.

✳ *Using What You Have Learned*

Using your knowledge of the weather associated with highs, lows, and fronts, sketch in the areas where you think precipitation is taking place on this map. Assuming that the fronts are moving 1000 km/day, what weather conditions do you predict for tomorrow at Denver and Chicago? When will the cold front reach Pittsburgh?

*H*IGHLIGHTS

The Big Idea

Weather is the state of the atmosphere at a given place and time. This is a simple enough statement on the surface, but how that state of the atmosphere is arrived at depends on the interaction of various elements. The way in which air masses, air pressure, winds, and many other conditions of the environment interact is what creates the weather. It can be described as a complex dance with many possible variations. How we use the information and signs of the weather doesn't affect or control it; it only makes it easier to understand.

For Your Journal

Look back at the ideas about weather and weather predictions you wrote in your journal at the beginning of the chapter. How have your ideas changed? Revise your journal to include any new ideas you have about these topics.

Connecting Ideas

The diagram below shows two different situations involving colliding air masses. Copy the diagram into your journal. Sketch in the fronts, the cloud types, and the weather conditions associated with each situation.

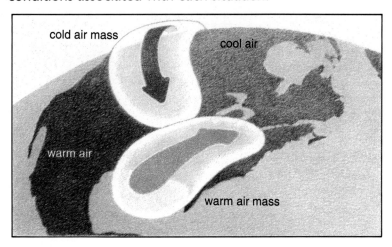

cold air mass

cool air

warm air

warm air mass

Understanding Vocabulary

1. For each set of terms, explain the similarities and differences in their meanings.
   a) air mass (370), front (371)
   b) thunderstorm (377), lightning (378)
   c) tornado (379), hurricane (381)
   d) weather (369), forecast (384)

Understanding Concepts

MULTIPLE CHOICE

2. A hurricane gets its initial energy from the heat released when water in the atmosphere
   a) evaporates.   c) precipitates.
   b) condenses.   d) diverges.

3. A continental Polar air mass contains
   a) moist, cold air.
   b) dry, warm air.
   c) moist, warm air.
   d) dry, cold air.

4. As a hurricane approaches, people are usually asked to evacuate all
   a) low-lying coastal areas.
   b) inland areas.
   c) urban areas.
   d) rural areas.

5. When the barometer reading falls, it usually indicates that
   a) stormy weather is approaching.
   b) the weather will improve.
   c) the weather will not change.
   d) air pressure is increasing.

6. When a low-pressure system develops,
   a) air moves away from the center of the system.
   b) air circulates in a counter-clockwise direction.
   c) widespread precipitation occurs ahead of the warm front.
   d) fair weather occurs along the cold front.

SHORT ANSWER

7. Describe the formation of a cold front.

8. Why are high places so dangerous during thunderstorms?

Interpreting Graphics

Use this weather map to answer the questions that follow.

9. Where is the stationary front?

10. Predict the weather conditions for New York City during the next 24 hours.

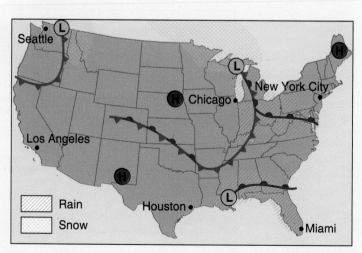

Reviewing Themes

11. *Environmental Interactions*
What causes an occluded front? What eventually happens to the occluded front?

12. *Energy*
Where does the energy for lightning come from?

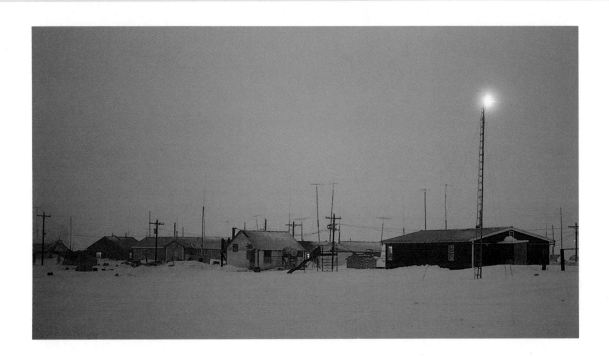

Thinking Critically

13. Cyclones typically form along a cold front more often in winter than in summer. Explain why.

14. Explain why thunderstorms occur more often in summer than in winter.

15. You are thinking about building a house on the coast of Florida. What weather-related risks should you consider?

16. Suppose a weather forecaster sees satellite photos of a storm developing near Puerto Rico. How would the forecaster know that the storm might develop into a hurricane and not a wave cyclone?

17. This photo shows a frozen landscape near Barrow, Alaska. Some scientists have predicted that the weather throughout North American might be like this if an asteroid collided with the earth. If you were to survive the asteroid itself, what problems might you encounter due to the severe weather?

Discovery Through Reading

Gannon, Robert. "The Lightning Stalker." *Popular Science* 242 (February 1993): 53–57, 96. This article provides interesting information about the work of Warren Faidley, a famous storm photographer.

CLIMATE

You might think the weather in your area changes from day to day, but the basic weather patterns of a region stay the same. Midwestern farmers learned a hard lesson in the 1930s, when they suffered a decade-long drought. Dry winds picked up the plowed earth and swept it around a vast "Dust Bowl" that stretched from Canada to Texas.

Here the diary of a Nebraska farm girl describes the conditions in 1934.

Last weekend was the worst dust storm we ever had. We've been having quite a bit of blowing dirt every year since the drouth [sic] started, not only here, but all over the Great Plains. Many days this spring the air is just full of dirt coming, literally, for hundreds of miles. It sifts into everything. After we wash the dishes and put them away, so much dust sifts into the cupboards we must wash them again before the next meal. Clothes in the closets are covered with dust.

Last weekend no one was taking an automobile out for fear of ruining the motor. I rode Roany [the author's horse] to Frank's place to return a gear. To find my way I had to ride right beside the fence, scarcely able to see from one fence post to the next.

Newspapers say the deaths of many babies and old people are attributed to breathing in so much dirt.

from **Dust Bowl Diary**
by Ann Marie Low

Different Places, Different Weather

Imagine that it is the middle of January and your class is trying to decide where to go for a week-long class trip. After some discussion, you split up into two groups. One group packs shorts, T-shirts, sandals, and bathing suits. The other group packs sweaters, down jackets, warm pants, and boots. Obviously, the two groups will not be spending their time at the same place. What inferences can you make about the weather conditions at the two vacation spots?

Figure 13–1. How does the weather of these places differ?

What Is Climate?

If someone were to ask you to describe the weather where you live, how would you answer? You probably would not describe the weather occurring at that moment. Instead, you would understand that the person is really asking about the climate where you live. **Climate** is an average of the weather conditions of a region over a long period. You could call the climate of a region the recent history of its weather patterns.

There are two major weather conditions that determine the climate of a region—temperature and precipitation. Recall that the sun heats the earth's surface unevenly. Because of the changing angle at which the sun's rays strike different parts of the surface, places near the equator are much warmer than places near the poles. Meteorologists have used the uneven heating of the earth's surface to divide it into climatic zones based on average temperature. Although latitude marks the boundary of each zone, the zones blend into each other.

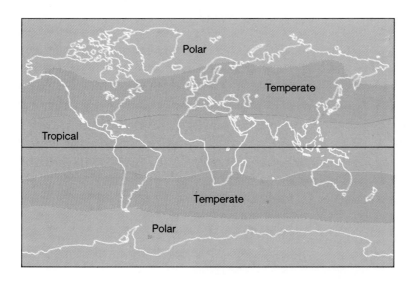

Figure 13–2. Major climatic zones of the earth

The region roughly between the latitudes of 30° N and 30° S has a tropical climate. A *tropical climate* is one in which average temperature of the coldest month is 18°C or higher. Regions with latitudes greater than 60° N and 60° S have polar climates. A *polar climate* is one in which the average temperature of the warmest month is lower than 10°C. As you might expect, regions in the climatic zones between 30° N and 60° S have temperatures that range between those found in polar and tropical zones. A *temperate climate* is one in which the average temperature of the coldest month is lower than 18°C and the average temperature of the warmest month is higher than 10°C.

The major climatic zones just described are based on temperature. However, climate is determined by temperature *and* precipitation. The most important factor affecting precipitation is distance from an ocean. Regions near an ocean generally have a *marine climate*. Inland regions generally have a *continental climate*. Places with a marine climate receive more yearly precipitation than places with a continental climate. A more complete map of world climates is shown on page 403.

Think about the two groups of vacationers described earlier. Where do you think each group is going? Obviously, one group is going to some place where the climate is warm; the other group is going someplace cold. It's possible, of course, that the first group is going to a resort near the equator, while the other group is traveling to the North Pole. However, it is more likely that both groups are going to places much closer to home.

As the map on the previous page shows, most of the contiguous United States has a temperate climate. Yet several variations in climate can be found within the boundaries of the United States. For example, in January, when the family members are going on vacation, parts of the country have warm, sunny weather, while other parts have cold, snowy conditions. By July, the conditions at both places will probably be more similar. Even though temperature and precipitation are the major determinants of climate, temperature and precipitation are influenced by many other factors as well.

Figure 13–3. These two photographs were taken on the same day in different parts of the United States.

 **ASK YOURSELF**

What are the major factors that determine the climate of a region?

Factors Affecting Temperature

The factors that influence temperature include how far north or south a location is, how high or low, how near or far from large bodies of water, and proximity to warm or cold currents. In other words, latitude, altitude, water, and ocean currents play a part in determining climate.

Latitude   The most important factor affecting temperature is latitude. As discussed earlier, the boundaries of the major climatic zones are based roughly on latitude. The warmest temperatures are found at places near the equator. The farther a place is from the equator, the lower the average yearly temperatures will be. In general, the coldest places in the world are the polar regions, which are at the highest latitudes.

Altitude   Altitude describes how high above sea level a place is. As altitude increases, air becomes thinner; fewer air molecules are present, and those molecules are spread farther apart. Thin air does not conduct heat very well, so the higher you go, the lower the temperature will be. This relationship between altitude and temperature explains why the peaks of high mountains located near the equator can be snow-covered.

Figure 13–4. Although parts of the Andes Mountains are on the equator, the high altitude keeps many peaks snow-capped year round.

Large Bodies of Water   Water and land heat and cool at different rates. Water heats up and cools down more slowly than does land. So the presence of an ocean or large lake has a moderating effect on the air temperature of nearby land areas. Places that are near water generally have cooler temperatures in summer and warmer temperatures in winter than do places that are not near water.

Ocean Currents   Planetary winds cause narrow "rivers" of water called *currents* to move across the surface of all the earth's oceans. Like rivers on land, each surface current follows a course. Ocean currents that move from equatorial regions toward the poles are warm-water currents. Those that move from the poles toward the equator are cold-water currents.

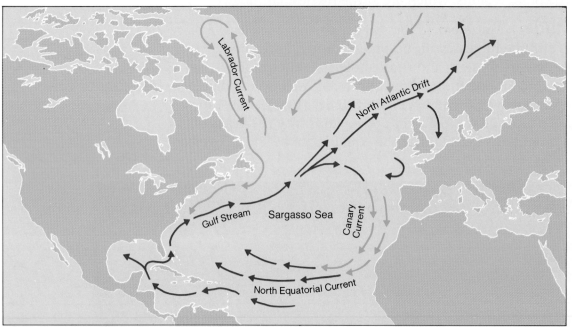

Figure 13–5. Major surface currents in the North Atlantic—warm currents are in red and cool currents are in blue

The temperature of ocean water affects the temperature of the air above it. Coastal areas located near warm-water currents are warmed by the air moving over these currents. For example, the Gulf Stream is a warm-water current that brings warm air to parts of the East Coast, such as Cape Hatteras. South of New England the Gulf Stream crosses the Atlantic Ocean, where it brings warm air to the British Isles and Norway as the North Atlantic Drift. Conversely, the cold California Current brings cool air to much of the West Coast of the United States.

 **ASK YOURSELF**

Why is temperature affected by altitude and latitude?

Factors Affecting Precipitation

The precipitation patterns, and therefore the climate, of a region can be affected by two factors—prevailing winds and mountain ranges.

Prevailing Winds   When the wind consistently blows more from one direction than from any other, it is called a *prevailing wind*. Frequently, the prevailing wind of a region is the most important factor in determining climate.

Recall that warm air can hold more moisture than cool air. Therefore, warm prevailing winds generally bring more moisture than cold prevailing winds. More importantly, however, is the surface over which the prevailing winds of a region blow. If the prevailing winds pass over water as they blow toward land, they will bring moisture and precipitation with them. For example, prevailing winds bring moisture from the Gulf of Mexico to the midsection of the United States.

If the prevailing winds blow across land, the air they bring into a region will be dry. Regions in the path of these dry winds will not receive much precipitation. The Sahara Desert in northern Africa is an excellent example of a region where the prevailing winds blow across a large landmass. As the map shows, the Atlantic Ocean lies just to the west of this desert. Yet, because the prevailing winds blow from east to west across the continent, the Sahara is one of the driest regions in the world.

Figure 13–6. Prevailing winds that originate over land are responsible for the dry conditions of the Sahara Desert.

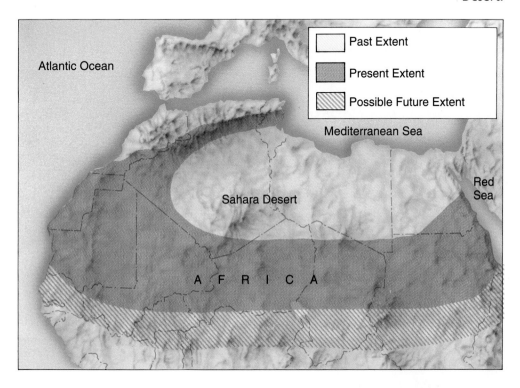

401

Mountain Ranges Mountain ranges can change the wind patterns of a region, thereby changing the patterns of precipitation. A mountain acts as an obstacle to prevailing winds, causing the moving air to rise up over the mountain. As the rising air cools, water vapor condenses out of the air, forming clouds and precipitation. As the diagram shows, the clouds and precipitation occur on the windward side of a mountain. The *windward side* is the side of a mountain that faces the prevailing wind. The windward side of a mountain has a wet climate.

Conditions on the other side of the mountain are quite different from those on the windward side. The side of a mountain that faces away from the prevailing wind is called the *leeward side*. By the time the rising air reaches the top of the mountain, it has lost most of the moisture it contained. As this air moves over the mountain and down the leeward side, it becomes warmer and drier, so the region on the leeward side of a mountain has a dry climate. Such regions are said to be in the **rain shadow** of the mountain. Deserts are often found in rain shadows.

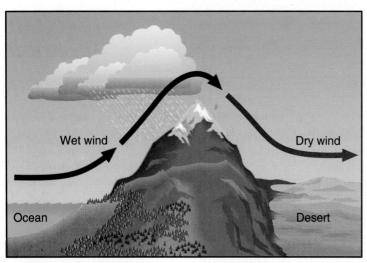

Wet wind

Dry wind

Ocean

Desert

Figure 13–7. Mountain ranges interrupt the movement of air, causing it to rise. This produces very different climates on opposite sides of the mountain.

 ASK YOURSELF

What is the factor most responsible for the climate in each of the following areas: Antarctica, the Amazon, the Sahara Desert?

Types of Climates

Remember, the climate of a given place depends on a combination of many different factors. Vacationers looking for that perfect spot need to take into consideration more than just temperature and precipitation. But additional information is available. Meteorologists have studied the climates of the world and classified them into a number of climatic types.

Every place in the world has its own climate. There are a number of different schemes for classifying climates. Table 13-1 shows the different climates of the continental United States. The map at right shows world climates.

Table 13-1 **Climates of the Continental United States**

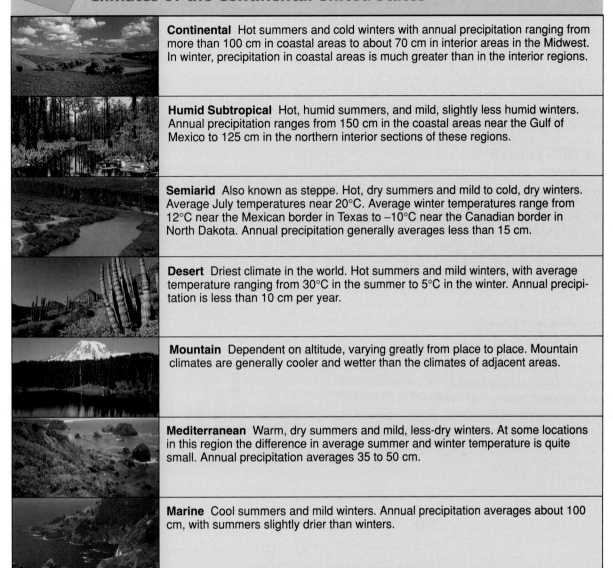

Continental Hot summers and cold winters with annual precipitation ranging from more than 100 cm in coastal areas to about 70 cm in interior areas in the Midwest. In winter, precipitation in coastal areas is much greater than in the interior regions.

Humid Subtropical Hot, humid summers, and mild, slightly less humid winters. Annual precipitation ranges from 150 cm in the coastal areas near the Gulf of Mexico to 125 cm in the northern interior sections of these regions.

Semiarid Also known as steppe. Hot, dry summers and mild to cold, dry winters. Average July temperatures near 20°C. Average winter temperatures range from 12°C near the Mexican border in Texas to –10°C near the Canadian border in North Dakota. Annual precipitation generally averages less than 15 cm.

Desert Driest climate in the world. Hot summers and mild winters, with average temperature ranging from 30°C in the summer to 5°C in the winter. Annual precipitation is less than 10 cm per year.

Mountain Dependent on altitude, varying greatly from place to place. Mountain climates are generally cooler and wetter than the climates of adjacent areas.

Mediterranean Warm, dry summers and mild, less-dry winters. At some locations in this region the difference in average summer and winter temperature is quite small. Annual precipitation averages 35 to 50 cm.

Marine Cool summers and mild winters. Annual precipitation averages about 100 cm, with summers slightly drier than winters.

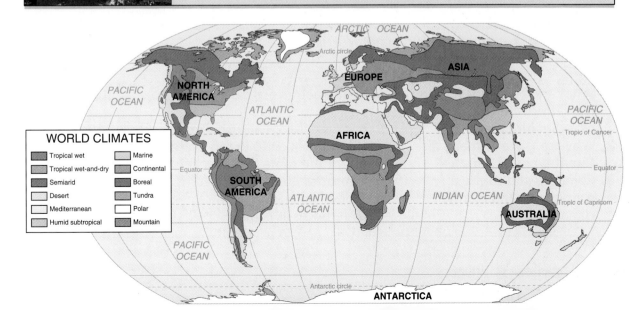

WORLD CLIMATES

- Tropical wet
- Tropical wet-and-dry
- Semiarid
- Desert
- Mediterranean
- Humid subtropical
- Marine
- Continental
- Boreal
- Tundra
- Polar
- Mountain

When considering the climate of a given location, you should keep in mind that the descriptions are general. Local conditions may differ somewhat. The following activity will help you study local climatic types.

ACTIVITY

How can you identify climates?

MATERIALS
paper, pencil

PROCEDURE
1. The combined graphs show average monthly temperature and precipitation for two different cities at the same latitude but in different parts of the United States. Study the graphs and answer the questions.

APPLICATION
1. What is the total annual precipitation of the city represented by red? by blue?
2. Which city has warmer summers? Which city has dryer summers?
3. Which city has the smallest temperature range between summer and winter?

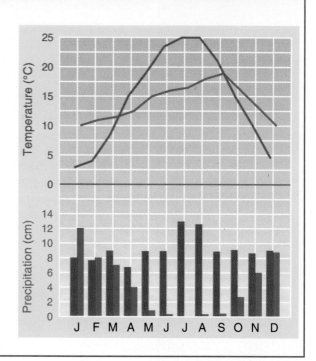

 ASK YOURSELF

What is the major difference between a continental climate and a humid subtropical climate?

SECTION 1 REVIEW AND APPLICATION

Reading Critically
1. What are the major factors that affect the temperature of a climatic zone?
2. Why are there so many different types of climates?
3. Compare and contrast continental and marine climates.

Thinking Critically
4. What factors are most responsible for the desert regions in the United States?
5. San Francisco, California, and Norfolk, Virginia, are two coastal cities in the United States. Both are at about the same latitude, yet the climates of the two cities are quite different. Identify the climatic type of each and the factors that account for the differences in climates.

SKILL Constructing Climatic Models

▶ **MATERIALS**
  ● paper ● colored pencils

▼ **PROCEDURE**

1. Models allow you to represent something that is too large to be observed all at once. For example, a map is a model of part or all of the earth's surface.

2. The first step in producing a model is to state the problem or identify the relationships that you wish to model. Then identify those details that may be relevant to the model. Let your imagination play with the details, experimenting with possible relationships, and then construct the model.

3. The figure below represents a large landmass with the boundaries of its climates indicated. The figure also shows lines of latitude and some surface features, such as mountains. Your task is to identify the different climates based on your knowledge of the factors that affect climate.

4. Copy the figure on a sheet of plain white paper.

5. Use different colors as shown in the key to show the climatic type associated with each region.

▶ **APPLICATION**
1. What do the black arrows on the figure represent?
2. What do the red and blue arrows on the figure represent?
3. Explain what reasoning you used in deciding what climatic type to associate with each region.

✳ *Using What You Have Learned*
Using the model you have constructed, make some predictions about local weather patterns in several climates on your model.

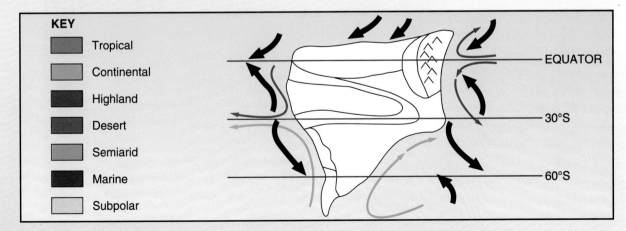

KEY
- Tropical
- Continental
- Highland
- Desert
- Semiarid
- Marine
- Subpolar

EQUATOR
30°S
60°S

Changes in Climate

Imagine that the meeting of your class to decide on a vacation had taken place some 18 000 years ago. How might the travel plans of the two groups have been different? The group heading for a warmer climate would have had to travel a great deal farther, but the cold-weather group might have been able to stay at home. The region of the world that is now the United States was in the grip of a long cold spell known as the Ice Age.

Natural Climatic Changes

You learned earlier that climate describes the average weather of a region over a long period. The question is what constitutes "a long period"? In general, the world's climatic regions seem to have remained relatively unchanged throughout recorded history. Recorded history is the time during which humans have made and recorded observations about the earth. That time span certainly could be considered to be "a long time." However, when compared to the age of the earth, the period that makes up recorded history is very short indeed.

Figure 13–8. As recently as 10 000 years ago, a huge ice sheet covered areas of North America.

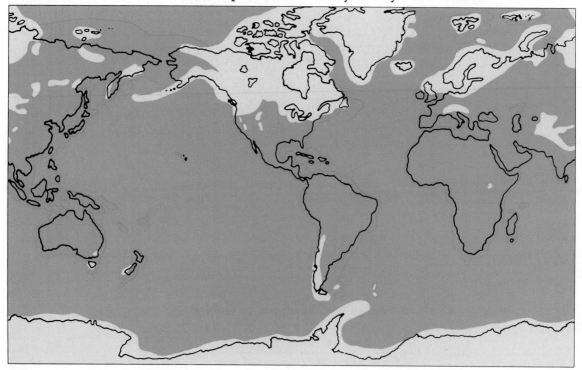

Considerable evidence has been collected indicating that major climatic changes have taken place throughout the earth's history. This evidence includes glacial features and formations, fossil remains, and changes in ocean currents. Many theories have been suggested to explain the causes of these changes.

Glacial Features and Formations   Today scientists study regions known to have been covered by ice at some time in the past to learn how glaciers change the land beneath them. Based on these studies, scientists have found evidence indicating that huge ice sheets have grown and moved over large regions of the earth several times. The Ice Age referred to earlier is the most recent period of major glaciation.

Imagine a huge sheet of ice, hundreds of meters thick and extending as far as the eye can see in all directions. Now try to think about what that sheet of ice would do to the land beneath it as the ice slid and scraped along the surface. Think also of the rivers of water that would flow over the land as the glacier melted. What other kinds of evidence of its existence would a glacier leave behind?

Figure 13–9. This landscape is the result of recent glaciation.

Scientists are not sure what caused the climatic changes that led to the formation and subsequent melting of these ice sheets. Many scientists hypothesize that variations in the relative positions of the sun and the earth may account for these changes. For example, a small variation in the size or shape of the earth's orbit or in the tilt of its axis could produce a change in the way the sun heats the earth.

Fossil Remains   Living things have inhabited this planet for billions of years. Evidence of this fact is found in the fossil record contained in the rocks of the earth's crust. Remember that fossils are the remains or traces of once-living things that have been preserved in rock. The next activity can help you relate fossil finds to climatic changes.

DISCOVER BY *Problem Solving*

Suppose you were to find the fossil of a fish in a rock formation near the top of a mountain. What might you infer from this discovery? What conclusions about climatic changes would you reach? In your journal, record your conclusions and the line of reasoning you followed to reach them. ✎

Figure 13–10. These fossil corals were found in the Arctic.

The fossil record can provide clues to what the earth's climate was like in the past. For example, fossils of tropical plants and animals have been found in arctic regions of the world. Assuming that these plants and animals lived and thrived in a tropical climate, one must conclude that these arctic regions must once have had a tropical climate. Other evidence of widespread climatic change includes indications of glacial activity in regions that have tropical climates today. Also, many scientists hypothesize that the mass extinction of the dinosaurs 65 million years ago was caused by a dramatic change in climate.

Throughout the earth's history, the major climatic zones have remained pretty much the same as they are today. Equatorial regions have been warmest; polar regions have been coldest. What, then, could cause such dramatic climatic changes as those described here? If a region near the South Pole were to be moved to the equator, the climate of that region would certainly be changed. According to the theory of plate tectonics, the landmasses of the world are constantly moving. Think about how the climate of a region would be affected by a change in position on the earth's surface.

Changes in Ocean Currents   You have learned that ocean currents and prevailing winds play important roles in determining climate. Warm ocean currents transfer heat to the atmosphere. In turn, this heat transfer helps to establish the circulation pattern of the jet streams. Any major change in this system would result in climatic changes.

For example, every few years a change takes place in the circulation of surface currents in the Pacific Ocean. This change involves the sudden appearance of a warm-water current, called *El Niño,* that flows from west to east across the Pacific near the equator. Normally, a cold-water current flows from east to west in this area of the ocean. The appearance of El Niño alters the circulation of the jet stream, which causes serious climatic changes in many parts of the world. In the early 1980s, the strongest El Niño in recent history caused severe droughts in some regions of the world and heavy rains and floods in others.

Figure 13–11. The climatic change that produced the drought conditions shown here was caused by El Niño.

 **ASK YOURSELF**

What are two examples of evidence indicating that major climatic changes have taken place in the past?

Human Activities Affect Climate

The climatic changes discussed so far have all been the result of natural causes. However, certain human activities also affect climate. In some cases, people actually create or control the climate in a very small area. For example, when you turn on an air conditioner or a furnace, you are controlling the climate of an enclosed area. You are creating a **microclimate,** the climate of a small area that differs from the surrounding climate.

Figure 13–12. The controlled microclimate of this shopping mall allows people to shop in comfort all year long.

One way that people create microclimates is to build cities. The climate of a city is different from the climate of the area that surrounds it. Tall buildings interfere with and change wind patterns. Concrete and asphalt absorb radiation and store heat better than trees, grass, and soil. Heat released by air conditioners, heating systems, and machinery tends to be trapped in the air over a city. Precipitation that falls on a city runs off paved surfaces and is transported away from the area by sewers. So, the air over a city is drier than air outside the city. It also contains more smoke and dust, which reduces sunlight.

Human activities can even affect climate on a large scale. For one thing, people add solid particles to the air, mostly in the form of smoke and dust. Any activity that adds particles to the air can reduce the amount of radiant energy that reaches the earth's surface. This, in turn, can produce a cooling effect. In the early 1990s, some people were concerned about the large number of oil well fires in the Middle East. They were worried that the smoke might lower air temperatures over a large area of the world. Fortunately, the fires were brought under control before any major climatic problems developed.

Figure 13–13. A forest fire like the one shown here adds tremendous amounts of solid particles and carbon dioxide to the atmosphere.

Other human activities increase the level of carbon dioxide in the atmosphere. These activities include burning fossil fuels and cutting down forests over large areas. The burning of fossil fuels releases carbon dioxide into the air. Trees remove carbon dioxide from the air as part of their photosynthetic process. So fewer trees means higher carbon dioxide levels remain in the atmosphere.

A dramatic increase in carbon dioxide levels in the atmosphere could lead to worldwide climatic change. Carbon dioxide absorbs heat and keeps it from escaping into space. The heat is held near the earth's surface, thereby raising air temperatures and producing an increased greenhouse effect. This *global warming,* a temperature increase of a few degrees Celsius, could cause all of the ice in glaciers and ice sheets to melt. Should this ever occur, sea levels would rise and coastal regions around the world would be flooded.

 ASK YOURSELF

How can human activities affect climates?

SECTION 2 *REVIEW AND APPLICATION*

Reading Critically

1. How might the large-scale removal of trees affect climate?

2. What evidence do scientists have that part of North America was once covered by an ice sheet?

Thinking Critically

3. Do you think there will be a major change in the climate of the region where you live in the next 100 years? In the next 10 million years? Explain your reasoning.

4. Imagine a large number of volcanic eruptions occurring in the eastern rim of the Pacific Ocean over a short time. How might these eruptions affect the climate where you live?

INVESTIGATION

*A*nalyzing Climatic Data

▶ MATERIALS
- graph paper (4 sheets) ● colored pencils

▼ PROCEDURE

TABLE 1: CLIMATIC DATA												
	Jan.	Feb.	Mar.	April	May	June	July	Aug.	Sept.	Oct.	Nov.	Dec.
A Temp (°C)	−2.1	0.9	4.7	9.9	14.7	19.4	24.7	23.6	18.3	11.5	3.4	−0.2
Precip (cm)	3.5	3.0	4.0	4.5	3.5	2.5	1.5	2.0	1.5	3.0	3.0	3.0
B Temp (°C)	−4.4	−2.2	4.4	10.6	16.7	21.7	23.9	22.7	18.3	11.7	3.8	−2.2
Precip (cm)	4.5	5.0	7.0	8.5	10.0	9.5	9.5	8.0	9.5	6.0	6.0	5.0
C Temp (°C)	25.6	25.6	24.4	25.0	24.4	23.3	23.3	24.4	24.4	25.0	25.6	25.6
Precip (cm)	26.0	25.0	31.0	16.5	25.5	19.0	17.0	11.5	22.0	18.5	21.5	29.0
D Temp. (°C)	24.0	23.0	21.0	17.0	14.0	11.0	10.0	12.0	14.0	16.0	20.0	22.0
Precip (cm)	10.5	8.0	12.0	9.0	8.0	7.0	6.0	7.5	8.0	10.0	9.0	8.5

1. On separate sheets of graph paper, prepare four graphs similar to the one shown. Label the left vertical axes "Temperature, (°C)," the right vertical axes "Precipitation, (cm)," and the bottom axes with the months.

2. For each location (A, B, C, D) in Table 1, prepare a temperature/precipitation graph like the sample.

▶ ANALYSES AND CONCLUSIONS
Use your completed graphs to answer the following questions:
1. Which location has the warmest average temperature all year long?
2. Which location, if any, is in the Southern Hemisphere? How do you know?
3. Which location has the greatest range between the coldest and warmest months?
4. Identify the climate at each location.

▶ APPLICATION
The following cities are represented by the graphs: Iquitos, Peru; Salt Lake City, Utah; Peoria, Illinois; and Buenos Aires, Argentina. Match each city with the appropriate graph.

※ *Discover More*
Using library resources, research the temperature and precipitation data over the last year for the area where you live.

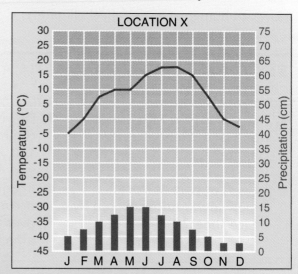

*H*IGHLIGHTS

The Big Idea

The climate of any one area depends on the complex interaction of many factors. Temperature and precipitation are subtly influenced by a variety of factors that, interacting together, create a unique climate. Now add to nature's interacting systems the factors of human intervention, and the delicate balances can change. Climates have changed in the past, as glacial features and fossil remains attest. How they change in the future must now include human influences, such as global warming.

For Your Journal

Look back at the ideas about climate that you wrote in your journal at the beginning of the chapter. Revise your journal to incorporate any changes in your thinking. Be sure to include information about factors that influence climate.

Connecting Ideas

Copy the concept map into your journal and complete it with the names of the factors that affect climate.

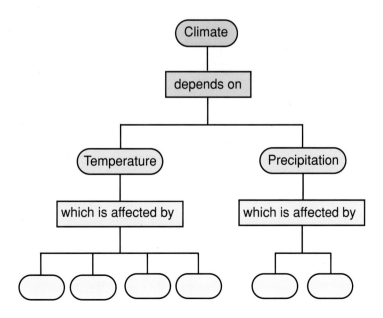

REVIEW

Understanding Vocabulary

1. Explain how the terms in each set are related.
   a) climate (396), microclimate (410)
   b) windward side (402), leeward side (402), rain shadow (402)

Understanding Concepts

MULTIPLE CHOICE

2. What are the two major factors that determine the climate of an area?
   a) winds and humidity
   b) temperature and precipitation
   c) terrain and soil
   d) seasons and tides

3. The climate of Portland, Oregon is
   a) humid subtropical.
   b) semiarid.
   c) continental.
   d) marine.

4. What kind of climate do the rainforests of Hawaii have?
   a) marine tropical
   b) continental temperate
   c) continental tropical
   d) marine temperate

5. What is the most important factor affecting the temperature of a location?
   a) longitude
   b) latitude
   c) large bodies of water nearby
   d) prevailing winds

6. Precipitation patterns of a region are least affected by
   a) warm prevailing winds.
   b) mountain ranges.
   c) latitude.
   d) prevailing winds from water.

SHORT ANSWER

7. What makes a tropical climate different from other climates?

8. How would you explain that some of the mountain peaks located near the equator are snow covered?

Interpreting Graphics

Use this map to answer the questions that follow.

9. The lines on this map connect locations that have the same annual temperatures. Using what you have learned in this chapter, explain why most of the lines bend northward as they approach the eastern coast of North America.

10. In what part of the world is temperature influenced almost entirely by latitude? Explain why you think this is so.

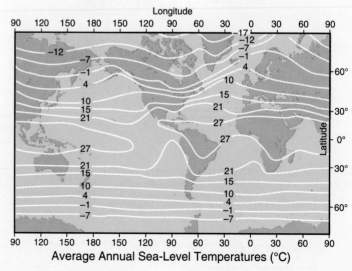

Average Annual Sea-Level Temperatures (°C)

Reviewing Themes

11. *Environmental Interactions*
Explain what El Niño is and how it affects climate.

12. *Changes Over Time*
If the overall climate in your area were to suddenly change, what immediate effects might it have on your life?

Thinking Critically

13. What evidence have scientists found to suggest that huge ice sheets have covered large regions of the earth several times in the past?

14. What effect are humans having on worldwide climate?

15. What effects might the climate of a country have on its long-range development?

16. Why is the downtown area of a large city, such as Chicago, considered a microclimate? How does a city's climate differ from that of the surrounding area?

Discovery Through Reading

Appenzeller, Tim. "What Drives Climate?" *Discover* 13 (November 1992): 64. This isn't the first time carbon dioxide has been blamed for global warming. This article tells why the same gas is blamed for global climatic shifts in the past, even before automobiles were invented!

Oliwenstein, Lori. "Climate Watch: Lava and Ice." *Discover* 13 (October 1992): 18. Seventy-four thousand years ago a volcano on the island of Sumatra might have triggered an ice age. This article discusses how the eruption accelerated glaciation by darkening the sun and creating a volcanic winter.

Hurricane Tracking

Hurricanes are nature's most devastating storms, as residents of South Carolina and Florida can testify. In 1989, Hurricane Hugo slammed into the South Carolina coast, crushing houses and boats. In 1992, Hurricane Andrew ripped across south Florida and completely leveled the town of Homestead.

The first effects of a powerful hurricane.

Hurricane Andrew leveled Homestead, Florida.

Hundred-Year Hurricanes

Both Hugo and Andrew were what meteorologists call "hundred-year hurricanes"—storms so powerful that forecasters predict they will happen only once every hundred years. When two" hundred-year hurricanes" struck the U.S. in just three years, meteorologists wondered if weather patterns might be producing a new wave of monster storms.

Some forecasters believe that global warming may be making hurricanes more severe. Global warming is the slow rise in global temperatures due to the burning of gasoline, coal, and other fossil fuels. If global warming has heated the Atlantic waters where hurricanes form, it may be increasing the power of the storms. Hurricanes feed off the water vapor above the ocean's surface, and warmer air holds more water vapor.

417

Contributing Factors

According to other scientists, residents on the eastern seaboard of North America may be suffering because of rains in Africa. A meteorologist has linked stronger hurricanes to the amount of rainfall in western Africa. The heavier rains reduce winds off the African coast, winds that normally retard hurricane formation.

The key to predicting the path of a hurricane lies in compli- cated computer programs that imitate conditions in and around a hurricane. With computers, meteorologists simulate winds, temperatures, and other weather conditions. They hope their computer simulations will tell them where a hurricane is going and how dangerous it will be to residents in its path. However, they're not always right. Hurricane Hugo surprised forecasters in 1989 by keeping its strength while travel- ing hundreds of miles inland. Despite computerized forecasting, meteorologists still have a lot to learn about hurricanes. ◆

Heavy rains off Africa may influence hurri- canes in the United States.

20N·025W 1535

10N·025W 1512

Computer simulations are used to predict hurricane movements.

HURRICANE HUGO
22 SEPTEMBER 1989
1201 AM EST

NOAA

Henry waited patiently
with his camera.

One Day in the Prairie

by Jean Craighead George

At 4:15 P.M. Henry Rush notices Red Dog's whiskers twitching, the signal he gives before he leaps. Henry waits without moving a finger. His camera is pointed and focused. The little rodent is sitting straight up on his haunches, his paws flailing, his teeth rattling. Something terrible is about to happen, his body language and those teeth sounds say.

The wasp senses this too. She races to finish her work before the disaster. She drags the tarantula, who is twelve times bigger than she, across the prairie dog town looking for a soft spot to bury him. Then she will lay an egg in him, and in spring the egg will hatch into a larva who will feed on the living but paralyzed tarantula.

The air pressure drops abruptly, warning the wasp to find shelter. She will not let go of her prize. She feels with her antennae for a place to dig.

Trotting through the meadow above Quanah Creek comes the herd of longhorns. They are loping toward the safety of the lowland, where prairie dog town lies. A red-shoul-dered hawk takes off from a dead tree along the creek. A flock of redheaded ducks who are migrating southward hastily leaves Quanah Lake, calling in alarm. The great blue heron hears them, stops stalking fish, and hunches down in the reeds.

At 4:45 P.M. the light is yellow-black and ominous. Red Dog whistles and . . . click . . . back flips.

"Got it," shouts Henry Rush. "Got . . .

"The buffalo are stampeding!" He dashes behind the tree.

The meadow by Quanah Lake is covered with billowing dust. Black heads and white horns show over the top of the swirling cloud churned up by the stampeding feet. The buf-falo gallop toward Henry like a roaring ava-lanche. A stick cracks. That is all it takes. The stampeding buffalo turn and thunder back toward Quanah Lake.

Prairie dog town is deserted. Everyone is underground. The vast miles of grass on the prairie flatten out under a powerful wind. Birds drop to the ground. Badgers

419

and skunks hide. The raging cloud spins into a funnel.

"Tornado," whispers Henry Rush.

"What a picture." He snaps the shutter.

"My gosh, it's coming this way!" He looks for a low spot in the land.

The buffalo turn at the stream edge. They snort in fear, panic, and trample a fallen cow. They thunder toward prairie dog town once more, bellowing as loudly as the wind that buffets their huge bodies.

As the black funnel descends, they drop to their knees and roll to the ground.

The buffalo were stampeding!

The tornado roared ten feet above his head.

Henry Rush sees them go down and drops to the ground too.

The tip of the funnel roars ten feet above his head like a thousand speeding freight trains. Rocks spin in the air, riding the powerful wind like paper. The sound is of screaming and roaring and all things exploding. The tornado touches down at the far edge of prairie dog town and wipes out everything — grass, trees, and creatures.

Henry holds his head and closes his eyes.

Rain gushes down as if released from a fire hydrant. In seconds prairie dog town is a lake. Only the cones stand above the flood, and the tunnels and bedrooms do not fill with water.

Within two minutes the terror is over. Rocks and trees and leaves fall to earth a mile away. There is silence in prairie dog town.

Presently Red Dog appears on his cone and barks, "All is well."

Henry looks up. The tornado is high in the sky, a small black funnel that looks as innocent as a teacup.

The torrential rain stops; the buffalo get to their feet. They shake the debris from their fur. A lark sings. The tarantula wasp lies miles to the east, dead beneath a fallen rock. Her egg is unlaid.

The bull elk stands calmly under a pecan tree. Not a hair on his body is ruffled. The tornado went to the south of him.

Henry Rush gets up and runs. He has had enough. He is lucky to be alive. ◆

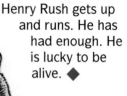

Red Dog signaled all was well.

Vilhelm Bjerknes (1862—1951)

One of the world's first meteorologists was Vilhelm Bjerknes. He predicted the weather based on the principles of physics and developed theories about the large-scale movements of the oceans and atmosphere.

Bjerknes was born in Oslo, Norway. As a youth he worked with his father, who was a mathematics professor. They studied the motions and actions of liquids under many temperature and pressure conditions.

When he was 38, Bjerknes began to work with Heinrich Hertz at the University of Bonn in Germany. He and Hertz studied the behavior of electrical discharges, such as lightning, through gases. Some of their findings were later used in the development of the radio.

Bjerknes became a physics professor at the University of Stockholm in Sweden in 1897. His students greatly respected and admired him. Many of his assistants joined him in 1904 in founding a program for weather prediction in Leipzig and Bergen, Germany. In 1919 his son Jacob was among several researchers who discovered how cyclones begin. ◆

Warren Washington (1936—)

Warren Washington is best known for his computer models that are used to study global warming. He has combined physics, mathematics, and computer science in the study of meteorology.

Warren M. Washington was born in Portland, Oregon. He showed an early interest in science, especially in physics and in the work of Albert Einstein. He received a bachelor's degree in physics from Oregon State University. Washington worked as a mathematician at Stanford University, where he solved mathematical problems related to the atmosphere.

In 1960, he returned to Oregon State for his master's degree. Now interested in the atmosphere, he studied at Pennsylvania State University, and in 1964 earned a doctorate in meteorology. Washington then began work at the National Center for Atmospheric Research (NCAR) in Boulder, Colorado, where he is coupling atmosphere–ocean–sea–ice models in the study of climate.

In 1977 President Jimmy Carter appointed Washington to serve on the National Advisory Committee on Oceans and Atmosphere. Today, Washington is the director of the Climate and Global Dynamics Division at the NCAR. ◆

Carolyn Kloth,
METEOROLOGIST

Growing up along the Gulf coast of Florida, Carolyn Kloth found both flying and the weather fascinating. In fact, while in the fourth grade, Kloth decided to become a "hurricane hunter." "As things turned out," she explains today, "instead of specializing in hurricanes, I ended up specializing in severe thunderstorms." Kloth is now a pilot and a meteorologist. She works at the National Severe Storms Forecast Center in Kansas City, Missouri, where she combines her interests in aviation and weather forecasting.

Every hour at the center, Kloth receives and analyzes weather data from all around the continent. She uses the data to predict where thunderstorms will form across the continental United States and its coastal waters. In addition to predicting where thunderstorms will develop, Kloth also tries to predict how intense each storm will be.

Kloth issues advisories that help pilots avoid thunderstorms by taking alternate routes. Kloth's advisories cover those thunderstorms that have winds over 26 m/s, hail greater than 19 mm in diameter, or that contain tornadoes. Only about 1 percent of all thunderstorms are this severe.

Kloth uses the information from six computers as she decides whether or not to issue advisories. She can view weather information collected by equipment at ground level, high in the atmosphere, or from space. "A lot of people tend to think that you need to be able to look out of a window to assess the weather," Kloth explains. "With the use of radar, weather satellites, and all of the other weather data available, you can do the job almost anywhere," she points out.

The kind of data that Kloth receives and analyzes includes pictures of cloud patterns, locations of cold and warm fronts, and lightning strikes. Measurements of air pressure, relative humidity, precipitation, and the temperature at the ground as well as at various levels of the atmosphere arrive in her office each hour.

Once all the weather data comes in, she uses her computers to analyze it. Kloth can construct a graphic on a video screen that contains surface observations that are taken every hour across the United States.

The speed with which Kloth can get "big-picture" views of the weather over the continent helps her make sound weather decisions. The frequency with which she can get weather data from so many observation points also aids Kloth in her forecasts. Every year, new advances in technology make the weather advisories Kloth provides for pilots more accurate. This means safer flights for everyone. ◆

Using Computers to Model Climate

Is the earth really heating up? What kind of effect will global warming have on our climate? Could industrial pollution slowly change Earth's climate?

Seeking the Answers

These are just a few of the common questions scientists at the National Center for Atmospheric Research (NCAR) in Boulder, Colorado, are trying to answer. Using computers, scientists are constructing models that show the effects of countless factors on climate. They hope in time to verify whether or not changes to Earth's climate are occurring. On a small scale, this kind of effort has had some success.

Meteorologists have used computer models to successfully predict the effects of the

Computers help project climatic changes.

Pacific's *El Niño* on winter weather. Advances such as this have encouraged scientists like Dr. Stephan H. Schneider at the NCAR to construct Global Circulation Models, or GCMs, using supercomputers. Schnei-

Weather patterns on a computer

der has programmed the computer with vast amounts of weather data. Using mathematics and principles of physics, he can simulate the effects of volcanic eruptions, industrial pollution, and even nuclear war on worldwide climatic patterns.

Worldwide Changes

If the GCM works, then scientists will be able to change any factor on the computer model and see how it could affect overall world climate. For example, they could increase the amount of carbon dioxide in the atmosphere, as pollution might, and see how it would affect temperature, rainfall, and winds around the planet.

Programing the billions of contributing factors that affect climate is not easy. Even when all the details of the GCM are finally programmed into the supercomputers, the results may not be conclusive. Some factors contribute more than others. There may even be factors that scientists are not yet aware of.

Despite the complexities involved, scientists hope that by using the GCM they can find out what kind of effects the accumulation of atmospheric carbon dioxide may have on Earth's future climate. ◆

UNIT 5

OCEANOGRAPHY

Every day the sun, the moon, and the earth play a game of tug of war with the oceans. Pulled by the force of gravity, ocean tides flow in and out of bays and inlets the world over.

Scientists have devised ways to harness this tidal energy to produce electricity. They have also begun to search into the ocean depths for other untapped resources.

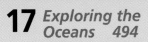

The Water Planet

The ancient Greek poet Homer knew about whirlpools. In the *Odyssey*, his hero Odysseus tells of sailing between Scylla, a rock monster, and Charybdis, a whirlpool that threatened to swallow his crew.

Thus we sailed up the straits, groaning in terror, for on the one side we had Scylla, while on the other the mysterious Charybdis sucked down the salt sea water in her dreadful way. When she vomited it up, she was stirred to her depths and seethed over like a cauldron on a blazing fire; and the spray she flung on high rained down on the tops of the crags at either side. But when she swallowed the salt water down, the whole interior of her vortex was exposed, the rocks re-echoed to her fearful roar, and the dark sands of the sea bottom came into view.

Odysseus' harrowing adventure was based on a real danger. In ancient times, whirlpools did form in the narrow Strait of Messina, between the mainland of Italy and the island of Sicily. The danger eased in 1908, when an earthquake rearranged the sea floor beneath the strait.

Since humans first sailed their planet's waters, they have told stories of the sea's mystery and terror. Some stories expressed sailors' vivid imaginations. Others embellished on real terrors of the deep. They told of sea creatures, violent storms, and whirlpools, which sucked whole ships into whirling funnels of water.

For Your Journal

- How large are the oceans and the seas and how deep are they?
- Why are the oceans salty and where does the salt come from?
- What is it like, under the sea?

Bodies of Water

Locate the oceans and seas of the earth, and **compare** their sizes and depths.

Describe some characteristics of the ocean floor, and **compare** the floors of the Atlantic and Pacific oceans.

Compare and contrast bodies of salt water and bodies of fresh water and the ways people travel on them.

People seem to have a fascination with the sea. Sailors have assigned the seas personalities. They talk about them as if they were living things. Why is this? Maybe it's because water covers 70 percent of the earth. Maybe it's because our own bodies are nearly the same percentage water. Maybe it's because we know so little about the seas. Or maybe it's just because of the feeling of beauty and mystery you have when you watch the sun set over the ocean.

Figure 14–1. Imagine standing at the edge of the ocean. There is water as far as you can see. The sun is setting—it looks like a huge, glowing disk, slowly melting into the horizon.

Oceans and Seas

All of the different bodies of water on the earth are composed of either fresh water or salt water. Bodies of fresh water include most rivers and lakes. Bodies of salt water include most oceans and seas. There are also many places, called *estuaries,* where fresh water mixes with salt water. Where a river flows into the ocean, the water changes gradually from fresh to salt. In the next section, you will find out why the ocean is salty. But first, let's explore some of the large bodies of salt water.

The surface of the earth is actually one huge ocean, with large islands—the continents—scattered here and there. The ocean is so large that you could sail on it for weeks and not see any land. While it is relatively easy to sail on top of the ocean, it is much more difficult to travel down into the ocean. Exploring the deep ocean is particularly difficult. Perhaps this is one reason why the oceans seem so mysterious.

People talk about "oceans" and "seas" as if they were the same things. What is the difference? Geographers usually call the largest areas of salt water **oceans.** The earth's oceans have surface areas of more than 12 million km². Even though the earth's ocean is really just one body of water, most geographers divide it into four oceans. Look at the map. Why do you think this is so?

Figure 14–2. The major bodies of salt water

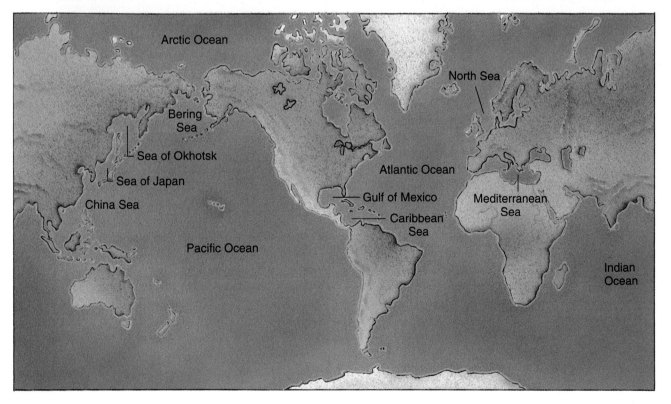

Also shown on the map are seas. Some people use the word *sea* to mean any body of salt water. But to geographers, **seas** are bodies of salt water with surface areas of less than 3.5 million km². Seas are also partially enclosed by land and are, therefore, somewhat separated from the oceans. Table 14-1 shows some comparisons among various bodies of salt water.

Table 14-1 **Major Bodies of Salt Water**

Name	Surface area (in millions of km²)	Volume (in millions of km³)	Deepest part (in meters)
Pacific Ocean	166.0	723.7	11 022 Mariana Trench
Atlantic Ocean	82.0	321.9	9144 South Sandwich Trench
Indian Ocean	73.6	292.1	7450 Java Trench
Arctic Ocean	12.2	13.5	5180 Fram Deep
China Sea	3.4	4.0	5016 China Sea Basin
Mediterranean Sea	2.9	3.8	5092 Helenic Trough
Caribbean Sea	2.6	6.4	7686 Cayman Trench
Bering Sea	2.3	3.7	4420
Gulf of Mexico	1.5	2.3	4029
Sea of Japan	1.0	1.7	4049
North Sea	0.6	0.054	700

The Caribbean Sea and the seas off eastern Asia are separated from the open oceans by chains of islands. Sea water passes freely between these islands. Their saltiness is similar to that of the oceans. Other seas, like the Mediterranean, have only a narrow connection to the ocean. The Mediterranean Sea is saltier than the ocean because of a high evaporation rate. Some seas are less salty than oceans due to runoff from numerous rivers.

Figure 14–3. The oceans and seas have always beckoned to daring explorers.

For thousands of years, the seas and oceans were the only means of travel between countries and continents. The late fifteenth and early sixteenth centuries were the "golden age" of exploration and discovery.

Suppose you were an underwater explorer, like Jules Verne's Captain Nemo in *20,000 Leagues Under the Sea*. You have many places you want to visit, but the only way you can travel is on or under the sea. Many questions come up regarding your travels. You will need to find out the answers before you go. The next activity can help you prepare for your journey.

Figure 14–4. To pilot Captain Nemo's submarine, the *Nautilus*, you must know the oceans and seas.

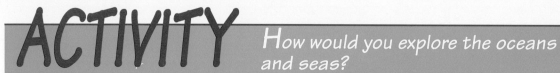

ACTIVITY

How would you explore the oceans and seas?

MATERIALS
paper and pencil

PROCEDURE

1. Copy the statements below into your journal.

2. Look at Figure 14-2, Table 14-1, and any maps or globes that you think you may need.

3. Complete the statements in your journal.

APPLICATION

You are in Spain and want to sail to Florida. The ocean you would cross is the _____ .

You are in Hawaii and want to sail to Japan. The ocean you would be sailing is the _____ .

You want to go to the deepest place in the Pacific Ocean. Your submarine must be able to withstand a depth of _____ .

You are a scientist, and you want to study the Java Trench. The ocean you need to go to is the _____ .

You have visited both the Atlantic and the Pacific oceans. You now want to go to the next deepest body of salt water. The ocean you would go to is the _____ .

You are in Australia and want to sail to Africa. The ocean you would cross is the _____ .

You are in California and want to sail to Australia. The ocean you would cross is the _____ .

You are in South America and want to sail to Africa. The ocean you would cross is the _____ .

You want to go to the ocean with the largest volume of water. The ocean you would go to is the _____ .

You want to go to the deepest place in the Caribbean Sea. The place you would go to is the _____ .

The activity gave you a chance to prepare for travel around some of the world's oceans. Now it's time to explore the ocean floor. In Chapter 17 you will learn about new technologies that scientists use to explore the ocean depths. But for now, let's climb into a submarine like the *Nautilus* and head out along the ocean bottom.

 **ASK YOURSELF**

Why are the continents on Earth like islands in a big lake?

The Ocean Floor

The diagram below shows the ocean floor off the coast of Virginia. Notice that just off the coast is the **continental shelf.** Most of the continents have shelves that surround them. As you head down the continental shelf, you notice something that looks like a river valley. How can there be a river valley on the ocean floor? Think about what could happen if, over time, sea levels changed.

In past ages, sea levels were much lower than they are now. This area used to be dry land. Rivers cut their beds into the land as they flowed to the ocean, farther to the east. As sea level rose, the river beds were flooded. This explains the signs of old rivers on the ocean floor. There are other signs showing that sea levels were once much lower. From your submarine's windows you see what appears to be an old beach.

As you head out to sea, the continental shelf begins to drop away. You have started down the **continental slope.** The depth increases quickly, and soon you have gone as far as your submarine can safely take you. You have to return to shore. If this is as deep as a submarine can go, how do scientists know what is in deeper water?

> Out of the sea floor the continents rise like mesas. Off many coastlines the edge of the continent is a sheer cliff dropping twenty to thirty thousand feet into troughs called trenches (the deepest parts of the ocean are all close to land). Slicing through the edge of the continents are submarine canyons. The Hudson Canyon becomes a colossal gorge wider and deeper than the Grand Canyon; the Congo Canyon is deeper still.
>
> from *Mysteries of the Deep*

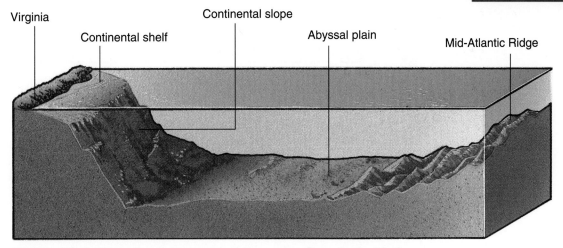

Figure 14–5. How do we know what the ocean floor looks like?

The answer has to do with sound. The technology is called *sound navigation and ranging,* or *sonar.* **Sonar** is the process of bouncing sound waves off a solid surface. It's not really new. Whales and dolphins have been using it for millenia to locate objects in the oceans. And you have probably experienced something similar when hearing your own voice echoing in an empty room.

A sonar instrument, as the diagram shows, sends a series of sound waves through the water. These sound waves strike the ocean floor and bounce back to the instrument on the ship. The instrument then "draws" a profile of the ocean floor based on the timing of the returning echoes. The deeper the water, the longer it takes for sound waves to return to the instrument.

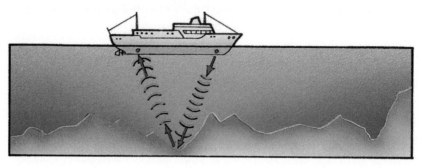

Figure 14–6. Sonar allows scientists to "see" the ocean floor.

To continue your exploration of the ocean floor, you will need to board a ship equipped with sonar. You have seen the continental shelf with your own eyes. Now you can "see" the rest of the ocean floor through sonar. You need to know the speed of sound traveling through water and the time it takes for sound to get from the sonar instrument to the ocean bottom and back to the instrument again. Then you can calculate the depth of the water. You can practice making these calculations in the next activity.

DISCOVER BY Calculating

Sound travels through water at about 1500 m/s, so in 3s sound would travel about 4500 m. If your sonar instrument sends a sound from your ship and the echo returns in 3s, how deep would the water be? Remember, the sound has to make a round trip in that time. Now do some more calculating. Copy the chart into your journal. Then calculate the depths, and fill in the blanks. Where do you suppose you were when you took the last sonar reading?

Time (seconds)	Depth (meters)
1.0	_____
2.3	_____
3.0	_____
6.7	_____
14.7	_____

The continental slope drops from the shelf to the deeper parts of the ocean. Much of the deep ocean floor is relatively flat. Since the deepest parts of the oceans are called the *abyss*, the smooth, deep ocean floors are called **abyssal plains.** As you continue to head out to sea, your sonar shows the abyssal plain extending for hundreds of kilometers from the continental slope. But after a while you start to see more mountainous terrain. The farther you go, the higher the mountains become.

You may have already studied plate tectonics. If so, you probably remember the relationship between the Mid-Atlantic Ridge and the two oceanic plates that are pulling away from each other. As the plates separate, lava escapes from deep within the earth, forming volcanic mountains. In some places, such as Iceland, these mountains reach above the surface of the ocean, forming islands.

As you can see on the map, the Mid-Atlantic Ridge is part of a very long mountain range that snakes through the earth's oceans. This range marks the boundaries where various plates are pulling apart, so there are many volcanic mountains along the ridges. However, because the plates are always moving, the mountains do not often get high enough to form islands. In fact, as they are pulled away from the hot center most of the mountains cool and actually shrink.

Since plates are moving away from each other in some places, they must be moving toward one another in other places. Your research ship now passes through the Panama Canal into the Pacific Ocean. As you can see from your sonar and ocean maps, the Pacific Ocean has a mid-ocean ridge, but it also has other mountains.

Figure 14–7. The ocean floor

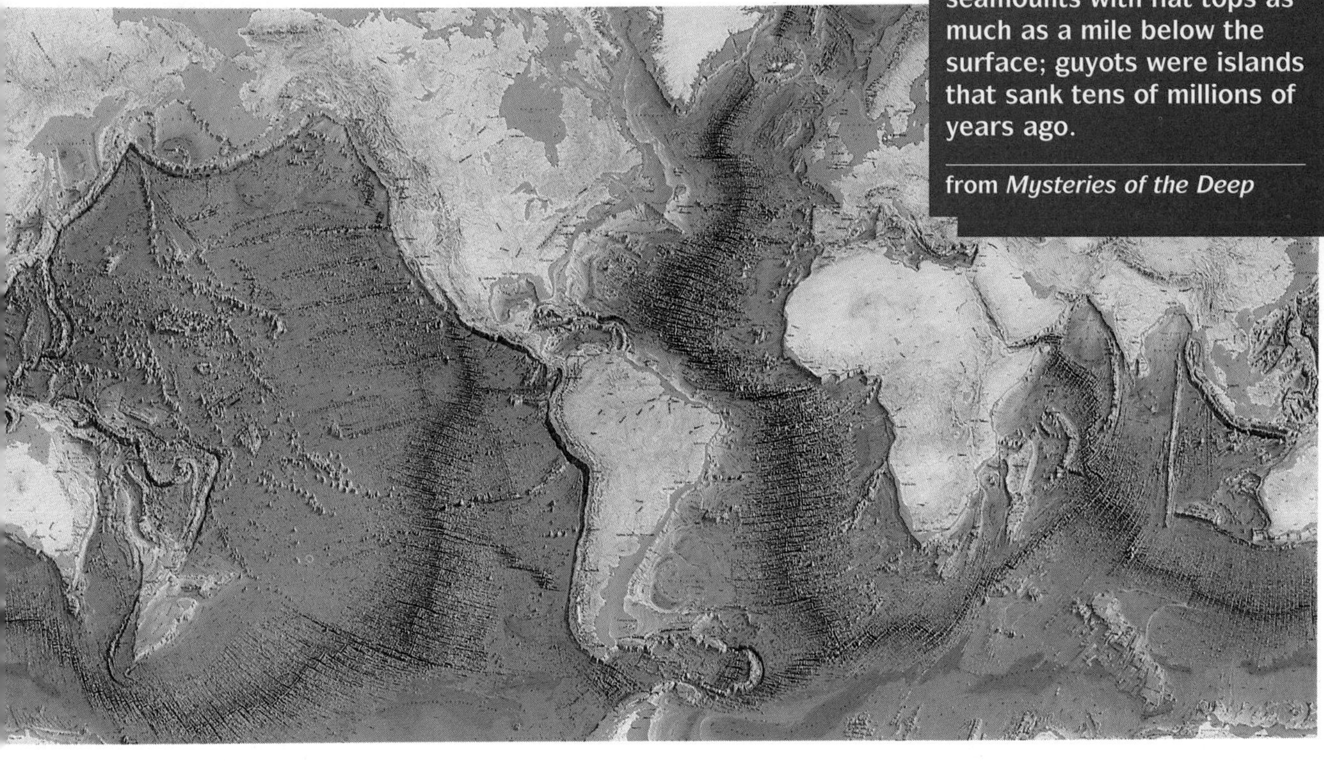

Far out from land, hills begin to sprout through the abyssal plain like islands from the ocean, rising and clustering into the foothills of the Mid-Atlantic Ridge. The ridge is a forest of volcanic mountains, a place of abrupt chasms, earthquake-shattered rocks, and fresh lava flows, the wildest of all the landscapes of the sea floor. From where it rises out of the abyssal plain on either side, it is over a thousand miles wide in the ocean between North America and North Africa. Its highest peaks rise within a few hundred feet of the surface, and sometimes above it. The ridge is constantly shaken by earthquakes and regularly builds up new peaks.

Submerged volcanoes pepper the ocean floor. Here and there are guyots, seamounts with flat tops as much as a mile below the surface; guyots were islands that sank tens of millions of years ago.

from *Mysteries of the Deep*

Remember, the deepest place in the oceans is in the Pacific. It is the Mariana Trench, south of Japan. The only vessel that has ever been down in the Mariana Trench is a submarine called *Trieste*. In a later chapter, you'll learn more about *Trieste* and other underwater research vessels.

Just as there is a link between mountain ranges and plates, there is also a relationship between plates and trenches. When two plates move toward each other, one of them usually slips under the other. Where one plate is pushed down, it forms a depression, or trench, in the ocean floor. But this is not a smooth motion. The plates scrape violently against each other, the friction producing earthquakes. Subduction produces enough heat to melt rocks and create magma. This magma may rise high enough to form islands, so many chains of volcanic islands are found along these plate boundaries.

Trenches also form where plates meet along the edges of some continents. As the sinking plate forms a trench, the rising plate forms a mountain range. Along the Pacific coast of South America, some of the earth's tallest mountains are found next to one of the deepest trenches in the oceans.

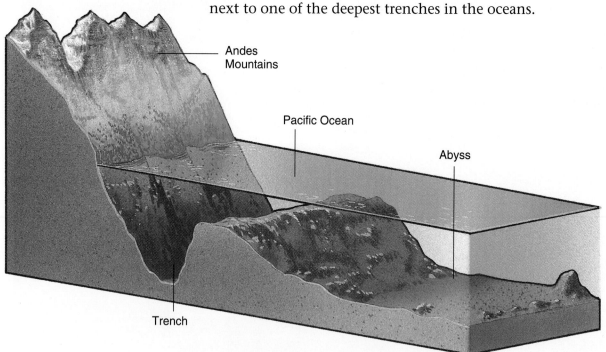

Figure 14–8. The distance from the bottom of this trench to the top of the nearby Andes mountains is the greatest difference in elevation on Earth.

 **ASK YOURSELF**

How do we know what the bottom of the ocean looks like?

Lakes and Rivers

The rain and snow that falls on land evaporates back into the atmosphere, soaks into the ground, or runs off on top of the ground into rivers and streams and then into the ocean. Sometimes this water sits for a while in ponds and lakes. All these streams, rivers, ponds, and lakes are bodies of fresh water. But do you know where most of the earth's fresh water is? Most of the fresh water on the earth is frozen in ice in the Arctic and Antarctica. What do you think would happen if all this ice melted?

Like the oceans and seas, lakes and rivers have been used for transportation for hundreds of years. To venture into unknown territories, early explorers used many kinds of boats, rafts, and canoes. As new areas were explored, settlements grew up on the shores of lakes and rivers. These bodies of water provided the easiest paths for transporting people and their tools, food, and other necessities of life. Look at Table 14–2. These rivers are all thousands of kilometers long. No wonder people have so often taken advantage of these natural highways.

Figure 14–9. Early explorers used rivers and lakes as natural highways into the interior of North America.

Table 14-2	**Some Major Rivers of the World**	
River	**Location**	**Length (km)**
Nile	Africa	6695
Amazon	South America	6276
Ob	Asia	5567
Chang Jiang	Asia	4989
Congo	Africa	4394
Mississippi	North America	3779
Missouri	North America	3726
Volga	Europe	3687
Yukon	North America	3185
Danube	Europe	2842

Figure 14–10. Large ocean-going vessels use locks to reach inland ports.

One of the problems with using rivers as highways is that they are not always smooth. As water flows off the land toward the oceans, it has to drop periodically. Explorers discovered this when they ran into rapids and waterfalls. They had to carry their canoes around these spots. Water that starts in a mountain lake 1000 m high has to drop 1000 m by the time it reaches the ocean. However, since rivers are just as important for transportation today as they were in the past, engineers have created ways to handle these drops.

The solution to the problem is to build dams and locks on the rivers. *Dams* keep the water level the same for long stretches, and *locks* raise and lower ships from one level, or *pool,* to another. Locks must be large enough to hold any of the ships that need to use them, yet small enough so that engineers can control the water levels in them. Look at the drawing of the system of locks that enables ships to reach the inland ports on Lake Ontario, Lake Erie, and Lake Huron.

The locks of the St. Lawrence Seaway allow large ships from any ocean in the world to travel nearly halfway into the middle of North America. The locks work by either raising a ship that is going up the seaway or lowering a ship that is going down the seaway. In the next activity you can find out about some of the U.S. and Canadian cities that have become international seaports along the Great Lakes and the St. Lawrence Seaway.

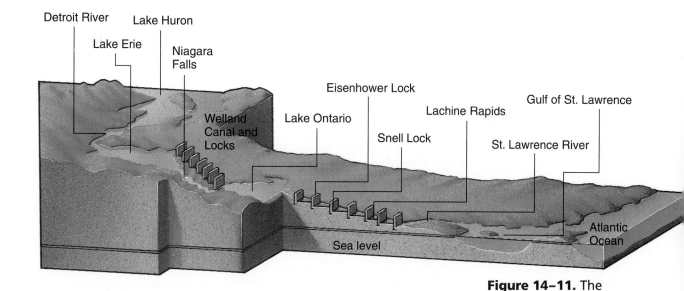

Detroit River

Lake Huron

Lake Erie

Niagara Falls

Welland Canal and Locks

Lake Ontario

Eisenhower Lock

Snell Lock

Lachine Rapids

St. Lawrence River

Gulf of St. Lawrence

Atlantic Ocean

Sea level

Figure 14–11. The drop from Lake Erie to the Atlantic Ocean is about 300 m.

ᴅⁱˢᶜᵒᵛᵉʳ ᴮʏ *Researching*

Look at a map or globe, and list in your journal at least five cities that are located along the Great Lakes or the St. Lawrence Seaway. Then use reference books in your school or community library to find out what materials are shipped to and from those cities to other countries. Find out the names of those countries if you can. ✐

▶ ASK YOURSELF

Why were lakes and rivers so important in the exploration and settling of North America?

SECTION 1   REVIEW AND APPLICATION

Reading Critically

1. Why are ocean trenches often located near chains of volcanic islands?

2. Why are some seas saltier than others?

Thinking Critically

3. If you know the surface area of an ocean (in km^2) and the volume of water (in km^3) in that ocean, how could you determine the average depth of that ocean?

4. Why is the Great Lakes area of North America sometimes referred to as the "north coast" of the United States?

SKILL Interpreting and Graphing Data

▶ **MATERIALS**
- graph paper • pencil

▼ PROCEDURE

Scientists use echo sounding, a type of sonar, to measure the depth of the ocean. Prepare a table like the one shown. Calculate and record the depth of the ocean bottom for each sounding listed.

The first sounding has been done for you.

$$\frac{1500 \text{ m/s} \times 1.2 \text{ s}}{2} = 900 \text{ m}$$

▶ **APPLICATION**

Use a piece of graph paper to draw the ocean bottom. Plot "ocean depth" on the vertical axis and "distance from origin" on the horizontal axis. Connect the plotted depths with a smooth line to diagram the features (mountains, valleys, and plains) of the ocean's bottom.

 Using What You Have Learned

How could you use soundings to locate a deep channel where a ship could safely sail?

TABLE 1: CALCULATING DEPTH			
Sounding	Distance from Origin (km)	Sounding time (seconds)	Calculated depth (m)
1	2	1.2	900
2	4	2.0	
3	6	3.6	
4	8	4.8	
5	10	5.3	
6	12	2.3	
7	14	3.1	
8	16	4.5	
9	18	5.0	
10	20	5.1	
11	22	4.9	
12	24	4.8	
13	26	6.7	
14	28	3.6	

Characteristics of Ocean Water

When you swim in the ocean, you can taste the salt. When you get out and the water evaporates from your skin, you can feel the salt that is left behind. If all the oceans evaporated, they would leave a layer of deposits 13 m thick. These deposits would include the salt that you can taste and many other dissolved minerals.

Why Is Ocean Water Salty?

Ocean water tastes salty because it contains compounds like sodium chloride. If you analyzed the material remaining after evaporating ocean water, over three-fourths of it would be sodium chloride. Sodium chloride is the salt you use on your food. You can also buy "sea salt" in some stores. It contains other compounds in addition to sodium chloride.

Almost all the elements of the earth's crust can be found in the oceans. You could, for example, get about 10 kg of gold from 1 km^3 of ocean water! Are you ready to go out and start mining the oceans? If so, you will have to think up a good way to do it. So far, the cost of extracting the gold from all that water has been much higher than the value of the gold itself.

Objectives

Explain why ocean water is salty.

Describe the effects of ocean water on the temperature of nearby land.

Correlate ocean depth with water pressure.

Figure 14–12. The salt that we use in our foods comes from deposits like this one on the island of Bonaire.

If you allow 1 kg of ocean water to evaporate, about 35 g of salt would be left over. That means the saltiness, or **salinity,** of ocean water is about 3.5 percent. Some bodies of water have a higher salinity. The Mediterranean Sea, for example, would leave behind about 40 grams of salt if you evaporated 1 kg of its water.

The salinity of the oceans has stayed nearly the same for thousands of years because salts and other minerals are removed from the oceans at about the same rate as new ones are added. All of the different salts and minerals dissolved in ocean waters are found in the rocks and sediments in the earth's crust. These materials are carried by rainwater to the oceans. The oceans also give back many of their minerals to the crust. You may recall that two kinds of sedimentary rocks—evaporites and precipitates—are formed from materials that were once dissolved in ocean waters.

 **ASK YOURSELF**

Why is the ocean salty?

How Does Ocean Water Affect Climate?

People tend to think of deserts, such as Death Valley, as hot. Most deserts are certainly hot during the day, but at night they cool off dramatically. The change in temperature from day to night is much greater in a desert than in a field or wooded area. The reason is the desert's lack of water. Water tends to keep the temperature of the air stable. Without the moderating effect of water, most of the earth would be a desert.

Figure 14–13. Death Valley

If the earth had no oceans, the air temperature would rise above 100°C during the day and plunge below –100°C at night. The oceans absorb a lot of heat energy from the sun during the day and release it slowly during the night. Breezes blowing from the oceans onto the land help cool the land during the day and warm it during the night. Oceans also distribute heat from the tropics to higher latitudes. You may want to review the section on oceans and climate in Chapter 13.

Figure 14–14. Salt is often spread on icy streets.

Ocean water has another effect on climate. Look at this photograph. If you live in a place that has snow during the winter, this is probably a familiar scene. Why do road crews put salt on icy streets? The next activity will give you the answer.

🔬 BY *Doing*

Fill two 50-mL plastic cups about halfway with water. Place 5 g of table salt in one of the cups, and stir the water until all the salt dissolves. Place a thermometer in each cup, and put both cups in a freezer for about 30 minutes. Is there any sign of freezing in either cup? Check the cups every half hour, and record your observations in your journal. Do this until the water in both of the cups starts to freeze. Note the temperatures and the times at which freezing begins in each cup. Now why do you think salt is placed on icy streets? ✐

Fresh water freezes at 0°C. Ocean water freezes at –1.9°C. This difference is due to the presence of salt in ocean water. Water that is saltier than the oceans freezes at even lower temperatures. Utah's Great Salt Lake, for example, freezes at –2°C.

In polar regions, ice forms on the surface of the ocean only. Some of the ice, like that in the picture, gets very thick, but the oceans never freeze solid. Why? The ice that forms on the oceans is made of fresh water—the salts stay behind, making the water even saltier. In addition, the ice acts as an insulating blanket over the rest of the water, keeping it from freezing. If the oceans were to freeze solid, Earth's climate would be much colder. Explain why.

Figure 14–15. The ocean never freezes solid.

Near the equator, the ocean's surface is warmed by the tropical sun. Surface temperatures can get as high as 30°C. But when you get below 1000 m, the temperature is usually below 5°C, even in the Caribbean Sea. Why isn't the ocean temperature uniform? You can find out why the ocean depths remain cold in the Investigation at the end of this section.

 ASK YOURSELF

Why do cities near an ocean have less temperature change from day to night than cities in the middle of a continent?

Why Does Water Exert Pressure?

You probably know that the atmosphere is held to the earth by gravity and that the atmosphere exerts pressure on everything. If you live at sea level, the air pressure is one atmosphere, but if you live on a mountain, the air pressure is less than one atmosphere because there is less air above you pressing down. Water is also held to the earth by gravity, so does it also exert pressure? You can find out more about water pressure in the next activity.

DISCOVER BY Doing

Take an empty glass, turn it upside down, and push it down into a tub of water. Now turn the glass over. What happens to the air in the glass? What happens to the water?

Since water is more dense than air, it exerts more pressure. If you dive 10 m into the ocean, you experience the equivalent of two atmospheres of air pressure—double the pressure at the surface. The deeper you dive, the more pressure you feel. Every 10 m of depth adds another atmosphere of pressure.

Unlike air, water under pressure does not compress very much. In fact, the density of sea water at the bottom of the Mariana Trench is almost the same as it is at sea level, even though the pressure is more than 1000 atmospheres! At those pressures, any ordinary vessel would be crushed as easily as you can crush a soft-drink can. That explains why the steel used in *Trieste* is 15 cm thick! You will learn more about the effects of pressure in Chapter 17.

Figure 14–16. Thick steel protects the crew of *Trieste* from a pressure of 1000 atmospheres.

ASK YOURSELF

Why are most submarines limited to certain depths?

SECTION 2 *REVIEW AND APPLICATION*

Reading Critically
1. How do large bodies of water affect the climate of nearby land?
2. How does salt change the freezing point of water?

Thinking Critically
3. What is the relationship between pressure and density in air? What is the relationship between pressure and density in water?
4. What is the relationship between pressure and volume in air? What is the relationship between pressure and volume in water?

INVESTIGATION

Demonstrating the Relationship Between Temperature and Density

▶ **MATERIALS**

- safety goggles ● laboratory apron ● beaker, 500-mL ● red and blue food coloring
- test tubes (2) ● crushed ice ● test-tube holder ● Bunsen burner

▼ **PROCEDURE**

1. **CAUTION: Put on safety goggles and a laboratory apron, and leave them on throughout this investigation.**
2. Fill the beaker nearly to the top with water at room temperature.
3. Place four or five drops of red food coloring in one test tube. Place four or five drops of blue food coloring in the other test tube. Add some crushed ice to the blue food coloring.
4. Fill both test tubes about two-thirds full with water.
5. Using the test-tube holder, heat the red water nearly to a boil with the Bunsen burner. Turn off the burner when you are done.
6. Slowly pour the contents of both test tubes, at opposite sides, into the beaker.

▶ **ANALYSES AND CONCLUSIONS**

1. What happened to the cold (blue) water?
2. What happened to the hot (red) water?
3. Did the colors mix or tend to stay separated? Explain.

▶ **APPLICATION**

How might this help explain why the deep waters in the oceans remain cold?

✳ *Discover More*

How could you modify these procedures to test the relationship between salinity and density?

HIGHLIGHTS

The Big Idea

The structure of the ocean floor is related to the interaction of the various plates that make up the earth's crust. Mid-ocean ridges, abyssal plains, trenches, and undersea volcanoes are created by the constant movement of these sea-floor plates. The oceans' salinity is also due to the interaction of several systems. The erosion of rocks and minerals from the land adds various salts to the ocean water, while factors such as evaporation and precipitation cause variations in salinity from place to place. The oceans, in turn, affect other systems. For example, the earth's moderate climate is due to the oceans' ability to store and slowly release the sun's heat.

For Your Journal

Now what do you know about the sizes and depths of the earth's oceans and seas? How have your ideas changed about why the oceans are salty and what the pressure is like under the ocean? Write your new ideas in your journal.

Connecting Ideas

It is easy to think of the oceans as stable bodies of water. But sea levels have changed a lot in recent geological time. Here is a chart listing events that could change sea level. Copy the chart into your journal. Then decide whether each event would cause sea level to rise or fall.

Event	Sea-level change
Glaciers around the world grow in size.	
Earth's climate gets warmer.	
Land rises.	
Earth's climate gets cooler.	
Erosion carries land into the sea.	
Glaciers around the world start to melt.	
Land begins to sink.	

REVIEW

Understanding Vocabulary

1. Explain how the terms in each set are related.
   a) oceans (429), seas (430)
   b) continental shelf (433), continental slope (433)
   c) abyssal plains (434), sonar (433)
   d) locks (438), dams (438)

Understanding Concepts

MULTIPLE CHOICE

2. The percentage of the earth's surface that is covered by water is about
   a) 25 percent.
   b) 50 percent.
   c) 10 percent.
   d) 70 percent.

3. The largest sea in the world is the
   a) Sea of Japan.
   b) Bering Sea.
   c) Mediterranean Sea.
   d) China Sea.

4. The longest river in the world is the
   a) Nile.
   b) Amazon.
   c) Chang Jiang.
   d) Congo.

5. The deepest ocean in the world is the
   a) Atlantic.
   b) Pacific.
   c) Indian.
   d) Arctic.

6. As the salt content of water increases, its freezing point
   a) goes up.
   b) goes down.
   c) varies.
   d) remains the same.

SHORT ANSWER

7. Explain how sonar is related to undersea exploration.

8. Name four features of the ocean floor.

9. Explain how the oceans help moderate the earth's climate.

Interpreting Graphics

10. Table 1 lists the average January and July temperatures for two American cities located at the same latitude. What is the difference between the January and July temperatures for both cities?

11. Kansas City is in the midwest, while San Francisco is on the west coast. Explain the greater differences in Kansas City temperatures.

TABLE 1: TEMPERATURES		
Temperature	San Francisco	Kansas City
July average	15°C	23°C
January average	10°C	–1°C

Reviewing Themes

12. *Environmental Interactions*
The oceans' various salinities are due to the interaction of several factors. Explain some of these interactions.

13. *Systems and Structures*
Mid-ocean ridges, trenches, and undersea volcanoes are all found on the ocean floor. Explain what these structures have in common.

Thinking Critically

14. A sounding from an oceanographic vessel takes 1.6 seconds to bounce off the ocean floor and return to the sonar instrument. A second sounding takes 2.4 seconds. A third sounding takes 4.8 seconds. Describe this section of ocean floor.

15. Suppose the average temperature of the earth rises a few degrees. How might this affect Baltimore? How would it affect Atlanta?

16. You might have seen a movie in which a submarine descends deeper and deeper as the crew looks on anxiously. What might the crew be worried about?

17. Air is much denser at sea level than it is on a mountaintop. However, water in the Mariana Trench is only slightly more dense than water at the ocean's surface. Explain this contradiction.

Discovery Through Reading

Kunzig, Robert. "Time Zero." *Discover* 13 (December 1992): 32–33. A new piece of the world was born in April 1991, on an undersea mountain range in the Pacific. Oceanographers were there to observe the event. This article discusses what they saw.

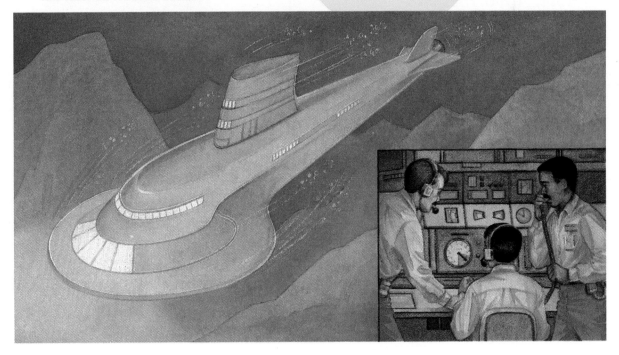

The Restless Oceans

The artists and poets of Japan have long been fascinated with the motions of the sea. The artist portrays the terrible force of the "Great Wave," which rises off Japan's stormy eastern coast, while the poet's refrain speaks of the sound and fury of the sea. Note the image of Mount Fujiyama in the waves, a reassuring symbol of nature's permanence.

Hokusai, 1818-1830, Block printing, *The Wave from 36 views of Mt. Fuji*

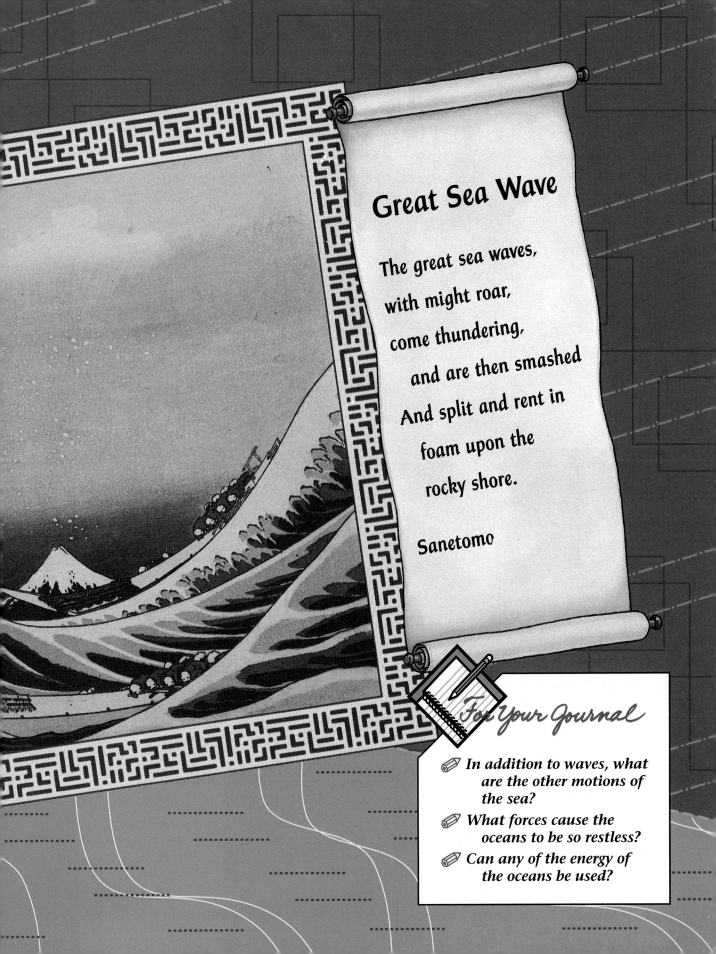

Great Sea Wave

The great sea waves,
with might roar,
come thundering,
and are then smashed
And split and rent in
foam upon the
rocky shore.

Sanetomo

For Your Journal

- In addition to waves, what are the other motions of the sea?
- What forces cause the oceans to be so restless?
- Can any of the energy of the oceans be used?

Waves

Describe how wind and waves affect coastlines.

Define an ocean wave's height, period, and wavelength.

Compare and contrast the flow of energy with the movement of water in an ocean swell.

Oceans seem to be alive—they are constantly moving. Oceans can gently caress the shore with tiny ripples or throw tons of water at the land in the form of huge breakers. Waves are fun to watch and fun for surfers to ride, but they can also be dangerous and destructive if they get too big.

Wind and Waves

Figure 15–1. Waves like the ones above can be a lot of fun. When waves get too big, like those on the right, they produce more energy than we can handle.

People often describe the size of waves by how high they are. Ocean waves can be a few centimeters high or a few meters high, like those in the picture on the left. Occasionally, ocean waves reach heights of more than 10 m, like those in the picture directly above, and come crashing into the shore, causing erosion and destruction.

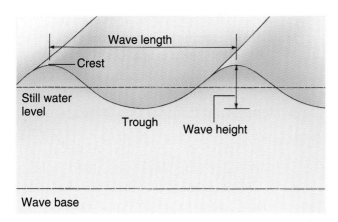

Figure 15–2. What determines the height of ocean waves?

Waves can also be described by how long they are and how fast they move. Look at the diagram. You can see that **wave height** is the distance between the top and the bottom of a wave. The highest point of a wave is called the *crest,* and the lowest point is called the *trough.* The diagram shows that **wavelength** is the distance between two consecutive crests. Common wavelengths in the ocean are 60 to 120 m. Finally, a **wave period** is the time between the passing of one crest and the passing of the next crest. If you were standing on a pier watching waves move below you, you could easily measure the period with a watch. In the ocean, wave periods are often six to nine seconds. In the following activity, you can make some waves yourself.

DISCOVER BY *Doing*

Half fill a shallow baking pan with water. Place your face near one end of the pan, and gently blow across the surface of the water. Observe what happens. Now blow a little harder. Note the relationship between how hard you blow and how much the water moves. ✎

When wind blows across water, friction between the air molecules and the water molecules starts the water moving in the same direction as the wind. Wind can also blow down into the water. This tends to push the water up, making the waves higher.

When the wind blows very hard, water can be picked up and blown. This is like the water that splashed out of the pan when you blew too hard. In the ocean, strong wind produces the whitecaps you see in a storm, where water is blown off the wave crests and flies through the air.

Figure 15–3. Swells on the ocean can look like rolling hills on land.

Choppy waves are created by short, sudden wind storms. These waves usually settle down as soon as the wind dies. There isn't time for much energy to be transferred from the wind to the water. When wind blows steadily across the open ocean, it can create waves that do not break. These rolling waves are called *swells*. Swells created by steady winds, such as trade winds, can travel great distances and keep much of their energy.

If the wind is strong, as with a tropical storm, wave heights and wavelengths can get to be very large. Swells created by a storm off Japan, for example, can move all the way across the Pacific Ocean and end up as good-sized waves for California surfers. These swells can have speeds of tens of kilometers per hour and wavelengths of hundreds of meters.

You know that the energy from storm waves can travel all the way across an ocean. But what about the water in the wave? How does it move? Is the water in a swell off Japan the same water that crashes onto a beach in California? In the next activity you can find the answers to these questions.

ACTIVITY
How does the water in a swell move?

MATERIALS
rope, 5 m; ribbon, 10 cm

PROCEDURE
1. Tie one end of the rope to a doorknob, a fence post, a tree, or another object about 1 m above the ground.
2. Tie the ribbon loosely to the middle of the rope. It should be loose enough to slide along the rope.
3. Hold the free end of the rope in one hand, and flip it up and down about once every second. Observe what happens to the rope and the ribbon.

APPLICATION
1. In what direction do the waves travel in the rope? How does the ribbon move?
2. How is the motion of the ribbon similar to the motion of the water in waves?

When you flip a rope, energy, in the form of waves, moves from your hand along the rope. The rope doesn't go anywhere. It moves up and down, but it does not move away from your hand. The energy in water waves can move all the way across an ocean, but the water itself doesn't go very far. It moves in circles, returning to where it started, as the diagram shows.

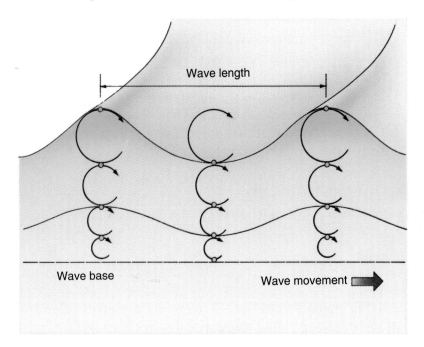

Wave length

Wave base

Wave movement

Figure 15–4. The water in an ocean swell moves in a circular motion.

Think of it this way; a drifting boat won't move across the ocean with the swells. The boat moves up and is pushed forward slightly on the crest, but then it moves down and slips back again in the trough. The boat stays with the water; only the energy of the wave is passed on.

When swells get close to shore, however, they start to change. Imagine hundreds of people walking, 2 m apart, in a straight line. Everything is fine as long as the ground is flat. But if the line reaches a hill, the people in front slow down as they start to climb it. The people in back start running into the slower ones in front of them. The people in front start to lose their balance and fall forward.

The same thing happens to waves. In the open ocean they can easily stay, say, 60 m apart. But as they approach land, where the water gets shallow, swells begin to slow down, as the people going uphill did. The bases of the waves start to drag on the ocean bottom, slowing them even more. Now the tops of the waves are moving faster than the bottoms. The waves start to bunch up and become unstable. The tops tip forward and eventually fall over, just as the people in the line did.

Figure 15–5. When the top of a swell moves too far past the base, the wave falls over, forming a breaker.

The most chilling of all the phenomena of the sea, to those who know its meaning, is the sudden withdrawal of the water from the shore. The sea drops far below the lowest tide level. Harbors dry out. Ships lie stranded. Land is exposed that no living person has ever seen before, and occasionally old wrecks. Sometimes people wander out curiously to explore, but they are never seen again. There is a hissing and sucking and rattling of stones as the water leaves. As the waves recede farther and farther, there is an eerie silence. When the sea returns, it comes in at a hundred or more miles an hour. Sometimes it simply rises, frothing and churning, dozens of feet higher than it ever has; and sometimes it comes as a cresting, towering wave, sweeping and crashing over everything.

from *Mysteries of the Deep*

Falling waves, which are called *breakers,* crash onto the beach. Beaches that face into the prevailing winds often have the high, curling breakers that are perfect for surfing. But are beaches the only place where waves break? Try solving this problem in the next activity.

DISCOVER BY *Problem Solving*

Imagine yourself walking along the shore on a windy day. Breakers are pounding on the beach. But when you look out over the ocean, you notice a separate line of breakers not too far off shore. What does this line of breakers tell you about the ocean bottom at that point? Draw a diagram to show what happens to create this other row of breakers. ✐

▼ ASK YOURSELF

How do energy waves move through the water?

Tsunamis and Storm Surges

Some of the biggest waves in the ocean are not even caused by wind. These waves, called **tsunamis** (SOO NAH meez), are caused by volcanic eruptions, earthquakes, or undersea landslides. Tsunamis have very long wavelengths, often more than 100 km. And they can travel at speeds of several hundred kilometers per hour. But their wave heights, in the open ocean, are often less than a meter. This means that a tsunami can pass under a boat without the people on board even being aware of it.

Figure 15–6. The waterfront of Hilo, Hawaii, was almost totally destroyed by a tsunami in 1946.

The problem is that tsunamis change as they get close to land. As the water gets shallow near the shore, a tsunami rides up the slope toward the beach, reaching a height of more than 30 m. Because tsunamis move so fast and carry so much energy, a tremendous amount of water is pushed very quickly onto the land. Few structures can survive such an unleashing of energy.

Tsunamis can also be caused by undersea landslides. Water suddenly displaced by the movement of large amounts of sediment or other material can send walls of water onto nearby shores. These waves strike without warning and can drown swimmers or people walking on the beach.

Hurricanes and other storms often produce large waves called *storm surges*. These are actually domes of water produced by the action of cyclonic winds in which the sea level can be up to 5 m higher than normal. If the crest of a storm surge should reach a shore in conjunction with a high tide, the damage can be much greater than that caused by wind-driven waves alone. In 1992 Hurricane Andrew flooded the island of Eleuthera in the Bahamas with a storm surge estimated to have been nearly 5 m high. On top of this were wind-driven waves more than 2 m high.

Figure 15–7. The storm surge and wind-driven waves associated with a hurricane washed completely over this island in the Bahamas in 1992.

 ASK YOURSELF

Why might a person in a boat at sea not notice a tsunami passing by?

SECTION 1 *REVIEW AND APPLICATION*

Reading Critically

1. What causes swells to become breakers?

2. What causes a tsunami?

Thinking Critically

3. How is the energy from the wind transferred to the sea to make waves?

4. Why do breakers form farther offshore in some places than in others?

INVESTIGATION

Modeling Waves

▶ MATERIALS
● shallow, rectangular pan ● water ● plastic straw ● food coloring

▼ PROCEDURE

1. Place the pan on a flat surface, and put about 1 cm of water in it.
2. Let the water settle until it is calm and the surface is flat.
3. Blow gently through the plastic straw across the surface of the water. Blow

from one end of the pan so that the waves move toward the other end.
4. Draw the pattern of the first waves you make.
5. Try blowing on the water from different directions and with varying strengths.

Draw the various wave patterns, and identify them.
6. After experimenting with different wind directions and velocities, add a few drops food coloring to the water and repeat step 3.

▶ ANALYSES AND CONCLUSIONS

1. How would you describe the pattern of the first waves you made? Try to describe the wave pattern.
2. How did the food coloring move when you blew on the water? Did the waves move faster than the colored water? Explain.

▶ APPLICATION

1. How is using a wave model similar to observing the actions of waves in the ocean?
2. When would scientists use a wave model instead of making observations of waves in the ocean?

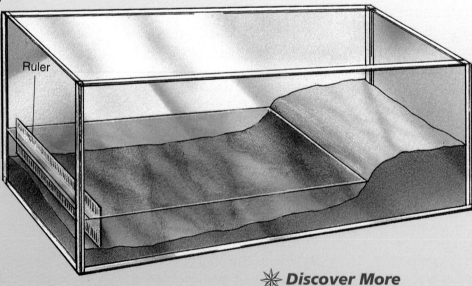

Ruler

✳ *Discover More*
Devise a wave-model experiment with the materials shown above that would simulate the effects of a tsunami or storm surge.

Currents and Tides

When early European explorers returned home from trips to the Americas, they discovered they couldn't sail back across the ocean along the same course they had come on. Even if the wind was right, the water seemed to be moving against them. It was like trying to paddle upstream in a fast-moving river. What are these "rivers" within the oceans, and what causes them?

Surface Currents

When wind blows along the surface of the land, it moves sand and other loose sediments. When wind blows along the surface of the ocean, it moves water. You may recall that it is friction between wind and water that causes water to move. In areas where the prevailing winds are strong and always from the same direction, this friction moves large quantities of water at the same time, creating surface currents. **Currents** are streams of water that move like rivers through the oceans.

Surface currents can transport millions of cubic meters of water per second. By comparison, the Amazon River, the largest in the world, moves only about 120 000 m³ of water per second. One of the reasons currents carry so much more water than rivers is that they can be hundreds of kilometers wide and hundreds of meters deep. The Amazon River averages only 10 km in width and about 12 m in depth. One of these currents, the Gulf Stream, is shown in the photograph. You can discover more about the Gulf Stream in the following activity.

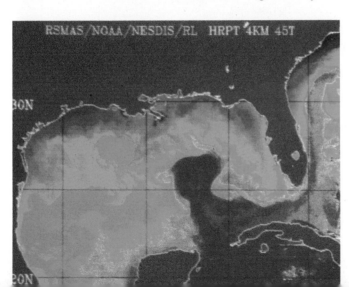

Figure 15–8. The red- and orange-colored water near Florida in this infrared satellite photograph is the Gulf Stream.

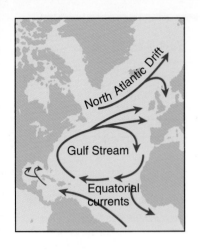

North Atlantic Drift

Gulf Stream

Equatorial currents

Look again at the satellite photo of the Gulf Stream, and compare it to this map of major North Atlantic currents. What is the source of the Gulf Stream? In what direction does the Gulf Stream flow? Do you think the waters of the Gulf Stream are warm or cold? How do you know? If you were sailing from Europe to North America and then back again, what routes would you take? Use the map to explain why Columbus first landed in the islands of the Bahamas. ✐

Besides helping explorers travel from the Americas to Europe, the Gulf Stream also affects the weather of Europe. The Gulf Stream helps move heat from the warm, tropical Caribbean to the cold North Atlantic. Although prevailing winds account for most heat transport, the contribution of currents is significant. This heat transport moderates coastal climates far from the equator.

Figure 15–9. In winter, the beaches of the Mediterranean are crowded, while those of New Jersey, at about the same latitude, are nearly deserted.

The Gulf Stream carries warm water north along the coast of the United States, then east across the Atlantic Ocean, giving Europe a more temperate climate than latitude alone would suggest. But there are other currents that have the opposite effect. You can learn more about surface currents in the next activity.

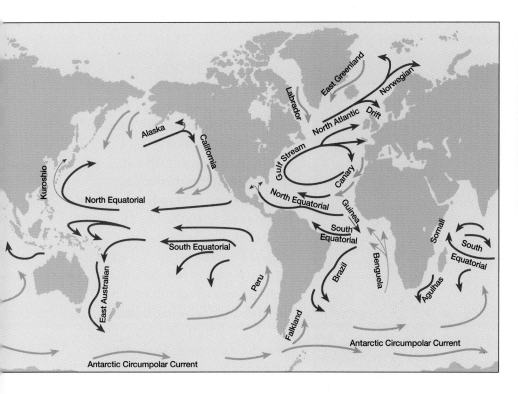

Figure 15–10. The major surface currents of the world's oceans.

Far from being a great sedate river confined as though between banks, the Gulf Stream meanders and eddies and shifts itself a hundred miles to one side or another. There are even cold streaks down the middle of it. The kinks in the Gulf Stream often break off from it, forming independent loops like miniature gyres that continue swirling about for weeks or months.

from *Mysteries of the Deep*

DISCOVER BY *Observing*

Look at the map of the major surface currents. Warm currents are shown in red, while cold currents are shown in blue. What current affects the western coast of the United States? Where do most cold currents begin? On which side of the oceans are most cold currents? What cold current is opposite the warm Gulf Stream? In what direction do the currents in the North Atlantic and North Pacific turn? In what direction do the currents in the South Atlantic and South Pacific turn? Recall the forces that drive the planetary winds. What prevailing winds create the equatorial currents in the Atlantic and Pacific oceans? Why do you think the ocean currents curve the way they do in the Northern Hemisphere? ✐

▶ ASK YOURSELF

What causes most surface currents?

Shoreline Currents

Figure 15–11. Why are there patches of white water near the arrows?

Imagine yourself standing on the beach near your home in San Diego on a windy day. You're facing the ocean, surfboard in tow. The wind is blowing in your face—directly onshore. You look at the surf and see that large breakers are dumping a great deal of water onto the beach. All the motion seems to be from the sea toward the land. But you know that all the water being thrown up on the beach can't stay there. It has to get back into the ocean. And on a day like this, that probably means breakouts, or *rip currents*. Rip currents carry water that has been blown onto the beach by the wind back to the ocean.

Rip currents are one type of shoreline current. **Shoreline currents** are local currents that run along coastlines. Shoreline currents may change from day to day, depending on the direction and strength of the wind. Look at the photograph of the beach. The rip currents are marked with arrows. Waves breaking on the beach stir up the sand, sometimes making the water brown. You know that rip currents flow away from the beach, while the water on either side of the rip current is blown toward and up onto the beach.

If you were swimming and got caught in a rip current, you would be carried out, away from the shore. Swimmers who don't understand rip currents sometimes try to swim back to shore against them. Some rip currents flow faster than people can swim. Swimmers get into trouble when they don't understand how these currents work. But, being an experienced surfer, you know how rip currents work. You know that if you swim parallel to the shore, you will quickly be out of the current. Then the waves will carry you back toward shore.

A second type of shore-line current has little to do with the size of waves. Suppose you've been invited by your cousin in Cocoa Beach, Florida, to visit and surf for a few days with her. You're standing on the shore of the Atlantic Ocean, just south of the Cocoa Beach pier. It is winter, and a strong nor'easter is blowing. Winter storm winds often blow from the northeast toward the south-west, hitting the shoreline at an angle. They blow water up onto the beach and then south along the coast, creating a shore-line current. The movement of this current—along the shore—gives it the name *longshore.*

Figure 15–12. On calm days, it is hard to see the effects of a longshore current.

Longshore currents can be very gentle on days when the wind is not blowing very hard. On calm days, they are hard to see and sometimes surprise swimmers. You remember a day last summer when you were here. You laid your towel on the sand, grabbed your board, and paddled out to ride the waves. The waves weren't too good, so after about half an hour you decided to return to the beach. You paddled straight into shore, but your towel wasn't there. What happened? Chances are a gentle long-shore current had carried you away from where you first went into the ocean. Think about this, and then try the next activity.

DISCOVER BY *Calculating*

Suppose you have been carried along the beach by a longshore current. You know your towel is somewhere near the pier. You've been in the water for about two hours—the waves are really great today. When you paddle into shore, you notice you are in front of lifeguard station 10. The lifeguard stations are numbered, in order, starting from the pier (station 0). The stations are about 200 m apart. What is the approximate speed of the longshore current that carried you away from the pier? ✎

◢ ASK YOURSELF

What is the difference between a rip current and a longshore current?

Tides and Tidal Currents

There is another kind of current that is different from surface currents and shoreline currents, a current caused not by the wind, but by the pull of the moon. Take a look at the diagram of the earth and the moon.

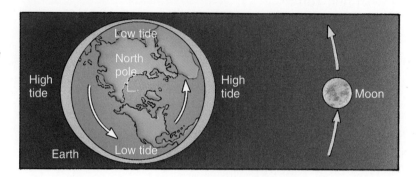

Figure 15–13. The moon's gravity pulls on the oceans of the earth, causing tides.

You probably know that both the earth and the moon have gravity. The gravity of each tends to pull on the other—the earth's gravity is one of the forces that keeps the moon orbiting around it. Although not as strong, the moon's gravity affects the earth, pulling up a bulge of water from the earth's oceans. You also know that the earth rotates on its axis. As the continents move toward this bulge in the ocean, the sea rises slightly onto the land. These changes in local sea level caused by the pull of the moon are called **tides.**

As tidal water moves in and out of rivers, bays, and harbors, *tidal currents* are produced. The direction of these currents depends on whether the river, bay, or harbor is filling with a rising tide or emptying with a falling tide. The change in sea level caused by tides varies a great deal in different parts of the world. On many coastlines, the difference between high tide and low tide is less than a meter. The tidal currents caused by these tides are weak. Other places have much greater tidal differences.

Figure 15–14. Imagine how fast the tidal current must be to change the water level this much.

In some places, the shape of the coastline causes the water from a rising tide to funnel through a narrow opening, producing large differences between low and high tides. Tidal currents produced under these conditions can be quite strong. The Bay of Fundy, in Nova Scotia, Canada, has some of the largest tidal changes on the earth. Sea level differences there may be as much as 17 m, causing tidal currents of 16 km per hour. Projects are underway to convert some of the energy of tidal currents into electricity. You can read more about these projects in the Science Parade at the end of this unit.

The sun's gravity also produces tides. However, because the sun is so far away, it affects tides only about half as much as the moon does. However, when the sun and the moon are in a straight line with the earth, the heights of the lunar and solar tides add up. The tides that result from this alignment are called *spring tides*. The name refers to the "springing up" of water, not to the time of year, as spring tides occur twice each month.

When the moon and the sun are at right angles with the earth, the resulting tides, called *neap tides*, are smaller than usual. Neap tides also occur twice each month.

Figure 15–15. As the examples show, the greatest tidal differences occur during a spring tide, while the smallest tidal differences occur during a neap tide.

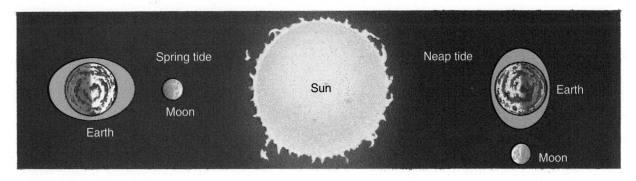

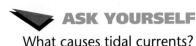 **ASK YOURSELF**

What causes tidal currents?

SECTION 2 *REVIEW AND APPLICATION*

Reading Critically

1. Why did early European sailors in the Caribbean use the Gulf Stream to take them home?

2. How are tides affected by the moon?

Thinking Critically

3. Can a beach have rip currents and longshore currents at the same time? Explain your answer.

4. How do ocean currents affect the climate of the British Isles?

SKILL *Interpreting Tables*

▶ **MATERIALS**
- paper • pencil

If you were the captain of a boat in the Bay of Fundy, you would certainly need to know the times of high and low tides. But boat captains and leisure sailors everywhere need to know about tides. This is why tide tables are published in local newspapers near coastal towns and cities.

▼ **PROCEDURE**

Here is part of a weekly tide table for Daytona Beach, Florida. Study the table, and answer the following questions:

TABLE 1: TIDES FOR PONCE INLET			
Low		**High**	
a.m.	**p.m.**	**a.m.**	**p.m.**
Sunday	2:41	3:16	9:06	9:31
Monday	3:28	4:08	9:55	10:19
Tuesday	4:14	4:59	10:45	11:06
Wednesday	5:04	5:54	11:36	11:58

1. What are the times for the high tides on Monday? What are the times for the low tides on Wednesday?
2. How many high tides are there each day? How many low tides?
3. About how many hours are there between high and low tides on the same day?
4. Look at the morning high tides. Do they get later or earlier each day? By about how many minutes do they change each day?

▶ **APPLICATION**

Suppose you own a boat with a 20-m mast. Near your dock, you must sail under a fixed bridge that has a clearance of 18.5 m at high tide. The tidal change in your area averages 2 m. On Wednesday, you sail out early in the morning, telling some friends you will meet them back at your dock at 5 p.m. How early or late will you be for your appointment?

✳ ***Using What You Have Learned***
Using the answers to the questions as your guide, continue this tide table to show the times for high and low tide for the rest of the week.

HIGHLIGHTS

The Big Idea

Most ocean waves are caused by the interaction of wind and water. Friction transfers the energy of the wind to the water, producing waves. While wave energy can travel across oceans, the water in waves moves in a circular pattern, ending up where it began.

Many ocean currents are also caused by the interaction of wind and water. Wind currents help to transfer the heat of the tropics to northern regions, while tidal currents can be used to produce electricity.

For Your Journal

Look again at your answers from the beginning of this chapter. What are the motions of the sea, and what causes them? How can we use the energy of these motions. How have your ideas changed?

Connecting Ideas

Look at the chart of places below. If you throw a bottle with a note in it into the ocean at these places, where might the ocean's currents carry it? Use the map on page 461 to help you make your decisions. Write your answers in your journal. Make sure you also name the currents that will carry your note.

Starting place	Ending place	Current
California		
Japan		
Florida		
Peru		
Alaska		
Northern Africa		
Australia		
Spain		

Understanding Vocabulary

1. Explain how these terms are related: wave height (453), wavelength (453), wave period (453).

2. For each set of terms, explain the similarities and differences in their meanings.
   a) crest (453), trough (453)
   b) swell (454), tsunami (456)
   c) rip current (462), longshore current (463), tidal current (464)

Understanding Concepts

MULTIPLE CHOICE

3. When the earth and the moon line up at right angles to the earth and sun, the tide is called a
   a) low tide.   c) spring tide.
   b) high tide.   d) neap tide.

4. Surface currents are caused by
   a) wind.   c) tides.
   b) waves.   d) oceans.

5. The time that elapses between the passing of two consecutive wave crests is the
   a) wavelength.   c) wave height.
   b) wave period.   d) trough.

6. Exceptionally high tides that occur bi-monthly are called
   a) spring tides.   c) tidal crests.
   b) neap tides.   d) seasonal tides.

7. Breakers form as waves
   a) become farther apart.
   b) enter shallow water.
   c) encounter tidal currents.
   d) become more stable.

8. Rip currents move
   a) along the shoreline.
   b) toward the shoreline.
   c) away from the shoreline.
   d) randomly.

9. Most currents in the Southern Hemisphere move
   a) clockwise.
   b) counterclockwise.
   c) west to east.
   d) south to north.

10. There is no circumpolar current in the Northern Hemisphere because
   a) the water is too cold.
   b) the earth rotates from east to west.
   c) tidal currents are too strong.
   d) continents are in the way.

SHORT ANSWER

11. Compare and contrast a tsunami and a storm surge.

12. Describe the movement of water in an ocean swell.

13. Why should swimmers always be aware of rip currents near shorelines?

Reviewing Themes

14. *Environmental Interactions*
Identify the factors that produce tides. How do these factors interact with each other?

15. *Energy*
What evidence is there that waves carry energy?

Interpreting Graphics

This diagram shows the positions of a water molecule in a wave at two different times.

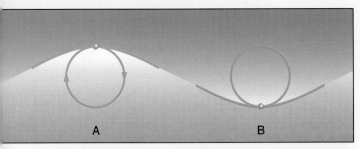

A   B

16. What part of the wave is passing at A? at B?

17. What does the vertical distance between the positions of the water molecule in the two diagrams represent?

18. What is the time span during which the water molecule makes a complete circle called?

19. What is the distance between the water molecule at A and one in a similar position called?

Thinking Critically

20. Explain why spring tides occur twice a month.

21. Why are tidal differences relatively small during a neap tide?

22. Describe the problems of docking a boat in the Bay of Fundy.

23. If the water in a wave does not move in the direction that the wave is moving, how do surfers ride waves?

24. Look at these photographs of the Arctic Ocean in winter and summer. Much of the winter ice melts in the summer because it averages only 3 m in thickness. Even at the North Pole, only a relatively thin layer of ocean is frozen. Ice is less dense than water, so ice floats on the surface, insulating the water below and preventing it from freezing solid. If ice were denser than liquid water, what would happen to the Arctic Ocean? What effect might this have on the earth's climate and on the organisms that live on the earth?

Discovery Through Reading

Lyon, Eugene. "Search for Columbus." *National Geographic* 181 (January 1992): 2-39. Columbus's early voyages, which prepared him to chart the way to the New World, are discussed in this article.

Ocean Resources

*T*he poet Walt Whitman loved the broad diversity of American life—city sidewalks where brawny laborers rubbed shoulders with ladies holding parasols, land-scapes that stretched from grand mountains to the seashore. Beneath the sea, Whitman imagined a world just as rich and varied—multi-colored forests, tribes of odd creatures, and vast riches.

Paul Klee, *Fish Magic*, 1925. Philadelphia Museum of Art: Louise and Walter Arensberg Collection.

The world below the brine;
Forests at the bottom of the sea—the branches and leaves,
Sea lettuce, vast lichens, strange flowers and seeds—the
    thick tangle, the openings, and the pink turf,
Different colors, pale gray and green, purple, white, and
    gold—the play of light through the water,
Dumb swimmers there among the rocks—coral, gluten, grass,
    rushes—and the aliment of the swimmers,
Sluggish existances grazing there, suspended, or slowly
    crawling close to the bottom,
The sperm whale at the surface, blowing air and spray, or
    disporting with his flukes,
The leaden-eyed shark, the walrus, the turtle, the hairy sea
    leopard, and the sting ray;
Passions there—wars, pursuits, tribes—sight in those ocean
    depths, breathing that thick-breathing air, as so many
    do. . .

from *The World Below the Brine*
by Walt Whitman

For Your Journal

✎ **What are the riches of the ocean?**
✎ **How important will the ocean's
    resources be in the future?**
✎ **How can we use those riches and
    still protect the oceans?**

Energy, Mineral, and Water Resources

Objectives

Explain the need for offshore oil drilling.

Describe the mineral and other nonliving resources of the oceans.

Evaluate the need for ocean resources.

Imagine that you are in Houston, Texas, standing under the whirling blades of a helicopter, ready to fly out to an oil-drilling platform in the Gulf of Mexico. This is your first assignment as a reporter for the biggest newspaper in the state. You are going to write a feature about ocean resources. You already know that more than 20 percent of all the oil we use comes from under the oceans. But you need to find out about the other resources of the oceans. You will be interviewing geologists, biologists, engineers, and possibly others. You should get quite a story!

Figure 16–1. Oil is only one of the resources found in or under the oceans.

Energy Resources

Thirty minutes later, your helicopter hovers and then lands on the chopper pad of the drilling platform. The platform is in 30 m of water off the coast of Texas. The first person you will talk to is the petroleum geologist. She will show you around the platform, including her geology lab.

Before you can ask what it is like to live on the platform, she asks you what you know about the formation of petroleum. Recalling your earlier studies in earth science, you tell her you know that petroleum is a fossil fuel, so it must have been formed from living organisms. And since it is found in rocks that are or were once under the oceans, you know that those organisms lived in the oceans.

Figure 16–2. Over millions of years, the organisms of the ancient seas were changed into petroleum.

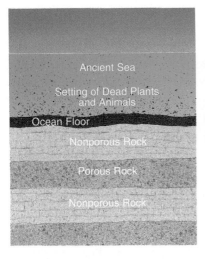
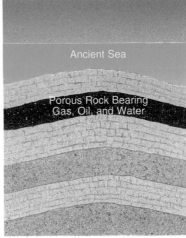
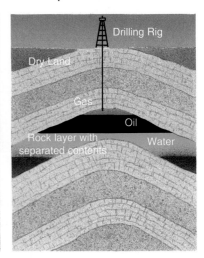

She congratulates you for having done your homework and fills you in on a few more details for your readers. In your notebook you write that the organisms were microscopic in size and lived in countless numbers in shallow seas between 100 million and 500 million years ago. As they died, they became part of the sediments on the sea floor. Over millions of years, heat and the pressure of overlying deposits changed the organically rich sediments to deposits of oil and natural gas. Drilling for them here is similar to drilling for them on land, except for the 30 m of water that covers the floor of the Gulf.

There are many offshore platforms in the Gulf of Mexico, and hundreds more in other places around the world. Many are off the coasts of California, Venezuela, central Africa, Alaska, Canada, and in the Mediterranean Sea, Red Sea, North Sea, and Persian Gulf. In fact, there is offshore drilling going on in more than 100 countries around the world.

Modern offshore platforms can be huge—the one you are visiting is more than 200 m high. The geologist explains that it is difficult and expensive to reach some petroleum deposits. But offshore drilling has become common as supplies of oil and gas become harder to find on land. You want to know why there is such a demand for oil and gas, so you decide to visit the platform's library for a little research.

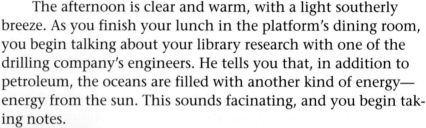
In your journal, list as many products made from petroleum as you can. Use books and encyclopedias in your school or public library to help you. Explain why each product is needed. For example, one of the products made from petroleum is gasoline. Gasoline is used for fuel in cars and trucks. Also describe what must be done to petroleum to make each product. For example, petroleum must be refined to make gasoline. Include information about any by-products or wastes that are produced when each product is manufactured or used. Gasoline, for example, produces pollution when it is burned in engines. Then compare your list with your classmates' lists, and answer the following questions: What are the three most common uses for petroleum? What other forms of energy could be used in place of petroleum for these uses? ✎

The afternoon is clear and warm, with a light southerly breeze. As you finish your lunch in the platform's dining room, you begin talking about your library research with one of the drilling company's engineers. He tells you that, in addition to petroleum, the oceans are filled with another kind of energy—energy from the sun. This sounds facinating, and you begin taking notes.

Figure 16–3. Ocean waves contain a tremendous amount of energy, some of which can be converted into electricity.

Heat from the sun is stored in the water. And sunlight indirectly causes the winds that produce waves and currents. The waves striking about 100 km of shoreline, for example, have enough energy to supply electricity to a million homes. He describes for you an experimental project that uses waves to generate electricity.

Tides can also be a source of energy. This you know from an article you read about a tidal energy plant in operation on the Rance River in France. However, all of this energy is spread through the entire ocean. It is technologically possible, but very expensive, to build the facilities to use this energy. It seems likely that for the time being, petroleum will remain the main source of energy from the oceans.

 **ASK YOURSELF**

How is the oil formed that is found beneath the ocean floor?

Mineral Resources

As you resume your interview with the geologist, you learn that there are other valuable resources in the oceans. While her major interest is petroleum, she tells you about resources such as iron, tin, and sulfur. Many of these resources originate on land. There they are weathered, and then carried to the oceans by rivers, where they are deposited as sediments. These sediments can be dredged up from the ocean floor and processed like those from deposits on land.

Figure 16–4. Vast accumulations of iron ore were deposited in ancient seas billions of years ago.

For example, about 3 billion years ago the ancient ocean floor accumulated layers of sediments rich in iron oxide. The most important deposits of this type are found around Lake Superior. Iron deposits are also found in eastern Canada, Russia, South America, and Australia. These deposits form the major iron resources of the world today.

Perhaps the most interesting deep-sea mineral resource is found lying on the ocean floor, in the form of a nodule. These nodules are usually from 3 cm to 6 cm in diameter, but they can be as large as 15 cm. No one is exactly sure how they formed. They are called *manganese nodules,* because manganese is often the major component. However, another common element in these nodules is iron.

Figure 16–5. Iron, manganese, nickel, and many other elements can be obtained from deep-sea nodules.

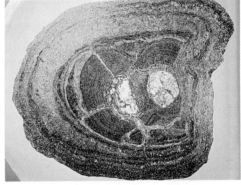

The geologist says that she has analyzed many nodules in her lab and has identified more than 10 other elements in significant concentrations. She estimates that manganese nodules cover more than 15 percent of the deep ocean floor, generally at depths of 4000 m or more. At that depth, mining them is a big problem. She asks you to see how creative you can be in solving this problem.

Problem Solving

If you were an ocean engineer, how would you mine manganese nodules? Remember that they are about the size of golf balls and baseballs. Remember, too, that they are found lying on the floor of the ocean, not buried in the sediments. Finally, remember that they are 4000 m underwater. In your journal, describe in words or draw pictures of the equipment you might use to get these nodules. Compare your ideas with those of your classmates. ✎

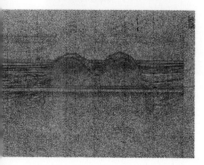

Figure 16–6. This is a sonar tracing of large salt domes in the Gulf of Mexico.

Other ocean resources the geologist tells you about are ones you had never thought about before: sand, gravel, and coal. The value of the sand and gravel taken from the sea is greater than the value of nearly any other type of ocean resource; only petroleum has a greater value. There are offshore coal deposits in the North Sea, but so far it is not practical to mine them.

One additional mineral resource mentioned by the geologist is salt. You read a little about salt before this trip. You know that about one-third of the salt used is collected in ponds from evaporated sea water. You also know that there are large salt mines in the neighboring state of Louisiana.

She explains that salt deposits often form large domes. These domes can be 3 km across and more than 10 km high. While most salt domes are buried, some rise above the surrounding sediments. A few of the salt domes in the Gulf of Mexico are like that. She even has a sonar tracing of two of these domes, known as the Sigsbee Knolls.

▶ **ASK YOURSELF**

What mineral resources can be obtained from the oceans?

Water as a Resource

At this point, you're satisfied with the amount of information you have on the nonliving resources of the oceans. Now you will talk with the platform's marine biologist to get information about the oceans' living resources. But before you leave, the geologist wants to talk to you about one more ocean resource—water.

Hundreds of millions of years ago, the earth had more than nine million billion metric tons of water. Almost all of that water is still available today. The same water molecule that

Curiously, the greatest energy potential of the ocean is salt, though as yet there is no technology for making use of the salinity differentials in the oceans. (The process could be like that of a battery, or might employ membranes, as in osmosis.) There is more potential energy in the salt of an offshore salt dome—salt domes are among the most common of oil-bearing geologic structures—than there is in its oil, by several hundred times, according to recent investigations.

from *Mysteries of the Deep*

dripped off a giant tree fern onto a dinosaur's head may pour into your drinking glass the next time you turn on the faucet. Water is a renewable resource—it is used over and over again, but it is never used up. Almost every drop is returned to the natural water cycle. In this cycle, water moves to the oceans, evaporates, and falls to the earth again as rain. The geologist takes you to her lab to show you a simple model of the water cycle.

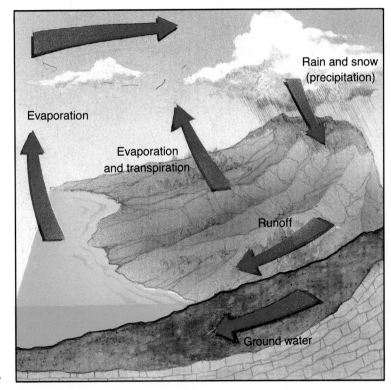

Figure 16–7. The water cycle

ACTIVITY

How can you make a model of the water cycle?

MATERIALS

safety goggles, 🥽 🔥
glass tubing, rubber tubing, plastic bag, string, ring stand, clamp, Bunsen burner, crushed ice, water

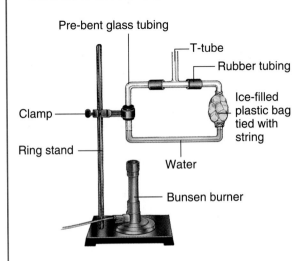

PROCEDURE

1. **CAUTION: Safety goggles must be worn at all times during this activity.** Set up the apparatus as shown, using the preshaped glass tubing.
2. Fill the bottom of the tube with water. Make sure that the plastic bag of ice is secured to the glass tubing with string.
3. Gently heat the water-filled part of the glass tube over a Bunsen-burner flame. Observe the results.

APPLICATION

1. Describe what you observed. How does this activity demonstrate the water cycle?
2. Explain how each part of the model represents a part of the water cycle diagramed above.

Figure 16–8. Fresh water can be made from sea water in a desalination plant.

The water cycle model was interesting, but you're still not sure why ocean water should be considered a resource; at least not in the same sense as petroleum or salt. The geologist explains that in some parts of the world, fresh water supplies are so limited that water is taken from the ocean. Salt can be removed from sea water by **desalination.** There are two basic methods; both leave water as fresh as that from a mountain spring.

In the first method, sea water is evaporated, leaving the salts behind. The vapor is then cooled and condensed back into a liquid. This is very similar to the way the natural water cycle works, you note. The second method uses a plastic film that allows pure water, but not dissolved salts, to pass through. This method is called *reverse osmosis*. Both methods are very expensive and can, at this time, produce only limited amounts of water.

Desalination is an important area of continuing research and development. As the world's population increases, water resources for agriculture and other human uses dwindle. Desalination could provide enough water for humans all over the world if a practical, inexpensive method can be developed.

 ASK YOURSELF

What is desalination?

SECTION 1   *REVIEW AND APPLICATION*

Reading Critically

1. What metal is often found with manganese in nodules?
2. Where do offshore petroleum deposits come from?

Thinking Critically

3. Why are oil and gas often found under the oceans?
4. Why are ocean deposits of sand and gravel so abundant?
5. How could human activities disrupt or contaminate the natural water cycle?

SKILL Analyzing Data with Models

The water cycle describes how water moves between the earth and the atmosphere. A water budget describes the credit (precipitation), storage, and debit (evaporation and runoff) of water at a particular location. Like a financial budget, a water budget will tell you whether there is a surplus or a deficit at any given time. The table shows a water budget for Savannah, Georgia.

▶ MATERIALS
● meter stick ● containers, 10 cm tall (2) ● small pan ● water

TABLE 1: WATER BUDGET FOR SAVANNAH, GEORGIA												
	Jan.	Feb.	Mar.	Apr.	May	June	July	Aug.	Sep.	Oct.	Nov.	Dec.
Credit (mm)	70	80	88	73	80	135	175	180	150	75	50	70
Debit (mm)	20	25	40	75	125	165	175	155	120	75	28	18
Surplus (mm)	50	55	48	0	0	0	0	25	30	0	22	52
Deficit (mm)	0	0	0	22	45	30	0	0	0	0	0	0

▼ PROCEDURE

1. Set one of the containers in the pan. This will represent Savannah's water supply. Fill the container to a depth of 70 mm. This represents the water credit for January.
2. Now remove 20 mm of water, which represents the water debit for the month. How much water remains in the budget?
3. Fill the second container to 80 mm to represent the water credit for February. How much water should you remove for the February debit? Pour the remainder into the first container. This is the surplus for the first two months. Any water that overflows into the pan represents water that cannot be stored. It runs off to another place.
4. Pour the water in the pan down a drain. Continue this procedure for the remaining months. If you have a deficit for any month, you must borrow the water you need from the surplus container.

▶ APPLICATION
1. Was there any month during the year when the surplus ran out?
2. Where would any runoff go in Savannah?

※ Using What You Have Learned
If the water budget shows a deficit for a given month, does that mean there is no water available for use? Explain your answer.

Living Resources

Categorize *ocean life forms by their places on the food pyramid.*

Describe *the major threats to ocean life.*

Evaluate *methods that reduce the threats to ocean life.*

You have pages of notes about the nonliving resources of the oceans. Now it's time to find out about the oceans' living resources. You've scheduled an interview with the platform's marine biologist. He says the best way to become familiar with ocean life is to dive beneath the platform. However, that will have to wait until morning, as it is already getting late. For now, he wants you to walk with him to see the spectacular sunset over the Gulf.

Ocean Life

As the sun sets, a shrimp boat comes into view. You remember the wonderful seafood lunch you had, and you think about all the fresh seafood available here on the Gulf Coast. You wonder how shrimp and other marine organisms rate as resources.

Darkness falls as you ponder this question, and the biologist steers you to the railing at the platform's edge. He wants you to see the Gulf at night. You wonder what you could possibly see at night. As you look down at the flat sea, you notice patches of water that glow with a greenish light.

Figure 16–9. A commercial shrimper at work

Plankton The light seems to come and go, creating a weird, but beautiful, show. The biologist explains that it is the plankton in the water that are glowing. You recall that plankton are so small you can't see them without a microscope, but at night they glow like fireflies. You realize that there must be millions of these tiny organisms in the water, drifting with the tides and ocean currents. The biologist invites you into his lab to have a closer look.

DISCOVER BY *Doing*

Place a small amount of diatomite on a microscope slide. Diatomite is the skeletal remains of diatoms—a type of algae. Add a drop of water and a coverslip, and observe under low power of a microscope. Draw and describe what you observe in your journal. ✎

But the greatest disappointment of recent ocean research concerns the often repeated promise that the oceans will feed humans when the land can do no more. Exploration has shown that the open ocean, the majority of the surface of the oceans, is a biological desert, in which little lives or grows.

from *Mysteries of the Deep*

While observing diatoms through the microscope, you remember the food pyramids you learned about in life science. Phytoplankton (plantlike plankton that make food by photosynthesis) form the base of the ocean food pyramid. Most of the larger animals of the oceans depend on phytoplankton as their source of energy. As you picture an ocean food pyramid in your mind, you remember that it takes a lot of plankton to support just one large fish.

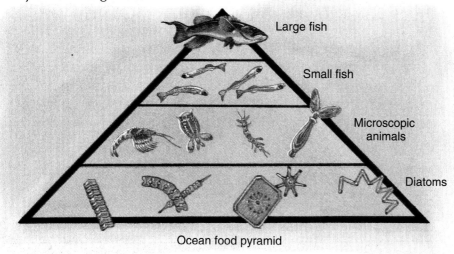

Large fish

Small fish

Microscopic animals

Diatoms

Ocean food pyramid

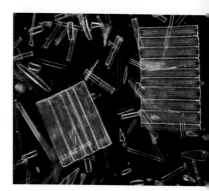

Figure 16–10. Diatoms form the base of most ocean food pyramids.

You also remember the engineer's talk about energy in the oceans. Phytoplankton get their energy directly from sunlight. But the sun's light is gone for this day, and it's time to turn in and get some sleep. You dream about the dive you will make tomorrow morning.

Figure 16–11. Fish are everywhere around the legs of the platform.

You've had a restful night, lulled by the sound of waves gently slapping the platform. The morning is brilliant with sunshine—a wonderful day for diving. After a breakfast of fresh orange juice, grits, toast, and freshly caught snapper—they surely do eat well on these oil rigs—you're ready to see the marine life you've heard so much about. Your partners are the marine biologist and the geologist you interviewed yesterday. The three of you wind down several long sets of stairs to reach the water. In a tiny room at sea level, fins, masks, air tanks, and all the necessary diving equipment are ready. You dress quickly, anxious to get into the water. After a final equipment check, you step into the Gulf.

Fish   The water is very clear. There are fish everywhere, attracted by the shelter of the platform. Right away you see the ocean's food pyramid in action. Large fish are eating the smaller fish. There are huge schools of the smaller fish, but only a few of the big fish. You see snapper, like the one you had for breakfast, grouper, and redfish, a good representation of the variety of fish in the Gulf of Mexico. If you were in the North Atlantic, you might see cod, herring, and haddock; or herring and salmon if you were in the Pacific. These fish are not just important to the food pyramid of the oceans, they are also important resources—part of the human food supply.

Algae   The biologist leads the group to one of the platform's legs for a closer inspection. There are small clumps of algae attached to the metal leg. These algae are not large, but you know that there are many kinds of algae that get very large. You remember the huge algae that grow off the coast of California. These organisms, called *kelp,* can grow to more than 30 m in length. If they were found here, they could be anchored to the bottom and still have their tops floating on the surface. They grow so fast, in fact, that they are harvested, just like fields of wheat on land.

Figure 16–12. This machine harvests kelp.

All too soon the geologist points to her watch, and gives the sign that it's time for the group to surface. Even though you know you are scheduled to dive again this afternoon, you're disappointed about having to leave this beautiful world.

Over a lunch of crab claws and hush puppies, you think again about kelp, so you ask the biologist why kelp is harvested. You figure they must do something with all that algae. The biologist asks you if you have ever eaten algae. Other than seaweed wrapped around some kinds of sushi, you tell him you don't think so. He says you might be suprised to find out just how much algae is in your diet.

Figure 16–13. How often do you eat algae?

One food additive, algin, comes from brown algae. Algin is used to make some jellies. Agar, made from red algae, is used as a food additive, and as a medium for growing bacteria in laboratories. Another red algae, carrageen, is also used in foods, as well as in hand creams and face lotions. The biologist invites you to try a little activity in the kitchen.

ACTIVITY *How do you like your seaweed?*

MATERIALS
various packaged foods

PROCEDURE
1. Check the ingredients on the labels of a number of popular food items in your kitchen. Look especially at foods such as puddings, salad dressings, condiments, ice cream, and yogurt.
2. Look specifically for algin (or alginic acid), agar (or agar-agar), and carrageen (or carrageenan).

3. Make a list in your journal of every product that contains one or more of these food additives.

APPLICATION
What do all of the products containing algae have in common? What role do you think algae have in producing this characteristic?

Other Organisms

After checking out the galley, your group prepares for the afternoon dive. Once in the water, your guide leads you down to the bottom of the platform's legs. The depth here is 30 m, just about the safe limit for sport divers.

Figure 16–14. The living resources of the oceans are almost too numerous to count.

On the ocean floor, you see crabs wandering around. Although not the same species as your lunch entree, they are important as part of the oceans' living resources. So, too, are shrimps and lobsters, which also live on the ocean floor, and clams and scallops, which burrow into the sand. Oysters, another resource, attach themselves to rocks and other solid objects, and most varieties of fish swim freely. You've seen so many examples of living ocean resources today that you begin to imagine the ocean as one big seafood buffet. You try a little mental activity to sort them all out, so you can include them in your article.

DISCOVER BY *Researching*

Below is a list of 10 kinds of seafood that some people like to eat. Copy the list into your journal. For each type of food, answer the following questions: What kind of animal is it? In what oceans of the world does the animal live? In what part of the ocean does the animal live?

tuna	salmon	crab	pollock	scallop
cod	flounder	shrimp	clam	lobster ✐

 **ASK YOURSELF**

Why are the oceans' living organisms considered to be resources?

Threats to Ocean Life

As you head back up to the surface, you notice an old net snagged on one of the platform's metal legs. It is obviously untended, because it is torn and ragged. But it still captures fish. Since no one is there to retrieve it, fish get caught and die. This is a waste of the oceans' resources. You swim over to the net and cut it loose from the platform. The only way to save fish from getting caught in it again is to take it out of the water.

Figure 16–15. Dead fish represent wasted resources.

You begin to think about all the things that can harm the ocean life you have seen. The abandoned net is only one hazard. What other hazards should you write about in your article, you wonder?

After you put all your gear away, you have a chance to ask the biologist about the abandoned net and other threats to ocean life. He tells you that the main threat to ocean life is people. There are billions of people in the world, and we all need food. In order to increase the amount of food we produce, we use chemical fertilizers and pesticides. Many of these chemicals end up in the oceans, killing ocean life and poisoning the water. We need to find ways to produce food without using so many chemicals.

Figure 16–16. Everyone wants a share of the oceans' resources.

We also use a lot of petroleum and buy things from companies that pollute the oceans. In many cases, we don't even know that we are part of the problem. We need to look for ways to become part of the solution.

Garbage   An especially big problem is garbage—what do we do with it? Most of us simply put our garbage in cans or take it to the curb where trucks pick it up. Then we never see it again. But what happens to all that garbage? It has to go somewhere.

Much of our garbage is dumped into the oceans. Some of it washes back up onto our beaches. A couple of years ago, hospital trash washed up on the beaches of New York and New Jersey. The trash included broken glass, needles, and other contaminated material. Swimmers complained about skin rashes. Some had trouble breathing after they had been in the water. Dead dolphins were found washing up on the shores. Could all of these incidents be related?

Figure 16–17. These barges are filled with garbage that will be dumped directly into the ocean. Some of it comes back to foul our beaches.

Plastic in our garbage is a real danger to ocean life. Plastic lasts a long time. It doesn't rust or break down, like most metals do. Many birds, turtles, and sea mammals have died from swallowing things like plastic bags, which they often mistake for jellyfish. Others get trapped in the plastic carriers from drink cans and suffocate or starve.

Figure 16–18. Plastic can be fatal to marine birds and mammals.

We can all work on solving the garbage problem. Recycling helps us reuse some of our garbage. Recycling saves resources and energy, too. Cutting down on all the paper, plastic, and boxes that package the things we buy will also help. Think of all the packaging you throw away just to get at the food you eat.

Pollution   Solving the problem of ocean pollution is more complicated. Have you ever thought about what happens to the water in your sink when you finish washing your hands? We all use a tremendous amount of water every day for cooking; cleaning; taking showers and baths; flushing toilets; and washing dishes, clothes, and cars. Perhaps you don't think much about water; it comes out of faucets whenever you need it, and then it disappears out of our homes. But where does all this waste water go? Let's do a little research to find out.

DISCOVER BY *Researching*

Find out where the waste water from your home or school goes. Does it go into a community sewer system? If so, find out how your community treats the water. Does it go into a septic tank near your home? If so, find out how a septic tank works. Take notes on your findings, and write them in your journal. Compare your notes with those of your classmates. ✎

Some communities dump waste water, or *sewage,* directly into the oceans. Most large cities have sewage treatment plants, but the treatment is sometimes inadequate. Other communities dump sewage into rivers, which flow into the oceans. A large percentage of ocean pollution is caused by improperly treated sewage. Many coastal areas have become unsuitable for recreation or commercial fishing because of sewage pollution.

Figure 16–19. Untreated sewage and chemical wastes pose a threat to ocean life.

Some factories and industries produce chemical wastes. Often these are dumped directly into rivers, lakes, and oceans. Industrial pollution kills ocean life or makes seafood unsafe to eat. Many harmful chemicals have been found in ocean water, in fish, and in shellfish. These chemicals become more concentrated as they move up the food pyramid. Some chemicals are not only harmful to ocean life, but they also cause health problems for people.

Figure 16–20. Sometimes birds and other animals are killed by the toxic effects of crude oil flowing from production platforms or leaking tankers.

The petroleum industry has also been responsible for polluting the oceans. Although petroleum platform operators try to be as careful as possible, accidents sometimes occur. Any leaks from an oil platform go directly into the sea. Spills from tanker accidents have also produced massive oil slicks that have coated beaches with oil and killed many fish, sea birds, and mammals. Huge tankers transport hundreds of thousands of metric tons of crude oil at a time.

In March 1989 the supertanker *Exxon Valdez* ran aground off the coast of Alaska. Forty-four million liters of crude oil poured into the water. The floating oil formed a slick covering 8000 km^2 of ocean, and washed up onto 1600 km of coastline. This was the largest oil spill to date in the United States. In January of 1993, an even larger tanker spill ocurred in the North Sea, fouling the water and coast of Scotland's Shetland Islands. Fortunately, rough seas dispersed the oil.

Figure 16–21. Some of the results of a super tanker spill.

As the earth becomes more crowded, people look to the oceans for more resources. But, as the human population grows, the problems of ocean pollution grow, too. Pollution must be controlled to protect the oceans' resources and our future.

It's time to leave the oil platform and head back to Houston. You say goodbye to your new friends and get into the helicopter. After listening to the engineer, the geologist and the marine biologist talk about the oceans' resources, energy, and pollution, you have lots of notes. Now you must decide what to tell your readers. You open your laptop computer and begin writing:

Figure 16–22. The oceans' resources are nearly as vast as their expanse.

Earth's human population increases every minute. Each new person requires a share of our vanishing resources. More fresh water is needed. More food is needed. More petroleum, minerals, and energy are needed. The oceans can help to meet these needs, but only if we treat them with respect. Pollution of the oceans is a huge problem that must be solved.

We are only now beginning to tap the vast resources of the oceans. In the future, these resources may be required to maintain human life on our planet. But where do we start? What do we do first? A friend of mine once asked me, "How would you eat a 50-pound tuna?" When I looked at him with a puzzled expression, he replied, "In small bites." ... Is anyone hungry?

 **ASK YOURSELF**

What are the biggest threats to ocean life?

SECTION 2 *REVIEW AND APPLICATION*

Reading Critically

1. Where do plankton fit on the ocean food pyramid?

2. How do you start to solve a big problem?

Thinking Critically

3. Why are there more small fish in the ocean than large fish?

4. How does the increasing number of people on the earth affect ocean resources?

INVESTIGATION

Cleaning Polluted Water

► **MATERIALS**
- hammer • nail • soup can • ring stand • clamp • beaker, 150 mL
- ruler • coarse gravel • fine gravel • muddy water • sand • crushed charcoal

▼ **PROCEDURE**

1. Using the hammer and nail, make about a dozen holes in the bottom of the can. Secure the can above the beaker as shown.
2. Make a 2-cm layer of fine gravel at the bottom of the can, then make a 2-cm layer of coarse gravel on top of the fine gravel.
3. Pour about 100 mL of muddy water into the can, and collect the filtered water. What does the water look like after filtering through the coarse and fine gravel?
4. Clean out the beaker and can, and try to improve your filter with the other materials you have.

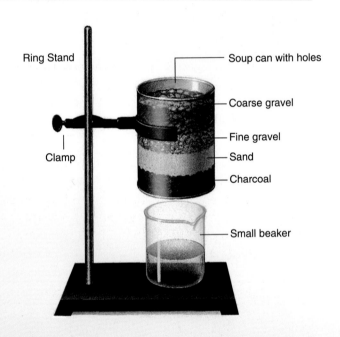

Ring Stand
Soup can with holes
Coarse gravel
Fine gravel
Clamp
Sand
Charcoal
Small beaker

► **ANALYSES AND CONCLUSIONS**
What combination of materials did the best job of filtering the muddy water?

► **APPLICATION**
How could this system be used on a large scale to clean polluted water?

✳ *Discover More*
Using the same materials, find the combination of materials that does the best job for the lowest cost. Assume that fine gravel costs twice as much as coarse gravel, sand costs twice as much as fine gravel, and crushed charcoal costs twice as much as sand.

The Big Idea

The human population of the earth is increasing. With this increase comes a need for more resources: food, water, minerals, petroleum, and energy. Increasingly, we look to the oceans to help meet those needs. But the oceans are in trouble. We must treat the oceans with respect—they hold the key to our future.

Look again at the way you answered the questions at the beginning of the chapter. How have your ideas changed now that you know more about the riches of the oceans? Be sure to write your new ideas in your journal.

Connecting Ideas

Copy the list of environmental problems that threaten the oceans into your journal. Next to each problem, describe some things that you could do to help solve the problem. The first two problems have some possible solutions suggested already.

PROBLEMS	SOLUTIONS
▶ discarded plastic harming wildlife	▶ recycle plastic ▶ don't buy so much plastic
▶ excess waste water flooding treatment plants	▶ take shorter showers ▶ put water restrictors on shower heads and toilets
▶ chemicals dumped into harbor killing plankton	▶ ▶
▶ fertilizer and pesticide runoff killing fish	▶ ▶
▶ oil spills fouling beaches	▶ ▶
▶ sewage dumped directly into oceans causing agal blooms	▶ ▶
▶ trash dumped into oceans littering beaches	▶ ▶
▶ overfishing depleting certain species	▶ ▶
▶ toxins in fish making people ill	▶ ▶
▶ fishing nets killing dolphins	▶ ▶

Understanding Vocabulary

1. Explain how the terms in each set are related.
   a) plankton (481), phytoplankton (481), food pyramid (481)
   b) manganese nodules (475), petroleum (473)

Understanding Concepts

MULTIPLE CHOICE

2. Algae are valuable ocean resources because they are widely used as
   a) fuel.   c) food additives.
   b) fertilizer.   d) sources of salt.

3. Plastic wastes are harmful to ocean animals because
   a) they decompose easily.
   b) the animals might swallow the plastic.
   c) they block out the sunlight.
   d) they sink to the ocean floor.

4. The base of the ocean food pyramid is formed by
   a) phytoplankton. c) ocean water.
   b) large fish.   d) clams.

5. The energy of ocean waves can be used to
   a) desalinate water.
   b) mine minerals.
   c) generate electricity.
   d) produce oil.

6. Many of the minerals in ocean water
   a) formed from microscopic organisms.
   b) were eroded from the land.
   c) are part of the food pyramid.
   d) are pollutants.

7. Life in the oceans is threatened by
   a) sewage.   c) fertilizers.
   b) oil spills.   d) all the above.

SHORT ANSWER

8. Why has offshore drilling become common even though it is expensive and difficult?

9. Describe two methods used to obtain fresh water from sea water.

Interpreting Graphics

Use the diagram below to answer the questions.

10. Which organisms in the food pyramid obtain energy directly from the sun?

11. Which use the sun's energy indirectly?

12. How do the sizes of the various populations change at each successive level of the food pyramid?

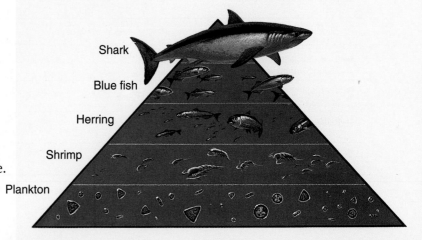

Shark

Blue fish

Herring

Shrimp

Plankton

Reviewing Themes

13. *Environmental Interactions*
How does water move from the oceans to the atmosphere and back to the oceans?

14. *Energy*
How do phytoplankton use the sun's energy?

Thinking Critically

15. Why are petroleum deposits found on land as well as under the oceans?

16. Why are harmful chemicals more concentrated in organisms that are at higher levels of food pyramids?

17. Give an example of an ocean food pyramid.

18. Why will ocean resources be increasingly important to humans in the future?

19. What can governments do to protect the oceans from pollution?

20. Most of the fresh water on the earth comes from the oceans without mechanical desalination. Explain this statement.

21. The salinity of the oceans varies in different locations. What processes might cause these differences?

22. The photographs show an underwater hotel in the Florida Keys, south of Miami. Here people vacation in the quiet solitude of the coral reef. Although this hotel is rather unusual now, it might one day be much more common. How else do you think the oceans might be used in the future?

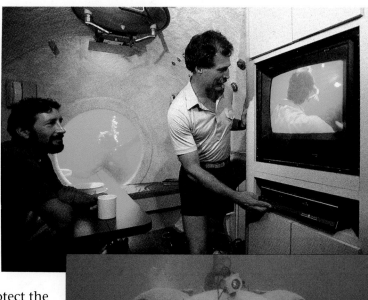

Discovery Through Reading

Bavendam, Fred. "Eye to Eye with the Giant Octopus." *National Geographic* 3 (March 1991): 86-97. The giant octopus once inspired tall tales of a monster from the deep. This article describes the true nature of this "beast."

CHAPTER 17

EXPLORING THE OCEANS

*I*t is early morning. I am alone on the deck of the research ship Calypso. The sea is flat and grayish blue. The air is hot and sultry. There is hardly any breeze. Overhead, heavy dark clouds are piling up into billowing pyramids. We are moving through a belt of water slightly north of the equator known as the doldrums. In the days of wind-driven sailing ships, mariners feared these waters. The doldrums is the region where the trade winds never blow. It is where ships used to becalm and where supplies would run out. But I hear the steady throbbing of Calypso's engines. I know this ship will pass through the doldrums without mishap.

*W*hat is it about the sea, I wonder as I stare into its depths, that has always lured humans on to explore its secrets? I think of the great voyages of the past—Ericson, Magellan, Vasco da Gama, Francis Drake. Their daring navigation told us much about the size and shape of the oceans of the world. Today we know the configuration of the surface of the sea with an accuracy they could not have dreamed of.

It is the underwater world we now explore. We are looking there for clues to our own distant past, for information about the formation of our planet and the origin of life. *Calypso* is crammed with equipment to aid us in that search. Our underwater observation chamber enables us to watch

and photograph marine creatures wherever we venture. Divers going down to depths of 200 feet for a personal look at the ocean's habitats enter and leave the water through our midships' diving-well. Our echo sounders help us map the ocean floor. Other research vessels have drills that bore into the earth's crust and bring up cylindrical samples of that ancient material. We are like detectives trying to unravel what happened millions and billions of years ago.

Somewhere on the ocean floor are bits and pieces telling the story of this planet's beginning. It is in the oceans, I believe, rather than in the skies that we will finally learn how earth and life formed. And the sea, which gave birth to us all, may shelter us once again. Looking down at the water I can imagine human-made islands of the future floating offshore and other settlements rising from the ocean floor. Before humans leave earth for another home I am sure they will retrace their steps through the sea.

from *The Ocean World of Jacques Cousteau*
by Jacques Cousteau

For Your Journal

- 🖊 *How have humans explored the oceans in the past?*
- 🖊 *What have we recently found out about the oceans?*
- 🖊 *What will the future of ocean exploration bring?*

Technology for Ocean Exploration

Objectives

Describe the technology scientists use to explore the oceans.

Determine the advantages and disadvantages of various submersibles.

Evaluate the need for teamwork in ocean exploration projects.

Y̶our first assignment as a reporter was a great success! The story you wrote on ocean resources was right on the money. Your editor loved it. In fact, she loved it so much she has given you a follow-up assignment. You already know quite a bit about ocean resources. Now it's time to find out how scientists learned about these resources. Your new assignment is the exploration of the oceans.

Submersibles

You fly from Houston to Melbourne, Florida, a short drive from Harbor Branch Oceanographic Institute in Fort Pierce. You think this might be a good place to start. Harbor Branch houses about 200 scientists, engineers, technicians, mariners, and support people. The institute's goal is to understand and protect the oceans and coastal regions. Harbor Branch is also the home of the Johnson-Sea-Link submersibles. **Submersibles** are vehicles that operate underwater.

Figure 17–1. A Johnson-Sea-Link and its support ship

You are greeted by a marine biologist who takes you to the Education and Conference Center. In the lobby she points to a large photograph of a Johnson-Sea-Link. You can see from the photograph how submersibles are launched from a support ship. One of the Johnson-Sea-Links is on assignment with its support ship in the Mediterranean Sea. But you can get a good idea of how submersibles operate by climbing into the other craft, which is currently at the institute for refurbishing.

The pilot and an observer sit in the big acrylic sphere. This "bubble" allows you to see in nearly every direction. It's almost like being in the water yourself. Your guide tells you that the Johnson-Sea-Links are categorized as midwater submersibles; they can go as deep as 900 m. You wonder how strong the walls are in that bubble. You feel better when she tells you that the acrylic in the sphere is more than 10 cm thick.

The Johnson-Sea-Link moves underwater by using nine electric motors. These "thrusters" get their power from 14 batteries. There is enough power in the batteries for scientists to make several three- to four-hour dives in a day.

One of the problems with many research submersibles is that the scientists can only look and take pictures. The Johnson-Sea-Link, however, is more versatile. It is equipped with manipulator arms, suction samplers, coring devices, and other equipment that can do just about everything that a human could. Using its powerful lights, the Johnson-Sea-Link can take both videotapes and still photographs.

 ASK YOURSELF

Why are submersibles like the Johnson-Sea-Link important to scientists?

Figure 17–2. Scuba allows nearly unlimited freedom but is restricted to shallow water.

Diving

After you've had a chance to examine the Johnson-Sea-Link thoroughly, your guide shows you the pool where divers check out new equipment. You ask her some questions about scuba diving. A certified diver yourself, your last dive was from the oil rig in the Gulf of Mexico. You had expected quite a lot of the exploration here at Harbor Branch to be done with scuba gear. She explains that scuba divers do most of the shallow-water exploration.

After talking with your guide about scuba, you realize that you have taken for granted all the conveniences that scuba offers. It has been around for only 50 years! You seem to recall that French ocean explorer Jacques Cousteau was one of the inventors of scuba. So you decide to do a little more research on scuba and Cousteau, and you find this quote, at left.

DISCOVER BY Researching

Use the resources of your school or public library to learn more about the development of scuba and other oceanographic work done by Jacques Cousteau. Record your findings in your journal, and discuss them with your classmates. ✐

The word *scuba* comes from the first letters of the words *s*elf-*c*ontained *u*nderwater *b*reathing *a*pparatus. Self-contained means that divers carry their air with them. Their air supply is compressed into aluminum or steel tanks they carry on their backs. The air is fed to the divers as they need it, but it is a limited supply. Scuba gives scientists a great deal of freedom, but it is limited to about 50 m. Below that, the pressure of the water causes problems.

Figure 17–3. This may be one of the earliest recorded explorations of the ocean.

While doing your research on Cousteau, you read about some early attempts at underwater exploration. Alexander the Great may have been the first underwater explorer. According to legend, in 322 B.C. he spent three days underwater in a glass barrel. You even found a picture depicting the event. What the picture doesn't show is how he managed to breathe underwater. The following activity will give you a clue.

DISCOVER BY Doing

Fill a pail nearly to the top with water. Then hold a clear plastic cup upside down, and push it into the water. Be sure to hold the cup straight. What do you observe? How might this explain how Alexander the Great was able to breathe underwater? ✐

The principle demonstrated in the activity was used for exploring the oceans during the 1800s. **Diving bells,** like Alexander the Great's and the one shown here, are chambers that allow people to work underwater for many hours. Air trapped in the diving bell keeps water out. However, explorers could only work inside the bell, where there was air.

As these early inventions show, Cousteau was certainly not the first person to think about breathing underwater. To breathe and move about underwater had long been a dream of many people. The invention of the diving suit brought them closer to their dreams. One of the first drawings of a diving suit, shown at right, would not have worked. Why not? Try the following activity to see for yourself.

Figure 17–4. A diving bell (left) and a 15th century drawing of a diving suit (above)

ACTIVITY

Why can't you snorkel in deep water?

MATERIALS
tall pail or plastic garbage can, water, rubber tube about 2 m long and 0.5 cm in diameter

PROCEDURE
1. Fill the container with water to 1.5 m in depth.
2. Place one end of the tube just below the surface of the water, and blow a few bubbles. Note how hard you have to blow to make bubbles.
3. Place the end of the tube 10 cm farther down into the water, and blow a few more bubbles. Note any difference in how hard you have to blow this time compared with last time. Record your observations in your journal.
4. Repeat step 3, putting the tube 20 cm into the water. Keep repeating step 3 until you reach the bottom of the container. Record your observations each time in your journal.

APPLICATION
1. Where is it easier to blow bubbles, near the surface or near the bottom? Explain your answer.
2. How is the work you have to do to blow bubbles similar to the work you would have to do to breathe through a tube of the same length?
3. Could you blow bubbles out of a tube that extended 10 m below the surface? Explain your answer.
4. Is it possible to breathe out of a 10 m-long tube? Why or why not?

Have you ever tried using a snorkel? A *snorkel* is a tube that goes from a diver's mouth to the surface. Most snorkels are about 40 cm long. Snorkels allow divers to float face down in the water without having to raise their heads to breathe. But why not make snorkels longer? In fact, why not use a garden hose and make a really long snorkel!

This idea is similar to what you tried in the activity. What happens if the snorkel is too long? In order to solve the problem, inventors discovered that they had to pump air under pressure to the diver. So, about 150 years ago a diving suit was designed with a hose connected to an air pump. Air pressure within the hose and helmet was greater than the water pressure outside the hose and helmet. Divers using these types of suits are often called *hard-hat divers*. This is because the helmets are usually made of hard metal or fiberglass.

A hard-hat diver can work at depths of several hundred meters, much deeper than scuba divers. But using hard-hat equipment requires much more training than using scuba. Hard-hat divers also require a surface pump to supply them with air. Look at the picture of the hard-hat diver. What problems might a diver have in this kind of equipment?

As you return to the harbor to continue your discussion with the institute's marine biologist, you pass several other scientists loading scuba gear into a boat. Diving provides a fairly easy way for scientists to get into the water and to the shallow reefs they often study. But when the water is too deep for scuba or even hard-hat divers, they sign up to use a Johnson- Sea-Link.

Figure 17–5. Why are these divers called hard-hats?

 ASK YOURSELF

Why are scuba divers limited as to the amount of work they can do underwater?

The Rest of the Team

One of the things you soon notice during your visit at Harbor Branch is that ocean research depends on a lot of people. One of the senior research scientists sums it up nicely: "Johnson-Sea-Link submersibles, developed for the most part right here at Harbor Branch, enable us to do things no one else can do. This is the result of unique cooperation among our scientists, engineers, machinists, and ship and sub crews. **Oceanography,** the study of the oceans, has always been a team effort. You just can't do it by yourself."

As you jot down his comments—they will make a good quote for your article—you realize that many different kinds of people are involved in the ocean research they are conducting here. During your day at Harbor Branch, you meet fish biologists, physical oceanographers, electrical and electronics engineers, marine chemists, ship's captains, ship's mates, ship's cooks, marine botanists, submersible pilots and technicians, museum curators, librarians, aquaculture scientists, and photographers. Each depends on the others to get his or her own job done and to keep the complex projects functioning. You make a note to yourself to be sure to let your readers know about the importance of teamwork.

As you walk around, you hear many of the scientists talking about Alvin. Who is Alvin? An oceanographer shows you a picture. *Alvin* is a submersible. It was built for the United States Navy in 1964, and is now operated by the Woods Hole Oceanographic Institute in Massachusetts. *Alvin* is a crewed submersible, made of titanium, one of the lightest and strongest metals known. Because of its strength, *Alvin* can carry a pair of scientists to a depth of about 4000 m. *Alvin*'s equipment includes measuring devices, cameras, and a pair of mechanical arms that can be used for collecting samples from the ocean floor.

Figure 17–6. Meet *Alvin*.

Ocean explorers have used dozens of different types of submersibles over the last 40 years. One of the earliest is also the one that has explored the deepest. Look at the picture of *Trieste II*. It has explored the deepest parts of the oceans. Do you remember where the deepest ocean trench is found? If you don't, look again at Table 14-1, on page 430.

Figure 17–7. *Trieste,* the deepest diver of them all.

Thirty-four thousand feet—no bottom . . . 35,000 feet, only water and more water . . . 36,000 feet, descending smoothly at sixty feet per minute. Now we were at the supposed depth of the Challenger Deep. Had we found a new hole or was our depth gauge in error? Then a wry thought—perhaps we'd missed the bottom!

from *Seven Miles Down*

Talking with scientists about their research makes you realize that the real problem in oceanography is getting deep enough. The deeper you go, the greater the pressure. One way around this problem is to design robots that can act as scientists' eyes and hands and operate underwater as freely as scuba divers. Robots are much less expensive to operate than submersibles such as *Trieste* and *Alvin* because they do not have to be designed to protect fragile people.

Your Harbor Branch guide takes you to a building where ROV SCOOP is being serviced. ROV SCOOP stands for Remotely Operated Vehicle and Sample Collecting and Oceanographic Observation Platform. It is a robot submersible. ROV SCOOP can be controlled safely from the deck of a ship; and it can explore where people cannot.

 ASK YOURSELF

When would you want to use a robot submersible instead of one with people in it?

SECTION 1 *REVIEW AND APPLICATION*

Reading Critically
1. What are the advantages of using robot submersibles?
2. What is the depth limit for scuba diving?

Thinking Critically
3. Why is teamwork important in exploring the oceans?
4. Why can't you breathe through a snorkel 30 m underwater?

SKILL Making Observations from Photographs

ANGUS

JASON

1. Observe the photograph of the robot submersible *ANGUS*. *ANGUS* has tools and instruments just like those of a crewed submersible.

2. You may already know that robot spacecraft use radio waves to send and receive information. But radio waves do not travel through water. How do you think a robot submersible and its equipment are controlled? Write some of your ideas in your journal.

3. *ANGUS* is connected to the command ship by a long cable called an *umbilical*. Based on this information, what can you infer about the function of the umbilical?

4. Now observe the photograph of the robot submersible *JASON*. What are the similarities and differences between *ANGUS* and *JASON*? Record your observations in your journal.

▶ **APPLICATION**

1. Are observations made from photographs always correct? Explain your answer in your journal.

2. What could you do to try to find out whether your observations are correct?

✳ ***Using What You Have Learned***
Are there some things that you infer that you cannot prove? Record your ideas in your journal, and discuss them with others in your class.

Aquanauts

Describe *some of the ways that sub-mersibles help make deep-ocean discoveries.*

Give examples *of future ocean exploration projects.*

Evaluate *the need for ocean explor-ation.*

Y ou had pinned this picture and poem to your bulletin board as you prepared for your trip to Harbor Branch. How does the poet's description of the ocean floor compare with what you think you would see if you could travel there? You might want to make your own drawing or write a description in your journal of what you think the ocean floor is like.

Children dear, was it yesterday
We heard the sweet bells over the bay?
In the caverns where we lay,
Through the surf and over the swell,
The far-off sounds of a silver bell?
Sand-strewn caverns, cool and deep,
Where the winds are all asleep;
Where the spent lights quiver and gleam,
Where the salt weed sways in the stream,
Where the sea-beasts, ranged all round,
Feed in the ooze of their pasture-ground;
Where the sea-snakes coil and twine,
Dry their mail and bask in the brine;
Where great whales come sailing by,
Sail and sail, with unshut eye,
Round the world forever and aye?
When did music come this way?
Children dear, was it yesterday?

From "The Forsaken Merman"
by Matthew Arnold

Robert Ballard

The idea of people working and living underwater has always been fascinating. As a science reporter, you have written about astronauts exploring space. But for this article, you need to find out about **aquanauts,** people who explore the sea. You already know something about Jacques Cousteau, but as you do your re-search, another name keeps coming up: Robert Ballard.

In 1977 Ballard and a team of scientists used *Alvin* to explore the floor of the Pacific Ocean near the Galapagos Islands, off the western coast of South America. Diving 2500 m into near-freezing water, Ballard watched in amazement as *Alvin's* temperature probe reached 17° C. Hot water at the bottom of the ocean? He had discovered hydrothermal vents in the ocean floor.

Figure 17–8. This underwater spring, or vent, is near the Galapagos Islands, off the coast of Ecuador.

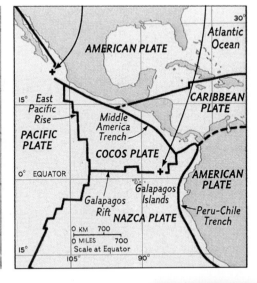

Even more astonishing than the warm water were the unusual life forms living in the dark environment near the vents. The scientists found 30 cm-long clams and giant, blood-red tube worms that had never been seen before. These organisms are more amazing than the "beasts" in the poem! The worms were more than 3.5 m in length. Hold your hands as far apart as you can. If you are of average size, your reach is probably less than 2 m.

Ballard's team also discovered masses of bacteria living near the hydrothermal vents. These bacteria make food from chemicals in the warm vent water, a process called *chemosynthesis*. You know that plants make food by photosynthesis. But plants can't grow in this world without sun. Sunlight doesn't penetrate water this deep. The chemosynthetic bacteria are the base of a food pyramid that includes the worms and all the other organisms living there.

Ballard and *Alvin* were also involved in locating and exploring the sunken ocean liner *Titanic*. You can read more about this adventure in the Science Parade that follows this chapter.

Figure 17–9. Animals that live near underwater vents depend on food produced by bacteria.

ASK YOURSELF

Why was *Alvin* used to explore the vents off the Galapagos Islands?

The Future of Ocean Exploration

You know that people have learned to live for extended periods in space. Can they also learn to live on the ocean floor? Your scientist guide explains to you that there are already examples of underwater habitats for humans. The picture shows a sea-floor home where divers can live and work for a long time. The home is called *Tektite*. There are other undersea habitats that have been built in addition to Tektite. In the next activity, you can find out more about them.

Figure 17–10. Much can be learned by living and working underwater for extended periods.

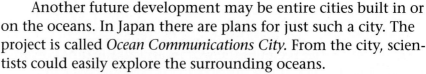

Researching

Use your school or community library to find material on undersea habitats. Select one of the underwater homes, and write about it in your journal. Compare information about your selection with that of your classmates. ✎

Another future development may be entire cities built in or on the oceans. In Japan there are plans for just such a city. The project is called *Ocean Communications City*. From the city, scientists could easily explore the surrounding oceans.

How else will future scientists explore the oceans? Remember, robot submersibles are often used to explore the more dangerous parts of the ocean. But these submersibles must remain attached to a command ship. Robot submersibles of the future may be computerized, so they can work without an umbilical. Such submersibles are under development right now. The one at left is called *Sea Otter*.

Figure 17–11. Robot submersibles of the future may not be tethered to a command ship.

Just as Alan Shepard and Yuri Gagarin were pioneers in the exploration of space, Robert Ballard and Jacques Cousteau were pioneers in the exploration of the oceans. Many others are sure to follow. But why, you wonder, is it important for humans to explore the oceans? Your guide reads you something written by Cousteau:

We must change our attitudes toward the ocean. We must regard it as no longer a mystery, a menace, something so vast and invulnerable that we need not concern ourselves with it, a dark and sinister abode of secrets and wonders. Nor do we want to follow the methods of the first scientists who sailed the seas to compile lists: lists of mammals, lists of seabirds, of jellyfish, of temperatures, of currents, of migratory patterns. Instead we want to explore the themes of the ocean's existence— how it moves and breathes, how it experiences dramas and seasons, how it nourishes its hosts of living things, how it harmonizes the physical and biological rhythms of the whole earth, what hurts it and what feeds it—not least of all, what are its stories.

 ASK YOURSELF

How will the future of ocean exploration be different from the present?

SECTION 2 *REVIEW AND APPLICATION*

Reading Critically

1. Why could the Johnson-Sea-Link not be used to explore the ocean vents that Ballard discovered?
2. Describe several ocean exploration projects.

Thinking Critically

3. How might Ballard's discoveries change what people think about the oceans?
4. Why did Jacques Cousteau say that we should not just be making lists of things when we explore the oceans?

INVESTIGATION

Mapping the Ocean Floor

▶ MATERIALS
- gravel ● pebbles ● small aquarium or glass bowl ● water ● graph paper
- ruler

▼ PROCEDURE

1. Arrange the gravel and pebbles in the aquarium to make a model of the ocean floor. Make your ocean floor uneven. Include some flat areas and some hills.
2. Fill the aquarium with water.
3. Label the left side of a sheet of graph paper "depth." Make marks at regular intervals to represent the depth (in centimeters) of your "ocean." Label the bottom of the graph paper "distance."

Make marks at regular intervals to represent the distance (in centimeters) across your "ocean."
4. Make depth measurements of your ocean. Start at one end of the aquarium. Use the ruler to measure the depth of the water.

5. Make an "x" on your graph paper above the first distance mark and across from the number that shows the depth of the water.
6. Repeat steps 5 and 6 from one end of the aquarium to the other.
7. When you finish your measurements, draw a line connecting the marks on your graph. This is your "ocean-floor" map.

▶ ANALYSES AND CONCLUSIONS
1. How does your map compare with your "ocean floor"? How could you improve your map?
2. Why is this technique impractical for mapping large areas of the oceans?

▶ APPLICATION
Why is a map of the ocean floor valuable? How might scientists use a map of the ocean floor?

✳ Discover More
Repeat the Investigation using a weighted string rather than a ruler. Does this affect your measurements? Why? Which procedure provides more accurate data? How might scientists use this procedure?

The Big Idea

Scientists use many different technologies to explore the oceans. Some involve using equipment that take the explorers themselves into the depths. Others use robots that send information back to the surface. All the technologies allow scientists to make observations, gather information, and make inferences about the oceans. As we explore, we must change our attitudes toward the oceans, and learn their stories.

For Your Journal

Refer back to your answers about ocean exploration. Based on what you have learned from the chapter, write down what you would change to make your answers more accurate.

Connecting Ideas

Copy the concept map into your journal, and fill in the missing pieces.

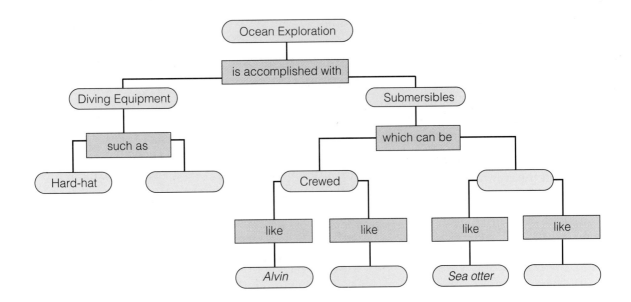

Understanding Vocabulary

1. Explain how these terms are related: oceanography (500), aquanauts (504), scuba (498), submersibles (496)

2. Define the word chemosynthesis (505).

Understanding Concepts

MULTIPLE CHOICE

3. Which of the following is *not* a reason why oceanography is often carried out by a team of many different kinds of scientists and professionals?
   a) Oceanographers tend to be competitive and cannot work together.
   b) Oceanography is a very complex science.
   c) Most oceanographers depend on other branches of science for information.
   d) Oceanographers might not have all the knowledge and experience to complete a research project without assistance from other scientists.

4. Scuba diving is usually limited to a depth of 50 m. Below 50 m,
   a) a diver's air supply runs out.
   b) water pressure causes problems for a diver.
   c) a diver becomes too cold.
   d) a diver cannot communicate with the crew on the surface.

5. The only submersible to reach the deepest part of the ocean was
   a) *ANGUS*
   b) Johnson-Sea-Link
   c) *Trieste II*
   d) *Alvin*

SHORT ANSWER

6. How does breathing in a hard-hat diving suit differ from breathing with scuba gear?

Interpreting Graphics

Use the diagram below to answer these questions.

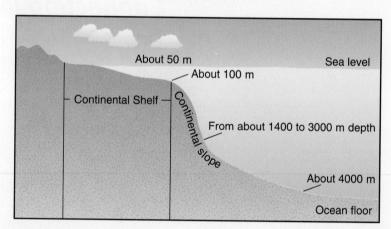

7. Holt Oceanographic Institute wants you to study corals along the Continental Shelf. Which of these methods—scuba, *Alvin*, Johnson-Sea-Link, or *ANGUS*,—would you use to reach your site? Explain your choice.

8. Suppose an oil company plans to explore the ocean floor. Which technology would you recommend? Explain why other methods wouldn't work.

Reviewing Themes

9. *Systems and Structures*
Why do you think the manufacturers of *Alvin* used titanium—one of the strongest and lightest of metals—in constructing its hull?

10. *Technology*
One recent advancement in ocean exploration is the use of computerized robot submersibles. Explain how these vehicles represent an improvement over older robot submersibles.

Thinking Critically

11. In what ways do you think future ocean exploration will differ from present operations?

12. Suppose you are on the planning committee at the Holt Oceanographic Institute. You have only a limited budget, but the institute wants you to carry out several deep-sea research projects. Based on what you have learned in this chapter, what one submersible would you choose? Explain your answer.

13. In spite of new technology, many oceanographers feel that research ships are still the most important instruments of study. Explain this position.

14. Since hydrothermal vents were found along the East Pacific Rise, near the Galapagos Islands, scientists have searched for vents in other areas. Why is there much interest in these vents from an earth scientist's point of view?

15. Some people think that expensive oceanographic research should be curtailed. Explain why oceanographic research is so important.

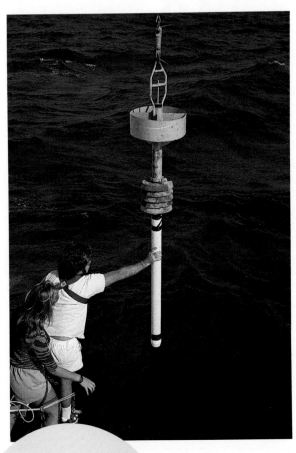

Discovery Through Reading

"Oceanography: The Undiscovered Country." *Discover* 3 (April 1992): 12. Research vessels have mapped about five percent of the world's oceans. In contrast, the navy satellite *Geosat* recently made radar measurements 1.6–3.2 km apart over most of the earth. Read this article to find out what the satellite discovered.

Science PARADE

Power from the Tides

*T*he dream of harnessing tidal power is a reality. On the Rance River near St. Malo, France, and near the city of Murmansk in Russia, tidal plants have been generating electricity from the oceans' tides since the late 1960s. In Canada, engineers are studying what would be the world's largest tidal power project.

France's Rance River tidal energy plant (above) and the energy turbine of a tidal plant (right)

Supertide

The Canadian study involves the Bay of Fundy. You may recall that the gravity of the sun and the moon cause the surface of the oceans to rise and fall. A supertide occurs where the tidal flow enters a long, narrow, funnel-shaped inlet. The tidal current sloshes up the neck of the funnel and back. If the bay is the right shape, the tidewater arrives back in the entrance just in time to get a push from the next high tide. The current is thus intensified, resulting in higher high tides and lower low tides. This is what happens in the Bay of Fundy. Tides here can rise and fall more than 15 meters.

A tidal power plant in the Bay of Fundy would take advantage of these supertides. The project would also surpass in magnitude any existing tidal power plants.

A small pilot project built across the Annapolis River is already in operation. It generates 20 MW of power and has provided scientists with preliminary data on some of the impact the larger project might have.

Environmental Concerns

The Bay of Fundy tidal project is an attempt to harness one of Earth's greatest untapped resources. Yet, it raises questions about how to balance the need for electrical power with the need to protect the natural environment.

Tidal energy plants create clean electric energy without consuming fossil fuels. There is no air pollution and there are no safety and waste disposal problems. However, tidal energy does have other problems. Impact studies at the Annapolis River pilot plant have indicated a possible danger to fish populations. The turbine used in the plant may harm the fish that pass through it. Further study is needed to ascertain the overall effect this might have on the fish population.

Tidal plants can work when the tide is rising, falling, or both. The Fundy plant would work only during the falling, or ebb, tide. This may not be the same time the demand for electricity is highest. Few people need to take a hot shower or bake a cake at 3 A.M. Also, tidal plants may be located hundreds of miles from the cities that use the most electricity.

Scientists also warn that the Bay of Fundy plant might change the very features that make the supertides possible. The bay's tidal current naturally regulates the settling of river sediments. A dam might alter this distribution. Instead of being continually flushed from the bay, these sediments could settle and fill the bay over time.

A final concern is cost. The Bay of Fundy project would require a dam about 15 stories high and would contain 42 turbines. It would cost over $2.6 billion. This cost, along with the other concerns, has warranted further study of the possible impact of such a project. ◆

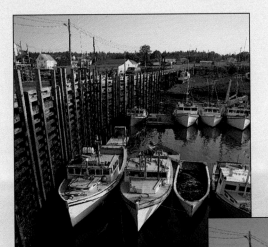

Tidal extremes in the Bay of Fundy

How it Works

The engineering principles of tidal power have been known for many years. Early tidal mills trapped the inflowing tide behind a dam, then allowed the water to flow out over a water wheel. Today's tidal plants have turbines that are turned by the tidal currents.

The Rance tidal plant has reversible turbines, so it operates on both incoming and outgoing tides. The Bay of Fundy turbines would turn only when the tide is flowing out. The total flow through the turbines would be at a rate of more than 13 000 m³/s, about three times the flow rate of Canada's largest river, the St. Lawrence. The water's speed would be controlled by a set of movable vanes, called *wicket gates*, which would open and close like Venetian blinds. The turbines would spin at 50 revolutions per minute.

There are a number of sites being considered for construction of the plant on the Bay of Fundy. Two of these sites are promising. One is in the Cumberland Basin. The second is across the Cobequid Bay inlet. Although both sites are feasible, the Cumberland Basin site is more likely. A plant there could generate an average of 1400 megawatts (MW) of power. By comparison, the nuclear plant at Indian Point, New York, produces only 873 MW.

Finding the Titanic

BY JONN C. HALTER FROM *BOYS' LIFE*

A SHIP OF DESTINY

On April 14, 1912, the pride of Britain's White Star Line was on her maiden voyage from Southampton, England, to New York. She was the most luxurious ship afloat.

And the biggest. The 46,000-ton *Titanic* was four blocks long and eleven stories high. When launched, she was the largest object ever moved.

Newspapers called her "the Wonder Ship," a crowning achievement of a century of rapid technological development. People looked at a floating palace like the *Titanic* and felt that, now, nothing was impossible.

She was the most glamorous of three sister vessels White Star hoped would dominate the profitable North Atlantic market. The *Olympic* was already in service; the *Britannic* was under construction.

The *Titanic's* crew of more than 900 promised a pleasant voyage. First-class passengers enjoyed an indoor swimming pool (the first on a passenger liner), a gymnasium, squash court, sauna, and library. They could eat in a 500-seat dining salon, a smaller private restaurant, or a French sidewalk cafe.

Even third-class passengers, used to cattle-car conditions in earlier ships, had four-bunk cabins, a dining hall, lounge, and promenade deck.

The 325 first-class passengers occupied 60 percent of the ship. Some were society celebrities; others were noted millionaires. Many brought personal servants. A one-way deluxe first-class suite cost $1,300—about $13,000 in today's money.

Several decks below the first-class accommodations, the 700 third-class passengers were confined to the ship's forward and rear areas. They had paid just $32 for their one-way tickets.

R.M.S. *Titanic*

THE "UNSINKABLE" SHIP

The *Titanic* had everything—except enough lifeboats.

The ship didn't need them, her designers said, because the *Titanic* was "practically un-sinkable." Her hull had a double bottom and 16 "watertight" compartments.

Her captain, the 62-year-old White Star veteran Edward J. Smith, had confidence in the new technology. "I cannot imagine any condition which would cause a ship to founder," he had said a few years earlier. "Modern shipbuilding has gone beyond that."

British regulations, last updated in 1898, said ships of more than 10,000 tons only needed lifeboats for 980 persons. No problem there. The *Titanic's* 20 boats would hold 1,178, almost 200 more than required.

But the ship held no lifeboat drills for passengers or crew. And most of the 1,320 passengers weren't aware she didn't have boats for everyone.

Four days out of Southhampton, however, the *Titanic* had experienced no major problems.

WARNING: ICE HAZARDS

An ice field lay ahead. Other ships had radioed alarms, and Captain Smith alerted his crew to keep a sharp eye out.

But he didn't slow the ship. No ship reduced speed until ice was sighted. Why should the *Titanic?*

The sun set, and the stars of a clear, moonless night came out. The outside temperature dropped steadily, to 43 degrees by 7 P.M., and passengers retreated inside.

By 11 P.M. the temperature was near freezing. In the radio room, the chief operator struggled to send out a pile of passenger ship-to-shore messages. In his haste, he brushed aside an interruption. The *Californian,* some 19 miles to the northwest, was reporting that ice had forced her to stop for the night.

Forty minutes later, lookout Fleet saw an iceberg looming ahead.

At his warning, the *Titanic* reversed engines and turned sharply to port.

Titanic's final moments

But not in time.

She scraped the giant iceberg, which stood as high as the top deck.

Below the surface, the rock-solid ice popped rivets and opened metal plate seams. Water poured into the first five compartments along a narrow, 300-foot long area.

The ship slowed to a stop, and her engines shut down. Crew members reassured passengers that nothing was wrong.

A DEADLY WOUND

The damage, however, was fatal.

The ship could float with her first four compartments flooded. But now water was filling the first five. As this weight pulled the bow down, water would soon spill into the remaining compartments. The ship would sink within hours.

The *Titanic* began sending the traditional CQD distress signal, as well as a brand new one —SOS. But on the *Californian,* just 20 miles away, the radio operator and captain had gone to bed. And her crew didn't understand why the *Titanic* was firing rockets.

Some 58 miles to the south, the 13,564-ton *Carpathia* picked up the SOS. She turned north, but at her 17-knot top speed, she would take four hours to reach the *Titanic.*

Captain Smith told the crew to begin loading the lifeboats. But—to avoid panic—he made no general announcement. The first boat—capacity 65—went into the water at 12:45 A.M. carrying only 28 persons. Passengers wouldn't leave the huge, brightly lit ship. They saw more danger in being lowered 60 feet in a wooden boat into the cold, dark sea.

And even though passengers became less hesitant as the ship's list increased, the inexperienced crew lowered many boats only half-full. Just seven carried more than 50 passengers and only three were filled to capacity.

Many third-class passengers never had a chance at a lifeboat. The crew restrained them below deck during the early loading. Then, as the boats dwindled, more passengers surged to the boat deck. Panic threatened the loading, and an officer held them back with a revolver.

DRAMA ON COLLAPSIBLES A AND B

By 2:05 A.M. only collapsible rafts A and B remained. Assistant radio operator Harold Bride joined some crewmen struggling to release the two boats.

Too late. Water swept the boat and men overboard. Collapsible A landed upright but half swamped. Collapsible B landed upside down near the *Titanic's* raised stern.

Harold Bride came to the surface underneath the overturned boat. More than three dozen other desperate swimmers struggled aboard or clung to the two collapsibles.

At 2:20 A.M. the *Titanic's* stern rose into the air. Then the darkened hull slid beneath the surface. The great ship was gone.

Two hours later, in the gray light of early dawn, 705 survivors in the lifeboats saw the *Carpathia* steam into view. A woman in the nearest lifeboat screamed the dreaded news: "The *Titanic* has gone down with everyone aboard!"

THEY FOUND THE TITANIC

Summer 1973: For decades the *Titanic* has rested two and a half miles down on the bottom of the North Atlantic. But with the passing years, her growing legend has captured the imagination of many people.

One who falls under her spell is professional oceanographer and former Life Scout Dr. Robert Ballard. He has long dreamed of finding the lost ship. Now, he begins organizing an expedition to find it.

August 1985: Dr. Ballard's team of scientists from the Deep Submergence Laboratory at the Woods Hole Oceanographic Institution in Massachusetts begins its search. The scientists join a French group already searching the ocean floor. The French use a powerful sonar they've developed. But its electronic sound waves bouncing off the ocean floor find nothing near the reported site of the sinking.

The searchers switch to video cameras on the Argo, a steel sled they tow by cable just above the ocean floor. They look for the thousands of small objects that fell from the *Titanic* as she sank.

Aug. 31, 1985: Only five days left before the expedition must return. Dr. Ballard feels "a rising panic." He begins to face "total defeat." Day after day, around the clock, the team sees "mud and more mud, endless miles of nothing" on the video monitors.

Just before 2 A.M.: "There's something," one of the crew says. On the screen, bits of wreckage stream past.

Deepsea submersible *Alvin*

And then a massive round object. It's one of the *Titanic's* giant boilers. The gigantic hull lies nearby.

Crew members cheer, shake hands, and pound each other on the back.

They've found the *Titanic!*

Despite rough seas, they carefully scan the hull with Argo and another sled, ANGUS, that takes still pictures. These reveal that the *Titanic's* stern is missing. So are her four enormous smokestacks.

Sept. 9, 1985: A cheering crowd welcomes the team back to Woods Hole. Once again the *Titanic* is front-page news around the world.

A CLOSER LOOK

July, 1986: The scientists return to visit the wreck. Crammed into their tiny titanium submarine, *Alvin,* "like three sardines in a spherical can," Dr. Ballard and two others descend into the freezing blackness. They listen to classical music during the two-and-a-half-hour dive to the bottom.

On the *Titanic,* they discover that deepsea worms have eaten away most of the ship's wood. They are surprised to find huge, bacteria-formed, powdery rust formations all over the hull. To Dr. Ballard they look like long icicles. He gives them the name "rusticles."

Their remote camera, a "swimming eyeball" called Jason Jr., or JJ, peers into the first-class cabins on the boat deck. It slithers down the grand stairway into the ship, where it photographs a light fixture still hanging from the ceiling.

Alvin also explores the 2,000 feet of debris between the bow and stern sections of the ship. Thousands of perfectly preserved objects lie on the ocean floor. The explorers see a rusty metal drinking cup sitting on top of a giant boiler.

The ship's stern section is badly damaged. Its huge propellers are buried in 45 feet of mud.

Before *Alvin* leaves for the last time, the scientists place a plaque on the *Titanic's* stern, in memory of the more than 1,500 persons who had died when she sank.

SOME ANSWERS AND SOME NEW QUESTIONS

The expeditions answered many questions about the *Titanic.*

They confirmed that the ship had broken in half while sinking (as many survivors had reported). They concluded that the iceberg had probably popped open steel plates rather than ripped a lengthy gash in the ship's hull (as long believed).

The scientists discovered that deep-sea organisms had removed most organic material, including wood, food, clothing, and human remains. And the ship's metal parts are slowly rusting away too.

Living survivors Ruth Blanchard and Eva Hart, two of the last people to see the *Titanic,* and Dr. Ballard, the first to see her again, all hope the world will leave the ship in peace, a memorial to those who died with her. They were saddened when a French expedition returned in 1987 to retrieve objects from the debris field.

We can still learn from the *Titanic* disaster, Dr. Ballard believes.

"Too much confidence was placed in technology," he says, "and the power of nature was overlooked."

THE END OF AN ERA

The *Titanic* disaster rocked society's belief in relentless (and often unplanned) progress. The ship was supposed to be a triumphant symbol of technology, but less than three hours after hitting the iceberg she was gone.

In our own time, the explosion of the space shuttle *Challenger* in 1986 had similar impact. A sudden, unexpected disaster exposed the flaws in our vaunted technology. Such events in history remind us that the price of progress is often tragically high. ◆

Matthew Fontaine Maury
(1806—1873)

Matthew Fontaine Maury has been called the "pathfinder of the seas." He contributed much to the science of oceanography. Following his voyages and the publication of many books, he organized the first international congress of oceanographers.

Maury was born near Fredericksburg, Virginia. When he was 12, Maury entered Harpeth Academy in the frontier town of Franklin, Tennessee. In 1825 he graduated from Harpeth Academy and became an officer in the United States Navy.

While in the navy, Maury helped develop the Naval Observatory and the Hydrographic Office. Maury was also in charge of the Navy Department's Depot of Charts and Instruments. Over the years he collected information and drew charts of the winds and ocean currents.

Maury's *Physical Geography of the Sea and Its Meteorology* was the first modern textbook on oceanography. His books contained information about all of the oceans of the world, including charts and maps of winds and ocean currents. Using Maury's charts, several shipping companies found that they could reduce their sailing time from New York to Rio de Janeiro by 10 to 15 days. ◆

Jacques-Yves Cousteau
(1910—)

Jacques-Yves Cousteau is best known for his television series, *The Undersea World of Jacques Cousteau*. Cousteau is also known for his books and for his undersea exploration techniques.

Cousteau was born in Sainte-Andre-de-Cubzac, France. As a child he spent much of his time at the beach, where he developed a love for the ocean. Cousteau attended the College Stanislas, in Paris. When he graduated, Cousteau entered the French naval academy at Brest. In 1933 he graduated second in his class and became an officer in the navy.

Cousteau is one of the inventors of the aqualung, a portable breathing device that made possible the Self-Contained Underwater Breathing Apparatus (SCUBA) in wide use today. He also developed a self-propelled underwater craft and a method for underwater filming.

Leaving the navy in 1956, Cousteau turned his full attention to ocean exploration and conservation. The expeditions of his research ship *Calypso* have earned him global recognition. His Conshelf Saturation Dive program has researchers considering the feasibility of permanent underwater communities. ◆

Thomasena Woods, SCIENCE SUPERVISOR

In the quiet of early morning, Thomasena Woods begins her work. She reads over her schedule for the day. It includes several phone calls, a morning planning meeting, two appointments, and an afternoon workshop. Does this sound like a busy day to you? Not to her. Thomasena Woods is science supervisor for the Newport News Public School System in Newport News, Virginia.

Before becoming a science supervisor, Woods was a science teacher. She taught biology, chemistry, and earth science for 16 years. As a science supervisor, Woods no longer teaches students, but is responsible for guiding the science program for grades K-12. A science supervisor works with all the science teachers in a school system. "I interact with all the science teachers and provide them with leadership, guidance, and special training to improve their teaching skills," explains Woods.

One of the ways she does this is by conducting teacher workshops. At the workshops, teachers learn about new ideas and ways to help students with science. This is often done with computers. Woods encourages teachers to use computers in their science classes. The workshops allow teachers to preview a variety of computer software.

Workshops also provide opportunities for teachers to increase their scientific awareness. The teachers do a variety of hands-on science activities. This gives them a chance to learn new and inventive ways of teaching.

When asked about her other responsibilities, Woods replied, "My job includes curriculum development and the selection of instructional materials such as textbooks, films, laboratory equipment, and other resources. I also coordinate summer science programs for students, science fairs, student visits, and tours to science museums and local research facilities." To accomplish these many tasks, Woods spends a great deal of time planning. She comes in contact with educational, public, and community organizations all the time. This requires good communication skills and the ability to work well with other people. To be a science supervisor, you need a science degree with additional training in supervision or administration as well as science-teaching experience. ◆

DISCOVER MORE

For more information about careers in teaching science, write to the

National Science Teachers Association
1742 Connecticut Avenue, N.W.
Washington, DC 20009

TOURIST SUBMERSIBLES

A tourist submersible

Ordinary people are venturing into the depths of the Pacific and Atlantic oceans in submersibles. Scores of tourists are taking advantage of a unique opportunity to discover for themselves the same breathtaking sites seen previously only by oceanographers such as Jacques Cousteau or Robert Ballard.

Recycled Submersibles

In the early 1980s, some oil industry executives made a decision that helped encourage the use of submersibles for tourism.

Rather than use submersibles crewed by humans to survey and do underwater work, the oil companies decided to switch to less expensive robot submersibles. As a result, several submersibles were pulled from service and sidelined in warehouses.

Businesses with an interest in ocean exploration then purchased the idle vehicles and refitted them. The idea was to lease them to scientists at an affordable rate. As it turned out, scientists were not the only ones interested in exploring the oceans. In places like Hawaii, Guam, Grand Cayman, Barbados, and St. Thomas, tourists were willing to pay for an excursion deep under the sea.

Safety First

Tourist submersibles must meet the strict safety requirements of the American Bureau of Shipping (ABS) for underwater systems and vehicles. In addition to the ABS safety requirements, tourist submersibles are also subject to Coast Guard safety regulations and enforcement. The submersibles must maintain continuous contact with a surface vessel when submerged, and they are routinely serviced and checked to ensure passenger safety.

Before each dive, passengers are fully briefed on the submersible's emergency system. For example, *Deep Explorer*, a sub-mersible in Grand Cayman, has five backup surfacing systems, carries scuba regulators for breathing in case of smoke, and carries enough air for five days underwater.

Photo Opportunity

Deep Explorer can take two passengers as deep as 300 meters. The vehicle uses floodlights so that passengers can view the spectacular colors, corals, and sea life present in the dark ocean depths.

Not all tourist submersibles dive as deep as *Deep Explorer*. Some, such as the *Atlantis,* carry up to 46 passengers to depths of 20 to 50 meters. *Atlantis* also makes night dives, on which tourists are awed by the array of night creatures visible under the vessel's floodlights.

Most tourist submersibles have a large, clear dome or portholes that allow passengers to watch and photograph what they see. Now ordinary people can observe the mysterious ocean depths for themselves, and experience the excitement and mystery once reserved for scientists. ◆

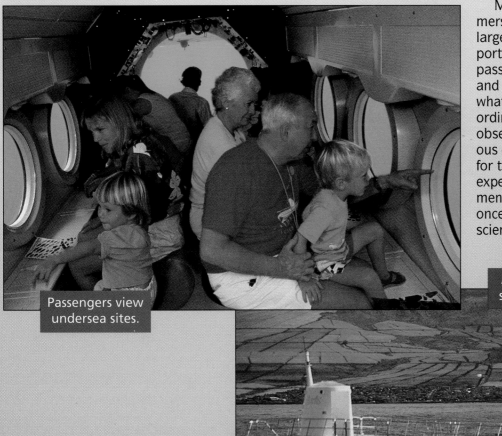

Passengers view undersea sites.

An *Atlantis* submersible

ASTRONOMY

People have been looking into the sky for centuries. Since the dawn of human history, people have desired to reach into the heavens.

This timeless quest has been helped by the tools of science. Astronomers have gazed beyond our moon into the outer limits of the universe. With space probes and spacecraft, humans have established a presence in space. We have touched the moon. Still, we continue to look into our solar system and beyond in search of life and knowledge.

21 *The Final Frontier   622*

Now we begin to venture out from our earth. Wherever we go in space, we must bring a part of Earth with us. How will we survive outside Earth's protective envelope?

Science
PARADE

OBSERVING THE SKY

People today find a rhythm to their lives in the hours ticking away on a clock, or the days marked off on a calendar. Ancient civilizations like the Maya of Central America measured time by observing the sky. Over thousands of years, Mayan priests and astronomers made observations and calculated the movements of the sun, the moon, and the stars.

The Maya had no concept of a round Earth. The movements of heavenly bodies were not thought of as revolutions, but as events that repeated themselves in a given pattern. You can imagine the mysterious power the priests must have had when they correctly predicted an eclipse of the sun! By mixing science with beliefs, Mayan astronomy mastered the awesome powers of the sky and gave order to time itself.

The Maya were extraordinary astronomers. It is no easy feat to make observations of sunrises and sunsets, eclipses, the movements of Venus, etc., in a country where it rains for nearly nine months of the year. It is also remarkable that the Maya were able to make astronomical observations with such minimal equipment. A pair of crossed sticks was set up inside the temple—perhaps this was a reason for building temples on top of high pyramids—as a fixed position on the horizon (some natural object) could be seen. Observations were made as the sun, moon, or planets repeated their movements in reference to these fixed points.

from *The Maya World*
by Elizabeth P. Benson

For Your Journal

🖋 **What instruments do astronomers use to observe the sky?**

🖋 **What do their observations tell us about the earth and other heavenly bodies?**

🖋 **What new technology will help astronomers in the future?**

Optical Astronomy

Describe *the accomplishments of ancient astronomers.*

Define *refraction and reflection.*

Compare and contrast *refracting telescopes and reflecting telescopes.*

When was the last time you looked at the sky to determine the time of day, the season, or the direction in which you were walking? You may never have. Why? Its just simpler to consult a watch, a calendar, or a compass. Thousands of years ago people were able to do all these things by simply looking to the sky. What did the sky tell them? How did early people manage to answer these questions without the modern technology we now take for granted?

Ancient Sky Watchers

All cultures have sought to understand the world in which they live. They have observed phenomena both on Earth and in the heavens and have determined the scientific laws that regulate the ways things work. People in ancient times were fascinated by the night sky, but they also found its apparent movements useful. The regular appearance of the sun and the moon led them to a system of time—days, months, and years. The appearance of certain stars or groups of stars told them it was time to plant, to harvest, or to prepare for the spring flooding. Sailors depended on stars to guide them on the oceans.

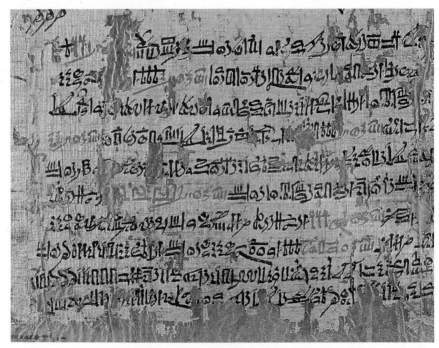

Figure 18–1. This portion of an Egyptian calendar dates from about 1230 B.C. The black figures represent what the Egyptians believed to be lucky days; the red figures, unlucky days.

It's in the Stone   Many records of the work of ancient astronomers remain today. From the Caracol, an ancient observatory at Chichén Itzá in Mexico's Yucatán peninsula, the Mayans made predictions of the motions of Venus that are the equal of computer-generated predictions made today. Stonehenge in southern England is an ancient astronomical instrument. Formed by a group of huge vertical stones, the arrangement of these stones suggests that the builders used the structure to plot the regular movements of the sun with great accuracy.

Figure 18–2. The Caracol (top) and Stonehenge (left) were ancient astronomical observatories.

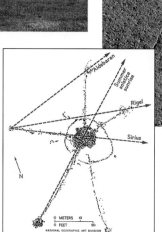

Native Americans also charted the heavens. At day's first light, sunlight shines through a window in an Anasazi dwelling in the American Southwest, lighting up a cut made in the wall. This cut lights up only at dawn on June 21, the summer solstice. The diagram shows a medicine wheel constructed by Native Americans in Wyoming. A central cairn (mound of stones, or raised marker) is surrounded by six other cairns. The colored lines show the rising of the sun at the summer and winter solstices, as well as the positions of three bright rising stars. Another medicine wheel found in Saskatchewan, Canada, dates from 600 B.C. and is the oldest astronomical observatory in the Americas.

Figure 18–3. Native Americans made meticulous observations of the stars using specially placed windows and medicine wheels.

In Sync with Nature   The ability to observe the heavens helped people in ancient times understand the cycles of the natural world and brought some constancy into their lives. Although they didn't understand the workings of everything they saw, they documented the appearance of all phenomena. What is truly amazing is that their observations were accomplished without modern technological help, not even simple corrective lenses for their eyes. Try the following activity to learn more about the accomplishments of our astronomical ancestors.

DISCOVER BY *Researching*

Use library resources to find examples of early calendars, star or navigational charts, or astronomical constructions. Choose one and write several paragraphs about it. Be sure to explain why it is an important example of the observational skill of ancient cultures. ✏

▶ **ASK YOURSELF**

What are some results of the observations of early sky watchers?

Visible Light and Lenses

The ancient sky watchers based their observations solely on what they could see. But the universe contains many things that cannot be seen with the unaided eye. It contains many forms of energy. One group of related energy forms is known as *electromagnetic radiation.* All matter in the universe gives off some form of electromagnetic radiation.

One form of electromagnetic radiation—the only form humans can see—is visible light. This is what the ancient sky watchers observed. Visible light is often referred to as white light. Although we perceive this light as white, it really contains a range of colors. If a beam of white light is passed through a prism, the light separates into all its colors.

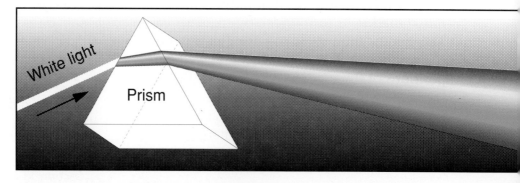

Figure 18–4. The prism reveals the colors that make up white light.

White light

Prism

Even though visible light represents only a small portion of all the electromagnetic energy, it is the portion through which we have gained the most information about the universe. Light behaves in certain predictable ways. It can be manipulated using refraction and reflection. **Refraction** means "to turn aside." If a beam of light hits a glass pane at a 90-degree angle, it will exit the glass in the same direction as it entered. If it strikes the glass at a lesser angle, some light will bend as it enters the glass and also as it leaves the glass. The resulting beam of light is parallel but offset from the original path. In the next activity, you can learn more about refraction.

Figure 18–5. Light that enters glass at a 90-degree angle passes through without refracting (left). If light enters the glass at a lesser angle, it is refracted from its original path (right).

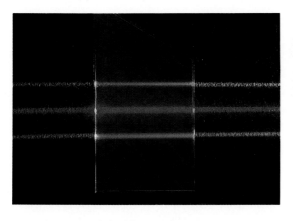

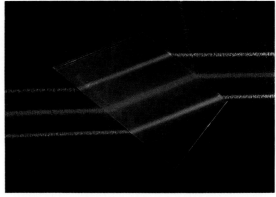

ACTIVITY

How can you show that white light is refracted as it passes through glass?

MATERIALS
glass blocks (2), flashlight, white construction paper

PROCEDURE
1. On a flat surface, stand one glass block on end, point the flashlight at the block, and turn the flashlight on.
2. Hold the white construction paper as shown, and move it around the glass block until the paper shows an image of the light coming through the glass block.
3. Position the second glass block between the paper and the first block, and note the angle of the light as it passes through both blocks.

APPLICATION
1. How is the white light changed as it passes though the first block? Give a specific description of the nature of the change.
2. What happens to the light passed through the first block when it is intercepted and passed through the second block?

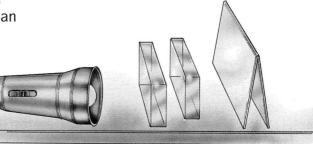

As you found in the activity, some light passes through an object even if it is refracted. **Reflection,** on the other hand, means to "turn back." Light reflects off many surfaces. If light hits such a surface at an angle, it bounces off at the same angle. The angle at which light is reflected is equal to the angle at which light hits a surface. Try the next activity to learn more about reflection.

DISCOVER BY *Observing*

Spend an afternoon looking around your home and community for good reflecting surfaces. As you walk down the street, look at store windows and see what conditions let you see yourself. Does a darkened store window appear to reflect better or worse than the window of a lighted store? Why do you think this is? When you look into a puddle or pond, is your reflection clearer if the water is still or moving? Make a list of the five best reflecting surfaces you find in your afternoon's travels. Compare your list with those of your classmates.

In the activity, you probably found that light reflects especially well off smooth, shiny surfaces. This property of reflected light enables anyone with a mirror to bounce light in practically any direction. Another example of reflection is seen in moonlight. The moon shines not by its own light but by the reflected light of the sun.

Figure 18–6. Our ability to see the moon is dependent on reflected light from the sun.

◤ ASK YOURSELF

Does a camera work on the principle of refraction or reflection? Explain your answer.

Optical Telescopes

The properties of refraction and reflection are useful in astronomy. Understanding how light behaves enables scientists to use a series of lenses to focus the light of distant stars. A piece of curved glass that is used to focus light is called a **lens.** Lenses, which were first developed in North Africa, form the basis for the most valuable tools used by astronomers: telescopes.

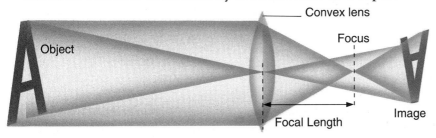

Convex lens

Focus

Object

Focal Length

Image

Figure 18–7. Light rays passing through a lens will converge at a focus.

If the surfaces of a lens are curved outward, all light rays will converge, or *focus,* at a single point. Lenses that cause light rays to converge are called *convex* lenses. If the surfaces of a lens are curved inward, all light rays will diverge, or separate. Lenses that cause light rays to diverge are called *concave* lenses.

As the diagram shows, a small image of the object being observed is formed at the focus. The distance from the lens to the focus is the *focal length*. The bending of the light as it enters the lens causes an image beyond the focus to appear upside down.

An *optical telescope* is an instrument with two basic purposes. First, it collects and focuses light, as your eyes do. Second, an optical telescope enlarges the image of distant objects. This enlargement factor is important for studying planets or the moon, but it has little effect on the appearance of stars. Stars are so far away that no amount of enlargement affects the way they appear to us. The two functions, light collection and enlargement, make a telescope an extension of your eyesight.

The first telescope was probably made by Hans Lippershey, a Dutch eyeglass maker. Based on Lippershey's model, Italian astronomer Galileo Galilei made a telescope based on refraction in 1609. Why does it make sense that the telescope was invented by an eyeglass maker?

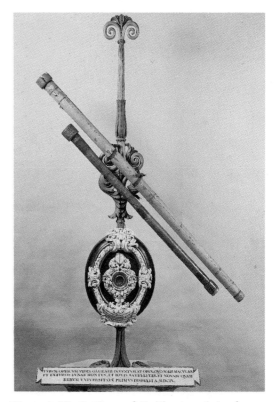

Figure 18–8. One of Galileo's original telescopes

Figure 18–9. Galileo made many drawings of the phases of the moon.

Although Galileo's telescope was primitive, he discovered many things with it: the craters and mountains on the moon, the phases of Venus, and the four large satellites of Jupiter. Each discovery contributed to the body of knowledge about the universe and allowed people to change their views of its makeup. The following activity will enable you to see what Galileo might have seen almost 400 years ago.

DISCOVER BY Observing

On a clear night, observe the moon with your unaided eyes and draw what you see in as much detail as possible. Then observe the moon with a telescope or binoculars. Again draw as much detail as you can. Describe the features that were not visible with the unaided eyes. If you have both a telescope and binoculars, make drawings based on each and compare them. When you look through binoculars you are seeing the moon at about the same enlargement that Galileo's telescope allowed. ✎

Figure 18–10. A modern refracting telescope has basically the same structure as the one that Galileo used in the early 1600s.

A "Refracted" View of the Universe

A *refracting telescope* is usually made with two tubes that fit together snugly. Light from the object being viewed passes first through a lens called the *objective lens*. The light comes to focus inside the tube and then passes through a second lens, called the *ocular lens,* to the observer's eye. The tubes can be adjusted to achieve better focus.

The brightness of the image you see in a telescope depends on the surface area of the objective lens. The larger the surface area of the objective lens, the greater the amount of light captured. Increasing the amount of light entering a telescope increases the brightness of the image. In the Skill on page 537, you can get an idea of how a refracting telescope works.

Refining the Basic Idea   Lensmakers improved the quality of refracting telescopes by making the lenses larger. The first large lenses were 38 cm in diameter and were made in Germany in the 1830s. The Yerkes Observatory in Williams Bay, Wisconsin, has the largest refracting telescope ever made. The objective lens of the Yerkes is about 1 m in diameter. Lenses larger than this probably will not be made in the future. Glass lenses must be absolutely smooth, and if they are made larger than 1 m they tend to develop air bubbles. Remember the activity in which you compared moving water with still water as a reflecting surface? What surface bubbles do to the reflecting ability of a pond, air bubbles do to the refracting ability of lenses. Also, a glass lens larger than 1 m is so heavy that it bends under its own weight, changing its shape and distorting the images.

Until the end of the last century, most astronomical observations were made with the human eye. Today most observations are made using photographic plates, which astronomers then study. Computers point the telescope, and motors on the telescope keep it focused on an object for hours or even days. This tracking system allows enough time for even very faint objects to leave images on film. In some cases, photographic cameras are replaced by electronic cameras. Electronic images can record the intensity of light more precisely than film.

Photographs taken of the same area of the sky at different times can be studied to see changes in stars. When an exploding star was observed in 1987, photos of the area taken before the explosion made it possible for astronomers to identify which star had exploded, since the star was simply no longer visible.

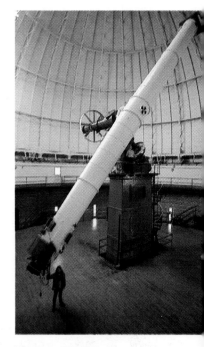

Figure 18–11. The Yerkes refracting telescope is the largest in the world.

Figure 18–12. Astronomers studied photos such as these to figure out which star had exploded.

Figure 18–13. This is a replica of the reflecting telescope made by Sir Isaac Newton.

A Reflection on the Universe

Sir Isaac Newton, an English scientist, made the first *reflecting telescope*, a model of which is shown in the photograph. Newton realized that using mirrors has several advantages over using lenses. First, mirrors reflect not refract; therefore, they produce no color distortion. Second, a good mirror absorbs less light than the most transparent glass, creating a brighter image. Third, a mirror is not as heavy as a lens of the same size. Therefore a mirror is easier to support and is less likely to bend. Much larger telescopes can be constructed using mirrors.

Reflecting telescopes have at least two mirrors. The large primary mirror receives the image and reflects it to a secondary mirror. The secondary mirror, placed in front of the focus, reflects the light through an ocular lens to the observer.

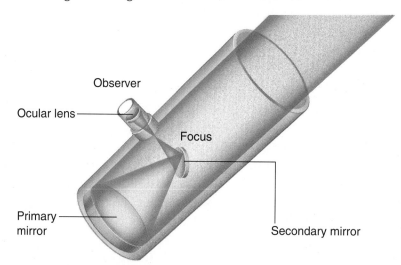

Figure 18–14. A reflecting telescope uses mirrors instead of lenses.

Observer

Ocular lens

Focus

Primary mirror

Secondary mirror

Figure 18–15. This 6-m telescope in Russia is one of the world's largest reflectors.

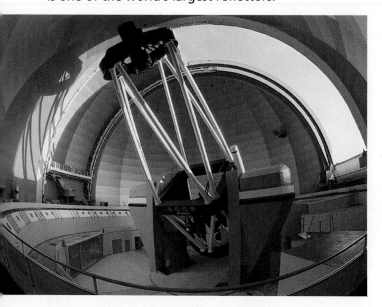

Most large telescopes used today are reflectors. The world's largest reflecting telescope is the Keck 1, which sits atop the Mauna Kea volcano in Hawaii. Scientists expect the Keck 1 to revolutionize ground-based optical astronomy. Its primary mirror, made up of 36 major segments, is 10 m in diameter. With it astronomers hope to see planets in orbit around nearby stars and to locate a black hole believed to lie at the center of our Milky Way Galaxy.

A View from the Outside

Although Earth's atmosphere lets in visible light from space, it also distorts that light. It is this distortion that causes stars to appear to twinkle. As a result, images produced by ground-based telescopes often appear a bit fuzzy. In April of 1990, after years of planning, the National Aeronautics and Space Administration (NASA) placed the Hubble Space Telescope into orbit. The size of a school bus and powered by solar panels, the Hubble telescope is expected to "see" light given off by stars near the edge of the universe.

Although its primary mirror was originally flawed, the repaired Hubble telescope has shown itself capable of seeing more clearly than any other telescope. It can make fine distinctions between distant objects not possible with ground-based telescopes. Just as Galileo's observations confirmed the idea that Earth was the center of the solar system, today's advanced observations may tell us we are not the only planetary system in the universe.

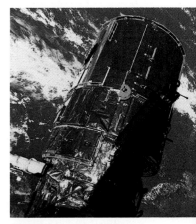

Figure 18–16. The Hubble Space Telescope, held here by a deployment arm, was transported to orbit by the space shuttle *Discovery*.

 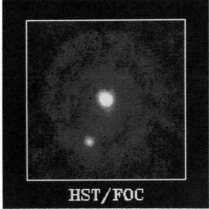

Ground Based HST/FOC

Figure 18–17. Even Hubble's pre-repair image of Pluto and its moon Charon (right) separates the two bodies much more clearly than the best ground-based image (left).

 ASK YOURSELF

What advantages do reflecting telescopes have over refracting telescopes?

SECTION 1 *REVIEW AND APPLICATION*

Reading Critically

1. How has the optical telescope advanced our knowledge of astronomy?
2. In what two ways does a telescope extend our sense of sight?

Thinking Critically

3. Why is the Hubble Space Telescope an even more important instrument today than it was when it was first launched into orbit?
4. Why do you think so many ground-based telescopes sit on the tops of mountains?

INVESTIGATION

Measuring Focal Length

► MATERIALS

● short candle ● matches ● Petri dish ● meter stick ● scissors ● white construction paper ● modeling clay ● convex lenses of varying thicknesses

▼ PROCEDURE

1. **CAUTION: An open flame is used in this investigation: be careful of your hair and clothing.** Set up the materials as shown in the illustration below.
2. Light the candle, and focus the image of the candle on the paper.
3. When the image is sharp,

measure and record the distance from the lens to the construction paper and from the lens to the candle. You can use this data to find the focal length of the lens.

4. Repeat steps 2 and 3 using different lenses and configurations. Keep a record of each distance measured.

5. The equation for determining focal length is

$$\frac{1}{f} = \frac{1}{d_o} + \frac{1}{d_i}$$

where f is the focal length, d_o is the object distance, and d_i is the image distance. Use your data and the formula to calculate the focal length of each lens.

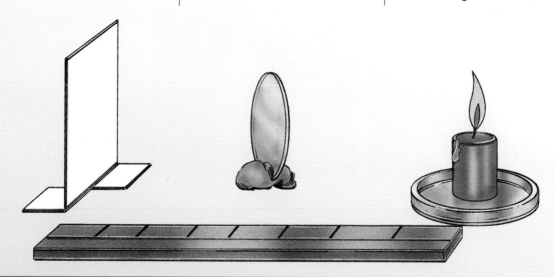

► ANALYSES AND CONCLUSIONS

1. Prepare a chart to compare and contrast the focal lengths of the lenses. Can you see any relationships among them? Explain.
2. How does the actual candle flame differ from its image on the construction paper? Be specific.

► APPLICATION

How could this procedure be used to find the focal lengths of telescopes, cameras, and other optical instruments?

✳ Discover More

Repeat this investigation using a concave lens. Can you locate an image on the construction paper? Do some research on the difference between a *real image* and a *virtual image*.

SKILL *Building a Telescope Model*

▶ **MATERIALS**
- paper tubes made from rolled construction paper, one with a slightly larger diameter than the other (2) ● modeling clay ● double concave lens ● double convex lenses (2) ● concave mirror ●flat mirror

▼ **PROCEDURE**

1. Fit together the two paper tubes. One should fit snugly into the other, but still be able to slide back and forth easily.
2. Into the end of the tube with the smaller diameter, insert the double concave lens. Secure the lens in place with modeling clay. This is the ocular lens.

3. Into the end of the larger tube, insert the double convex lens, also using clay to secure it. This is the objective lens.
4. Slide the tubes back and forth to focus on distant objects.

▶ **APPLICATION**

1. Draw to scale several objects as they appear both with the unaided eye and through your telescope.
2. Use your telescope to view the moon. What features can you see with the telescope that you could not see before?

✳ *Using What You Have Learned*

1. When you have mastered the use of this telescope, try making and using a simple reflecting telescope.
2. Find information in the library about eyeglasses and contact lenses and how they are made. What principles are similar in making telescopes and corrective lenses for eyes?

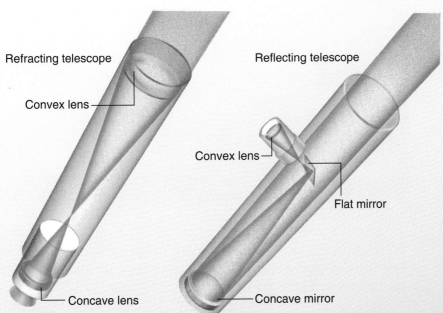

Refracting telescope

Convex lens

Concave lens

Reflecting telescope

Convex lens

Flat mirror

Concave mirror

Radio Astronomy

Perhaps you have played the game in which one person is blindfolded and given objects to identify. For example, you can probably identify a soccer ball by feeling its shape. You can even distinguish between an orange and an apple without touching them simply by smelling each one. Just as you depend on senses other than sight for information, there are many ways to learn about the universe in addition to your sense of sight. There are many technologies that astronomers use to gather information.

The Unseen Universe

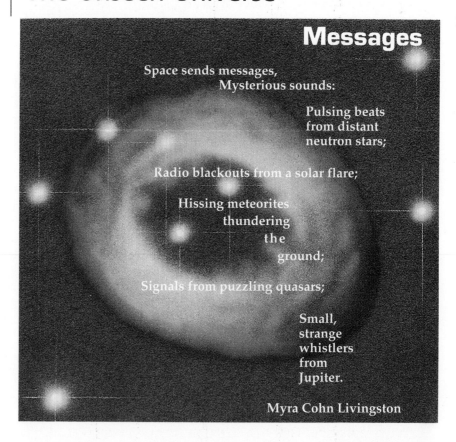

Messages

Space sends messages,
Mysterious sounds:

Pulsing beats
from distant
neutron stars;

Radio blackouts from a solar flare;

Hissing meteorites
thundering
the
ground;

Signals from puzzling quasars;

Small,
strange
whistlers
from
Jupiter.

Myra Cohn Livingston

As this poem implies, there are many kinds of information coming from space. You learned in Section 1 that visible light is one form of energy through which we have learned much about our universe. There are other forms.

Stars, galaxies, and other objects, including some on Earth, give off many kinds of electromagnetic energy. X-rays, infrared radiation, ultraviolet radiation, gamma rays, microwaves, and radio waves are all forms of electromagnetic radiation. Scientists have learned to gather, measure, and interpret these forms of energy to learn about the universe and its evolution.

The illustration below shows the range, or *spectrum,* of electromagnetic energy forms. Notice that visible light represents only a small portion of the entire spectrum. Although we cannot see the other forms diagramed here, they can be detected in other ways. For instance, you can feel infrared radiation as heat, as from a fire or the sun. You feel the effects of ultraviolet radiation as a sunburn. Microwaves can cook food. The programs you hear on the radio or see on TV are decoded radio waves.

Figure 18–18. Other forms of electromagnetic radiation can provide information about the universe.

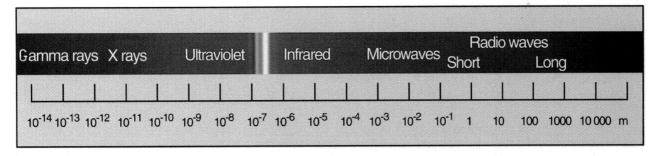

| Gamma rays | X rays | | Ultraviolet | | Infrared | | Microwaves | | Radio waves | |
| | | | | | | | | Short | Long | |

10^{-14} 10^{-13} 10^{-12} 10^{-11} 10^{-10} 10^{-9} 10^{-8} 10^{-7} 10^{-6} 10^{-5} 10^{-4} 10^{-3} 10^{-2} 10^{-1} 1 10 100 1000 10 000 m

Look at the spectrum again. Notice that the names of some forms of energy have the word *wave* in them. Also notice the wavy lines below. Light and other forms of electromagnetic energy travel through space as waves. You can see that radio waves are very long. They can be measured in meters. Light waves, on the other hand, are very short.

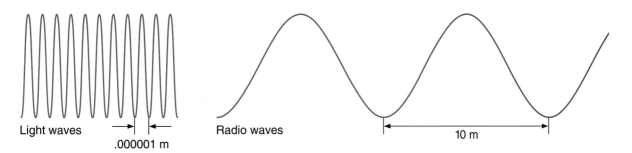

Light waves → .000001 m ←

Radio waves ← 10 m →

Try to remember that the different forms of energy in the spectrum are really very similar. They travel at the same speed, and they bring us similar kinds of information. The only difference is in their wavelengths, which means that they must be detected and measured with different kinds of technology.

Figure 18–19. Radio waves are very long, while light waves are very short.

Figure 18–20. You can measure the frequency of ocean waves by counting each wave as it crests.

It's a Matter of Frequency

Imagine that you are watching waves at the seashore. Certain groups of waves have long periods between their crests. You might count only 10 waves in a minute's time. Other groups of waves pile up on each other so fast that you might have difficulty counting them at all.

Electromagnetic waves can be counted in the same way as ocean waves—by counting each curve, or crest. When you have counted the number of crests in a given time period, you have determined the **frequency** of the waves.

When people refer to a certain radio frequency or TV channel, they mean the device is set to detect and decode that number of radio waves in a given period of time.

Although radio waves had been discovered coming from the sun in 1890, American Karl Jansky was the first to detect other sources. In 1931 Jansky, a communications specialist, was trying to pinpoint the source of static that interfered with short-wave radio reception. Jansky detected a regular hissing noise and eventually determined that it was coming from the center of the Milky Way Galaxy.

Since radio waves are not absorbed by gases and dust, they can travel from the farthest corners of the universe to Earth without interruption. This enables astronomers to create "pictures" from radio waves. Radio wave "pictures" of the heavens are different from optical photos of the same areas. For instance,

Figure 18–21. Karl Jansky and his strange-looking radio antenna.

radio emissions from visible objects such as stars are often very weak. Stronger emissions come from the centers of galaxies, including our own. Radio waves also reveal the presence of huge hydrogen clouds throughout the universe, which do not show up as visible light. Astronomers have long suspected that there was much unseen matter in the universe—radio astronomy has confirmed it. Try the following activity and see how many sources of radio waves you can find.

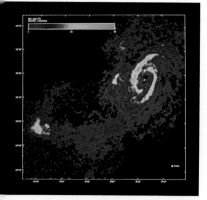

Figure 18–22. The bright areas mark the centers of galaxies with a large hydrogen cloud between them.

In your home, locate all the objects that produce or receive radio waves. After you have identified several sources, do some library research to determine the frequency that each produces or receives. ✑

Radio astronomy and optical astronomy have one major element in common—both can be practiced on Earth. The other forms of electromagnetic radiation are generally absorbed by various elements in Earth's atmosphere and so do not reach the ground. It is only when we journey above the earth's atmosphere that these forms of radiation can be detected.

X-rays from places other than Earth were first detected by a V-2 rocket in 1948. They came from the sun and affected the photographic process used on the rocket. Other X-ray sources, such as extremely hot stars and the remains of exploded stars, were identified in the following decade. But it was not until 1970 that technology allowed us to put an X-ray satellite into orbit.

Infrared rays are found just beyond the red end of the visible spectrum. This form of cosmic radiation was discovered in 1800 by British astronomer Sir William Herschel. Most objects on Earth give off infrared rays that can be detected, but our atmosphere absorbs most infrared rays from space.

Look at the infrared image of a man drinking a cup of hot coffee. Infrared detectors record heat sources: white and yellow represent the hottest areas, blue and green the coolest. Look at the face and the cup. Is the hotter area represented by yellow or white? How do you know? Why is the area at the back of the man's head cooler than his forehead and face?

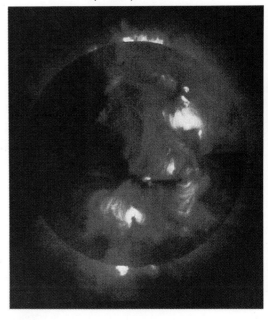

Figure 18–23. X-rays reveal a universe not seen by visible light. This photo shows areas of violent upheaval on the sun, none of which would show on an optical photo.

Figure 18–24. In an infrared image, white and yellow represent the hottest areas.

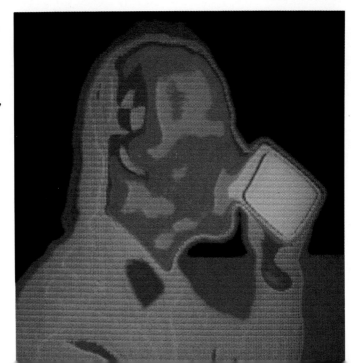

541

Figure 18–25. Arno Penzias and Robert Wilson, with the antenna that recorded cosmic background radiation.

Beaming to You from Outer Space Other forms of electromagnetic energy from space are gamma rays, microwaves, and ultraviolet rays. Gamma rays are emitted by exploding stars and other objects. Microwave radiation from space was discovered in 1965 by American astronomers Arno Penzias and Robert Wilson. Their instruments recorded constant background hissing coming from all parts of the sky. Today we know this to be background radiation, perhaps left over from the origin of the universe. Penzias and Wilson won the Nobel Prize in physics in 1978 for their discovery.

Ultraviolet (UV) radiation is the energy that tans or burns your skin. It is the energy just beyond the violet end of the visible spectrum. UV radiation was discovered in 1801 by the German physicist Johann Ritter. One source of UV radiation is new stars. In the next activity you can learn about some of the sources and uses of electromagnetic radiation on Earth.

DISCOVER BY *Researching*

Use library resources to find several Earth-based sources of and uses for the invisible forms of radiation. For what are microwaves used, for instance, or X-rays? What are some of the dangers of using these forms of energy on Earth? ✍

▼ ASK YOURSELF

Why does astronomy based on forms of electromagnetic radiation other than visible light have to be practiced mostly above Earth's atmosphere?

Figure 18–26. The radio telescope at Arecibo, Puerto Rico, the largest on Earth.

Recording the Unseen Universe

Astronomers capture radio waves from space with instruments called *radio telescopes*. Since the early 1940s, radio telescopes worldwide have been detecting objects in the universe. Some radio telescopes look like large TV satellite dishes. Others look like the one in the photograph.

Radio waves are reflected by metallic surfaces just as light is reflected by a mirror.

Like optical telescopes, most radio telescopes can be pointed in different directions. The largest movable radio telescope is 100 m in diameter and is located in Effelsberg, Germany.

The largest nonmoving radio telescope is near Arecibo, Puerto Rico. Built into a deep natural depression in the earth, the dish is 305 m in diameter. The rotation of the earth enables this telescope to scan different parts of the sky regularly. The radio signals collected in the dish are then fed into the antenna above it.

After the radio signals are fed into the antenna, they are changed into electrical signals. These signals are amplified and analyzed by a computer, which can then make a "picture" of the sky, just as a camera can make a photograph of the same area.

Figure 18–27. The workings of a radio telescope

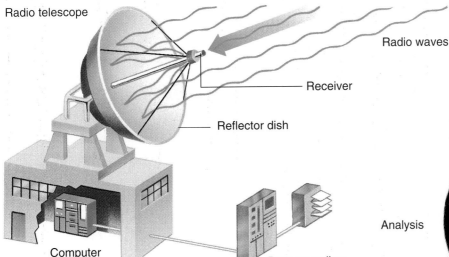

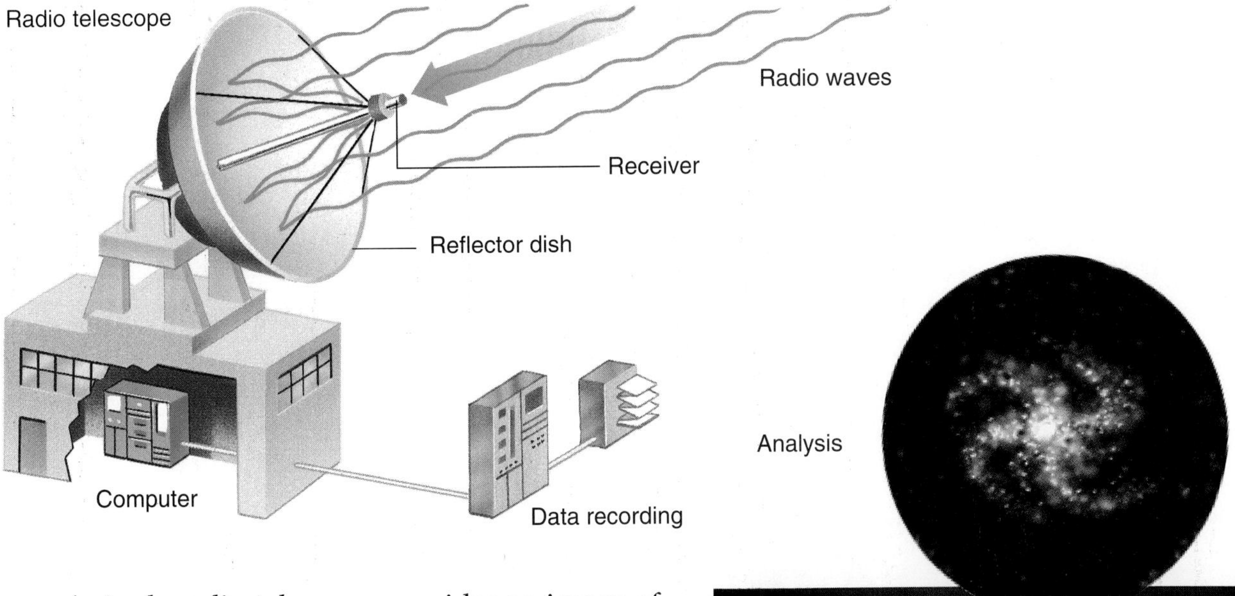

Radio telescope

Radio waves

Receiver

Reflector dish

Computer

Data recording

Analysis

A single radio telescope provides an image of the sky that is less precise than the image provided by an optical telescope. Radio telescopes can be linked together, however, so that the image received by one can be compared and mixed with the image received by another. Combining images from several instruments can produce a much clearer radio image of the sky.

The radio telescopes in the photograph are in the desert near Socorro, New Mexico. Called the Very Large Array (VLA), the 27-dish setup creates an image of the sky that compares favorably with the image provided by the best optical telescopes. You can read more about radio telescopes in the Science Parade on page 658.

Figure 18–28. The VLA has 27 radio telescopes, each about 25 m across. Its combined images compare favorably with optical photos.

Starting with Karl Jansky's discovery that radio-wave sources exist at the center of our galaxy, astronomers have determined that many galaxies contain these sources. Radio astronomy has also revealed objects called *pulsars,* the rapidly spinning remains of huge collapsed stars, and *quasars,* or quasi stellar objects. Quasars are believed to be very active nuclei of galaxies or black holes. You will learn more about galaxies, pulsars, and quasars in a later chapter.

Figure 18–29. The pulsar in the Crab nebula flashes (on/off, left/right) thirty times each second.

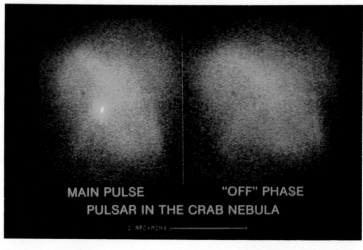

MAIN PULSE   "OFF" PHASE
PULSAR IN THE CRAB NEBULA

Figure 18–30. A technician balances on cables leading from support towers on mountains surrounding the Arecibo antenna.

Radio astronomy, unlike optical astronomy, can work both ways. Radio telescopes can not only receive radio waves from space, they can also transmit them. Human beings have long wondered whether we share the universe with others like ourselves. In 1960 American astronomer Frank Drake tuned a government radio telescope to two nearby stars to listen for signs of intelligent life. In 1974 he sent a coded message, shown in the diagram on the next page, from the Arecibo telescope out into space. The search for extraterrestrial intelligence (SETI) has been going on ever since. The Arecibo telescope regularly sends messages about life on Earth into deep space.

Though government funding of a large SETI project has been cancelled, radio telescopes will keep up the search at a slower pace. Astronomers, computer experts, and others will continue to monitor radio channels to discover whether we have neighbors in space and, if so, where. Now you can use what you know about radio astronomy to complete the next activity.

ᴅⁱˢᶜᵒᵛᵉʳ ʙʏ *Calculating*

If anyone is out there listening to Earth, about how far away would they have to be in order to hear the 1974 message right now? Remember that all radio and television waves travel at the speed of light. Would radio waves transmitted during the early years of television have traveled farther than the 1974 message? Explain. Why does it make more sense for astronomers to listen on a grand scale than to transmit on a grand scale? ✎

Satellite Central   In the last 20 years, many nations have worked together to study different forms of energy coming from space. Many joint satellite projects have been launched in the search for information. Table 18-1 summarizes many of these projects. The next pages present a brief pictorial view.

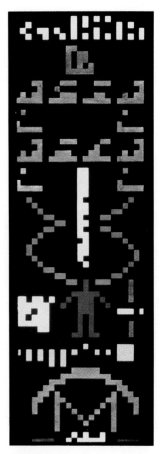

Figure 18–31. Frank Drake's message

Table 18-1	**Some recent satellites**	
Satellite	**Type of Radiation Detected**	**Task/Accomplishment**
Uhuru	X-rays	Gave early astronauts first X-ray maps of the sky
Exosat	X-rays	Detected new X-ray sources and measured their various stengths
ROSAT	X-rays	Documents X-ray sources
IRAS	Infrared	Found enormous souces of infrared emissions
IUE	Ultraviolet light	Probes UV characteristics of planets and moons and detects stars forming in Orion nebula
GRO	Gamma rays	Records gamma ray bursts, solar flares
COBE	Microwave	Detects microwave radiation that supports big-bang theory

Exosat (left) being tested before launch; an X-ray map of the Milky Way Galaxy (right).

An artist's conception of IRAS in orbit.

The center of the Milky Way Galaxy, shown by the line in the bottom photo, is obscured by large dust clouds.

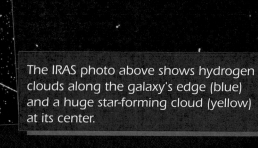

The IRAS photo above shows hydrogen clouds along the galaxy's edge (blue) and a huge star-forming cloud (yellow) at its center.

The IUE (left) has shown us the ultraviolet universe (right).

Scientists once thought gamma ray bursts came from our Milky Way Galaxy. This gamma ray map shows that they originate fairly regularly from all over the sky (white dots), both inside and outside of the galaxy.

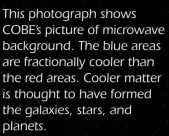

This photograph shows COBE's picture of microwave background. The blue areas are fractionally cooler than the red areas. Cooler matter is thought to have formed the galaxies, stars, and planets.

The unseen universe is dynamic; even violent. It continues to evolve each second, with new stars forming and old stars dying. As scientists learn more about evolution and change in the universe, we can see what might be in store for our sun and planet Earth.

▼ **ASK YOURSELF**

Why it is useful to compare the images of the same area of the universe made by satellites collecting different frequencies of electromagnetic radiation?

SECTION 2 *REVIEW AND APPLICATION*

Reading Critically

1. How is visible light different from other forms of electromagnetic radiation?

2. Why are invisible forms of energy from space important to astronomers?

Thinking Critically

3. Why do many Earth-based observatories contain both optical telescopes and radio telescopes?

4. Compare the optical (left) and X-ray (right) images shown. Which phenomena show on both images? What phenomena appear on the X-ray image that do not show up in the optical image? Why do you think these phenomena are invisible to an optical camera?

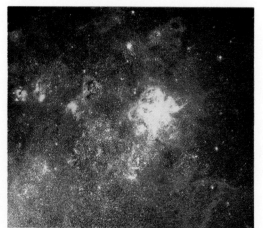

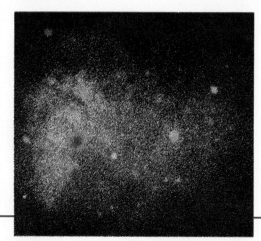

The Big Idea

People have always looked to the heavens. In ancient times humans did not know about the structure or workings of the universe, nor did they have the technology to help them find out. However, they were fine observers and used their observations of heavenly bodies to bring order and constancy into their lives.

With the advent of the telescope, humans began to change their views of the universe in response to their discoveries. They found that the universe was an evolving system that could, over time, be observed and increasingly understood. We have moved from an Earth-centered view to a sun-centered view, and from there to infinity. We know now that our sun and our planet, as special as they are to us, are but one tiny part of an immense universe that will continue to reveal itself over time.

For Your Journal

Look again at your answers to the questions at the beginning of this chapter. How have your ideas changed about the instruments that astronomers use? What have astronomers learned from their observations? Write your answers in your journal.

Connecting Ideas

Copy this concept map into your journal and complete it.

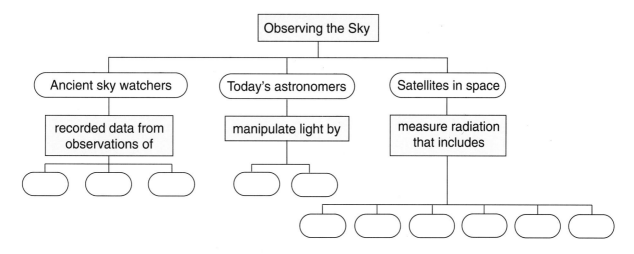

REVIEW

Understanding Vocabulary

1. Explain how the terms in each set are related.
   a) refraction (529), reflection (530)
   b) lens (531), convex (531), concave (531)
   c) spectrum (539), frequency (540)

Understanding Concepts

MULTIPLE CHOICE

2. Increasing the surface area of a lens in a refracting telescope will
   a) brighten the image.
   b) darken the image.
   c) sharpen the image.
   d) shrink the image.

3. The form of electromagnetic radiation with longer wavelengths than visible light is
   a) gamma rays.   c) radio waves.
   b) X-rays.   d) ultraviolet rays.

4. If a light ray hits glass at less than a 90° angle, the light ray
   a) increases in wavelength.
   b) refracts.
   c) speeds up.
   d) passes through without refracting.

5. Electromagnetic radiation is emitted by
   a) stars.   c) television sets.
   b) galaxies.   d) all of these.

6. The enlargement capability of a telescope would be most useful in studying
   a) Earth.   c) the Big Dipper.
   b) the moon.   d) the Milky Way Galaxy.

7. Which of the following is not an advantage of mirrors over lenses?
   a) Mirrors produce a brighter image than lenses do.
   b) Mirrors produce no distortion.
   c) Mirrors can be smaller than lenses.
   d) Mirrors are not as heavy as lenses of the same size.

8. The type of electromagnetic radiation we feel as heat is
   a) gamma ray.
   b) X-ray.
   c) ultraviolet radiation.
   d) infrared radiation.

SHORT ANSWER

9. Compare and contrast the operation of refracting telescopes and reflecting telescopes.

10. How are the forms of electromagnetic energy similar? How do they differ?

Interpreting Graphics

Use the diagram of the electromagnetic spectrum to answer these questions.

11. Which form of energy on the electromagnetic spectrum falls between 10^9 Hz and 10^{12} Hz in frequency? between 10^{-9} m and 10^{-10} m in wavelength?

12. Which energy form in the spectrum has the longest wavelength? the highest frequency?

13. Which statement below is true, based on the diagram?
    a) The longer the wavelength, the lower the frequency.
    b) The shorter the wavelength, the lower the frequency.

Reviewing Themes

14. *Technology*
Why is the ability of a telescope to magnify objects in the universe limited?

15. *Energy*
How has non-optical astronomy revealed information about the structure of the universe?

Electromagnetic spectrum		Frequency (f) in hertz	Wavelength (λ) in meters	
Gamma rays		10^{23}	10^{-15}	
		10^{22}	10^{-14}	
		10^{21}	10^{-13}	
		10^{20}	10^{-12}	
		10^{19}	10^{-11}	
X rays		10^{18}	10^{-10}	
		10^{17}	10^{-9}	
Ultraviolet radiation		10^{16}	10^{-8}	
		10^{15}	10^{-7}	
Visible light				
		10^{14}	10^{-6}	
Infrared radiation		10^{13}	10^{-5}	
		10^{12}	10^{-4}	
		10^{11}	10^{-3}	
Microwaves		10^{10}	10^{-2}	
	UHF	10^{9}	10^{-1}	
	VHF	10^{8}	1	
	Short wave	10^{7}	10	
Radio waves	Medium wave	10^{6}	10^{2}	
	Long wave	10^{5}	10^{3}	
		10^{4}	10^{4}	
		10^{3}	10^{5}	

Thinking Critically

16. What were the strengths of the ancient sky watchers? What were their weaknesses?

17. How does the technology used in modern astronomy extend the human senses?

18. This array of radio telescopes is "eavesdropping" on the universe, listening for signals that might indicate the presence of other life forms. If other life forms are listening to us, what kinds of things do you think they have learned about humans in the last 10 years?

Discovery Through Reading

"Stunning Images from an Ailing Space Telescope." *Smithsonian* 22 (March 1992): 98–101. See what the Hubble Space Telescope has observed—even with its "bad eye"—and imagine what it will be capable of "seeing" after repairs are made.

The Solar System

The pine, the spruce, the hemlock, the fir—all those conifers that know no leafless season—have been held in special favor when man would have symbols of life that outlast all winters.

Since man was first aware of the changing seasons, the winter solstice has been occasion for awe and wonder and a challenge to faith. Hope and belief are easy in a warm, green world, but when the cold days come and the sun edges farther and farther south, . . . utter darkness and oblivion are at hand. Then the sun stands still. The turn comes. The crisis passes and the sun slowly climbs the sky once more, reaching toward another spring, another summer. It was, and it still is, an annual miracle.

from Hal Borland's Twelve Moons of the Year
by Hal Borland

*W*hy, in northern regions, is late December a time for hanging evergreen wreaths and decorating pine trees? The origin of these winter traditions lies in the rhythms of the solar system. As one nature writer explains, people bring evergreen branches into their houses to remind them that the sun will rise again and that spring will soon follow.

You may not think of the movements of the solar system as a miracle, but you probably still hang an evergreen wreath in December. And the sun climbs higher, and spring returns.

Motions of Earth and Its Moon

Objectives

Compare and contrast rotation and revolution as they relate to Earth's path through space.

Explain how the tilt of Earth's axis affects the seasons.

Describe the causes of solar and lunar eclipses.

GRAND TOUR—PLANETS AND MOONS OF THE SOLAR SYSTEM! MAKE RESERVATIONS NOW! SEE THE RINGS OF SATURN, URANUS—THE SIDEWAYS PLANET, AND TRITON—NEPTUNE'S CANTALOUPE MOON! DON'T DELAY! FIRST STOP, EARTH'S MOON!

Suppose that you saw such an advertisement in the not-so-distant future. Who wouldn't want to sign up? Although the nine planets are part of one solar system, each planet and moon is strikingly different. Our own planet and moon are the first stop on the tour.

Figure 19–1. A skater's period of rotation is very short.

Rotation

Imagine that you start your tour looking at Earth from its moon. You see our planet turning on its axis, as all planets do. An *axis* is an imaginary line that passes from pole to pole through the center of a planet. The turning of a planet on its axis is called **rotation.** Earth rotates much like the ice skater in the picture, who wraps her arms around her body and spins in place like a top. Unlike the skater, however, it takes Earth 24 hours to make one complete rotation. This is called Earth's *period of rotation*. The direction of Earth's rotation is counterclockwise as seen from above the North Pole.

With a telescope, you and your friends on the tour can see North America. As Earth turns, you watch the sun's light move across the continent as yet another day dawns. The first part of the continent to receive the sun's light is the east coast of Canada. At this time, the United States is still in darkness. As Earth rotates, the light moves westward, until all of North America is in sunlight. This process takes about six hours.

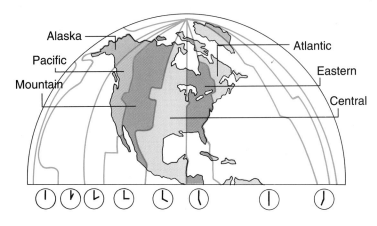

Figure 19–2. The rotation of Earth led to the formation of different time zones. The time zones of North America are shown here.

For sunrise to occur at approximately the same time all across North America, six different time zones have been established. These time zones change roughly every 15° of longitude. However, time zones are not straight lines, as lines of longitude are. Time zones have been drawn so that they do not pass through and divide cities. Complete the following activity in order to learn more about time zones.

ᴰⁱˢᶜᵒᵛᵉ ᴿ ᴮᵧ *Researching*

Use library resources to find information on the history of time zones. How were schedules made before time zones were standardized? What problems did people have due to the absence of time zones? What segments of the population were particularly interested in standardizing time zones? Why? Record your answers in your journal. ✎

▶ ASK YOURSELF

According to the diagram, which state fills a whole time zone?

Revolution

From the moon, you and the tour members also observe a little of a second motion of Earth. You learned in an earlier chapter that gravity is the major force in the universe that pulls objects toward one another. The sun exerts gravity on the Earth and all the planets, so that in addition to rotating on its axis, Earth moves around the sun. The movement of a planet around the sun is known as **revolution.** To compare rotation with revolution, try the next activity.

DISCOVER BY Doing

You can demonstrate both rotation and revolution in your classroom or at home. Remove the lampshade from a lamp placed in the center of the room. Then turn your body slowly in place (rotate). Now, as you are turning, or rotating, gradually move in a wide circle (revolve) around the lamp. Notice that during part of each rotation the lamp shines on your face, and during the other part of your rotation it shines on the back of your head. Which part of your movement represents the passage of a day's time? Which part of your movement represents the passage of a year's time? ✐

It takes Earth one year, or 365.25 days, to complete one revolution, so its *period of revolution* is one year. The path that Earth follows as it revolves around the sun is called its **orbit.**

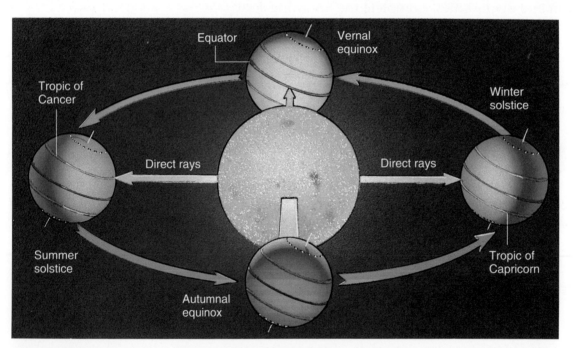

Figure 19–3. Earth revolves around the sun every 365.25 days, the length of an Earth year.

Summer or Winter?

Look at the diagram. As Earth revolves around the sun, its axis remains in a fixed position. It always points in the same direction. Also notice that the axis is tilted. This tilt, or *inclination,* is 23.5° in relation to a line perpendicular to the plane of Earth's orbit. The tilt causes first the Northern Hemisphere and then the Southern Hemisphere to lean toward the sun during the course of one revolution. This shift creates seasonal changes in climate. It also causes changes in the relative amounts of daylight and darkness during the year.

When the Northern or Southern Hemisphere is tilted toward the sun, that hemisphere receives very direct rays from the sun. During this summer season there is sunlight for long periods of time. When the same hemisphere is tilted away from the sun, that hemisphere receives less direct sunlight. During this winter season, there are fewer hours of sunlight.

In the Tropics

On June 20 or 21 the tilt of the Northern Hemisphere toward the sun is greatest. The sun is directly over 23.5° north latitude—the Tropic of Cancer. In the Northern Hemisphere this date is the *summer solstice*, the day of the year with the most hours of daylight.

As Earth's revolution around the sun continues, the sun is directly over the equator on September 22 or 23. In the Northern Hemisphere this is the *autumnal equinox*. On this date the periods of daylight and darkness are about equal. The word *equinox* comes from two Latin words meaning "equal" and "night." What else might people have called this date?

As Earth continues its revolution, the daylight hours in the Northern Hemisphere grow shorter until the *winter solstice* on December 21 or 22. On this date the sun is directly over the Tropic of Capricorn, at 23.5° south latitude. This is the day with the least daylight in the Northern Hemisphere.

On March 20 or 21 the sun is directly over the equator again, and the hours of daylight and darkness are nearly equal. In the Northern Hemisphere this is known as the *vernal equinox*. As Earth completes its revolution around the sun, the hours of daylight continue to lengthen in the Northern Hemisphere until it is once again the summer solstice. In the Southern Hemisphere the summer and winter solstices and the seasons are the opposite of those in the Northern Hemisphere. Why is this so? Complete the following activity to keep track of the passing of the seasons in your area.

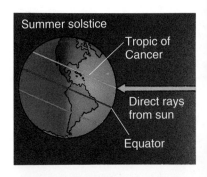

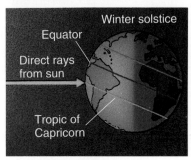

Figure 19–4. Our summer solstice (top) occurs when the sun is directly over the Tropic of Cancer. Our winter solstice (bottom) occurs when the sun is directly over the Tropic of Capricorn.

Figure 19–5. The seasons of the year are determined by the tilt of the earth in relation to the sun's rays. In temperate climates, shown here, four seasons occur each year.

ACTIVITY

How can you observe the revolution of Earth?

MATERIAL
drawing paper, pencil, calendar

PROCEDURE
1. Find the highest location from which it is convenient for you to make regular observations toward the west. Locations might include a west window in a high-rise, an attic window in a house, or a hilltop area near your home.
2. Draw as accurate a representation of the western horizon as you can, noting such landmarks as big buildings, trees, mountains, and utility towers.
3. Every day for a month, note the position of the setting sun on your drawing.

APPLICATION
1. Imagine you are a time traveler and you crash-land on Earth. You are all alone. You have food and shelter but no idea what time of year it is because you are in a climate that is much the same year-round. How could you determine what season it is?
2. How could early civilizations have used this activity to create a calendar of the year?
3. Imagine you live in Australia or Argentina. What are the months of each season?

Now that you have observed some basic movements of your own planet, it's time to take a look at Earth's companion in space, its moon. To do that, you need to change your viewing perspective.

◤ ASK YOURSELF

Does the sun ever shine directly overhead where you live? Why or why not?

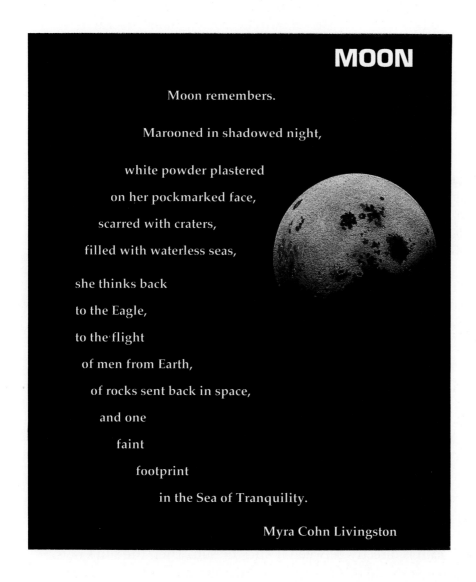

MOON

Moon remembers.

Marooned in shadowed night,

white powder plastered

on her pockmarked face,

scarred with craters,

filled with waterless seas,

she thinks back

to the Eagle,

to the flight

of men from Earth,

of rocks sent back in space,

and one

faint

footprint

in the Sea of Tranquility.

Myra Cohn Livingston

Motions of the Moon

After reading this poem, you decide to catch a shuttle flight to Earth to observe the moon. Studying the moon from Earth is like studying Earth from the moon. Earth and its moon together form a system that revolves around the sun.

Like all smaller bodies in the universe, the moon has movements that are similar to the rotation and revolution of Earth. The moon rotates on its axis as it revolves around Earth. However, since the moon's periods of rotation and revolution are the same, the same side of the moon always faces Earth. The changes in the appearance of the moon as it revolves around Earth are called the *phases of the moon.* The next activity can help you learn about the moon's phases.

Keep a record of the phases of the moon for two weeks, longer if you live in an area with a lot of cloudy evenings. Make drawings of each phase, and record the time the moon rises and sets. Consult a newspaper to find out when to watch. ✎

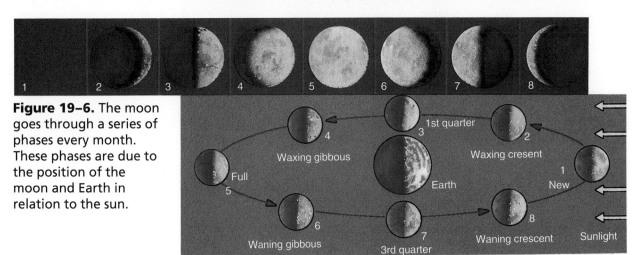

Figure 19–6. The moon goes through a series of phases every month. These phases are due to the position of the moon and Earth in relation to the sun.

Although the moon appears to shine on its own, moonlight is really reflected sunlight. When the entire side of the moon that faces Earth is lighted, you see a *full moon.* When the side of the moon facing Earth is not lighted, the phase is called a *new moon.*

Occasionally the moon moves into a position in which Earth is directly between it and the sun. Then the sun's rays, which would ordinarily fall on the moon's surface, are temporarily blocked by Earth. When Earth's shadow falls on all or part of the moon's surface, a **lunar eclipse** occurs. Because of the relative sizes of Earth and the sun, two shadow zones form during a lunar eclipse. The zone of complete shadow is called the *umbra.* The umbra is surrounded by a zone of partial shadow called the *penumbra.*

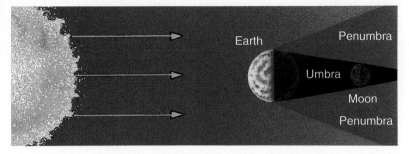

Figure 19–7. A lunar eclipse (above) occurs when Earth casts a shadow on the surface of the moon (right).

During a total lunar eclipse, the moon passes through the penumbra and into the umbra. You can watch the moon pass through a region of dim light until it crosses a distinct boundary into relative darkness. In a partial lunar eclipse, all or part of the moon passes through the penumbra but not the umbra, so the moon never completely darkens.

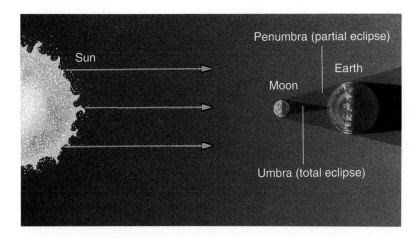

Figure 19–8. A solar eclipse (below) occurs when the moon passes between Earth and the sun (left).

Sometimes the moon passes between the sun and Earth, causing a **solar eclipse.** A partial solar eclipse occurs when only part of the sun is blocked by the moon. During a total solar eclipse, only the corona of the sun is still visible. The moon's shadow moves rapidly across the surface of Earth. In fact, a total solar eclipse can be seen from any one position on Earth for at most about 7 minutes. **CAUTION: Never look directly at the sun during an eclipse. You could be blinded.**

Now that your tour group has seen the relationship between our planet and its moon, it's time to take a look at the rest of the solar system. Pack your bags again.

 ASK YOURSELF

When would observers on the moon see an eclipse of Earth?

SECTION 1 *REVIEW AND APPLICATION*

Reading Critically

1. Compare and contrast the revolution and rotation of Earth.

2. Describe the relative positions of the sun, moon, and Earth during a full moon.

Thinking Critically

3. If you travel west in a jet plane, does the time get later or earlier? What if you travel east?

4. Why is it always hot at the equator?

5. Why is the far side of the moon never seen from Earth?

The Inner Planets and Asteroids

Objectives

Name *the inner planets, and* ***describe*** *their arrangement and orbits in the solar system.*

Compare and Contrast *Earth with the other terrestrial planets.*

Explain *why the inner planets and the moon have different surface ages.*

Your tour ship would have to travel far into space to see all the planets in orbit around the sun. Since this would take much too long, let's imagine that we will be observing a slide show on the solar system in the ship's lounge. This will give us the perspective we need while the ship is on course to the first stop on the inner planets tour. As the lights dim, your guide reads you this poem to get you in the proper frame of mind.

JEWELS

Space blazes with jewels,
          a shimmering ice
          of billions of diamonds

     dazzle

          the Milky Way:

               Jupiter, a giant agate,

          Uranus, a ball of jade,

          Pluto, a luminescent pearl,

     Saturn, a halo of rings.

                    A slice
          of moon, a crescent brooch.

     Bright rubies splay
     Antares

               in this
                    midnight
                    masquerade.

                              Myra Cohn Livingston

A View of the Solar System

Earth and its moon are only part of a complex system of orbiting bodies that make up the solar system. The solar system is named for the sun, *sol,* its most important body. About 5 billion years ago, a massive cloud of dust and gas was pulled together by gravity. The more massive this collection of dust and gas became, the more matter it could attract to itself through the force of gravity. Eventually the mass began to spin, and the pressure and heat increased enough for the sun to ignite. Other masses of gas and dust were pulled together to form the planets and other bodies of the solar system.

Our solar system contains nine known planets. Astronomers are examining the possibility of a tenth planet, although it will be some time before they reach any conclusion. The word *planet* comes from the Greek word for "wanderer." Ancient people were puzzled by the planets. While the sun, moon, and stars had regular movements, other heavenly bodies seemed to wander against the background of stars. The ancients named these wandering bodies *planets.*

Planets are the main bodies that revolve around the sun. They all orbit in the same direction—counterclockwise, if viewed from above the plane of the solar system. The gravity of the massive sun keeps the planets in orbit. The nine planets, in order of average distance from the sun, are named *Mercury, Venus, Earth, Mars, Jupiter, Saturn, Uranus, Neptune,* and *Pluto.* Some clever person invented a mnemonic—a memory device— to help people remember their order. It goes like this: *My Very Energetic Mother Just Sent Us Nine Pizzas.* Perhaps you can think of a different mnemonic to help you remember the names of the planets in order.

Figure 19–9. The nine planets in our solar system are shown (not to scale).

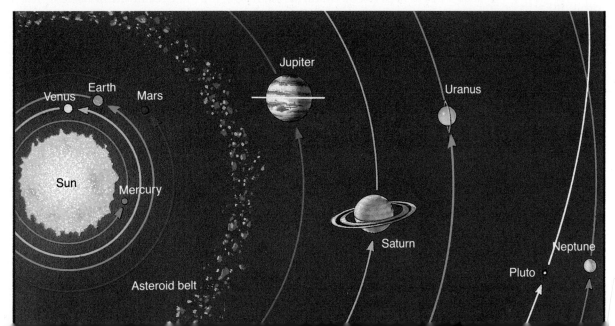

Accurate mathematical observations of some of the planets were made in the late 1500s by a Danish astronomer, Tycho Brahe. For many years Brahe observed the positions of the planets. Upon his death he left his notes to his assistant, Johannes Kepler. Kepler, a German, continued to study the planets, using Brahe's data, and plotted their orbits.

DISCOVER BY *Researching*

Use library resources to find information about both Tycho Brahe and Johannes Kepler. Choose one major observation of Brahe and one law of Kepler and explain each in your own words. Write your findings in your journal. ✎

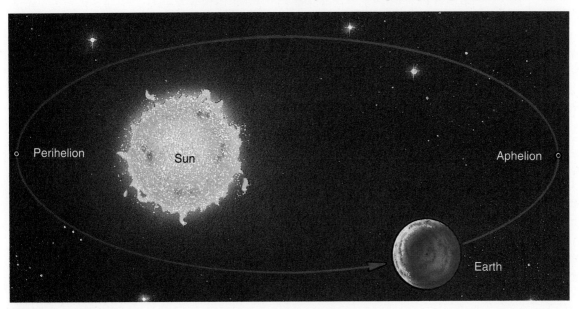

Figure 19–10. The orbit of Earth or any planet is not circular. Rather, the path a planet takes around the sun forms a flattened circle, called an *ellipse*. The farthest point in the orbit is called *aphelion*; the nearest is called *perihelion*.

A planet's distance from the sun is measured in astronomical units. An *astronomical unit* (AU) is the average distance between Earth and the sun, or about 150 million km. The distances in the solar system are so vast that astronomers often find it more convenient to measure them in terms of AUs than kilometers.

Your tour now begins at the planet nearest the sun. As you travel, you might wish to consult Table 19–1. It shows a picture of each planet in the solar system and gives a brief description of each one.

 **ASK YOURSELF**

Why do the planets orbit the sun instead of the sun orbiting one of the planets, say Earth?

| Table 19-1 | **The Planets of the Solar System** |

Mercury, with a diameter of about 4900 km, is about 58 million km (0.39 AU) from the sun. A year on Mercury is 88 Earth days, and a day is 58.67 Earth days. Mercury has practically no atmosphere. Its surface is rocky and old, with many craters and steep cliffs. Temperatures on Mercury range from −170°C to more than 400°C. Mercury has no satellites.

Venus, with a diameter of more than 12 000 km, is about 108 million km (0.72 AU) from the sun. A year on Venus is 225 Earth days, and a day is 243 Earth days. The planet's thick atmosphere is mostly carbon dioxide and sulfuric acid. The surface of Venus has vast plains, a few high mountains, much evidence of former volcanic action, and few craters. Temperatures on Venus average about 500°C. Venus has no satellites.

Earth, with a diameter of about 12 760 km, is about 150 million km (1.00 AU) from the sun. A year on Earth is 365.26 days, and a day is 24 hours. Earth's atmosphere is mostly nitrogen and oxygen. The surface of Earth is 70 percent water. Temperatures on Earth range from −90°C to about 60°C. Earth has one satellite, the moon.

Mars, with a diameter of nearly 6800 km, is about 228 million km (1.52 AU) from the sun. A year on Mars is 1.9 Earth years, and a day is 24.4 Earth hours. The thin atmosphere of Mars is mostly carbon dioxide and nitrogen. The surface of the planet has polar ice caps, dry riverbeds, and the largest volcano and canyon in the solar system. Temperatures on Mars range from −100°C to about −30°C. Mars has two satellites.

Jupiter, with a diameter of nearly 143 000 km, is about 778 million km (5.20 AU) from the sun. A year on Jupiter is 11.9 Earth years, and a day is 10 Earth hours. Jupiter's atmosphere is mostly hydrogen and helium. Jupiter has no solid surface. Temperatures range from −125°C in the atmosphere to nearly 30 000°C in its core. Jupiter has 16 satellites, including the four large Galilean moons.

Saturn, with a diameter of about 120 000 km, is about 1427 million km (9.54 AU) from the sun. A year on Saturn is 29.5 Earth years, and a day is 10.5 Earth hours. Saturn's atmosphere is mostly hydrogen and helium. Saturn has no solid surface. Temperatures in the atmosphere average about −180°C. Saturn has 18 confirmed satellites and 4 more whose orbits are not yet identified.

Uranus, with a diameter of over 52 000 km, is about 2871 million km (19 AU) from the sun. A year on Uranus is 84 Earth years, and a day is 17 Earth hours. The planet's atmosphere is mostly hydrogen, helium, and methane. Uranus has no solid surface. Temperatures in the atmosphere average about −225°C. Uranus has 15 satellites.

Neptune, with a diameter of about 50 000 km, is about 4497 million km (30.3 AU) from the sun. A year on Neptune is 165 Earth years, and a day is 16 Earth hours. Neptune's atmosphere is mostly hydrogen, helium, and methane. Neptune has no solid surface. Temperatures in the atmosphere average about −220°C. Neptune has 8 satellites, including a large moon, Triton, which has current volcanic activity.

Pluto, with a diameter of about 2300 km, is about 5913 million km (39.5 AU) from the sun. A year on Pluto is 249 Earth years, and a day is 6 Earth days. Pluto's atmosphere is mostly methane. The surface of Pluto is solid with unknown features. Temperatures on Pluto average about −230°C. Pluto has one satellite.

Mercury and Venus

Your spacecraft is flying over Mercury at present, the first of the inner planets. The inner planets are often called terrestrial, or earthlike, planets because they have some characteristics in common with Earth. The inner planets are mostly giant balls of rock, because most of their lighter gaseous elements have been obliterated by the sun's heat and gravitation. Only solid elements, such as iron, nickel, and silicon, remain.

Mercury is the smallest of the inner planets and the one closest to the sun. From your privileged point of view you can see Mercury as it looks in the larger photograph. But from Earth it is a different matter. Even through a telescope, Mercury appears as an indistinct object just above the horizon in the morning and evening sky.

Figure 19–11. Mercury is named after the fleet-footed messenger of the Roman gods. The name is appropriate because Mercury has such a rapid period of revolution.

Like all the planets, Mercury underwent massive bombardment by meteorites and other objects during the early period of the formation of the solar system. This period was known as the *impact period*. You can see the results in the larger photograph—crater upon crater covers the planet's surface. Because Mercury is so small, it has little gravity—certainly not enough to hold onto much of an atmosphere or a moon. And without a substantial atmosphere, little weathering can occur to wear away the craters. Also, Mercury has had no recent volcanic activity to alter its surface.

Scientists say that Mercury has a very old surface. They mean that its surface exists today pretty much as it did billions of years ago, because little has happened to change it. Based on this information, do you think Earth has a young or an old surface? Explain your answer.

The density of Mercury is calculated to be similar to that of Earth. This indicates that it may have an iron-nickel core as Earth has. Unlike Earth, however, Mercury's magnetic field, is very weak. Scientists think this may be because the planet rotates too slowly for electrical currents to develop in its core.

The second planet from the sun is Venus, named for the Roman goddess of love. Venus is often thought of as Earth's sister planet because the two have similar masses, densities, and sizes. Often visible in the evening or morning, Venus appears to have phases similar to those of Earth's moon.

Venus rotates very slowly, taking 243 Earth days to turn once on its axis. Unlike Earth and most of the other planets, Venus rotates east to west, or clockwise. Our sister planet takes 225 Earth days to orbit the sun, so a day on Venus is longer than its year! Complete the following activity to compare your age on Earth and on Venus.

DISCOVER BY *Calculating*

How may Venus years have passed since you were born? Multiply your age by 365 plus the number of days in your current uncompleted year. This total gives you the total number of days old you are on Earth. On Venus, a year is 225 Earth days. Divide the total number of days in your Earth age by 225. How old would you be in Venusian years? ✎

As your ship cruises over Venus, you will observe a scene like that in the photograph at right. Unlike Mercury, Venus is large and dense enough to hold an atmosphere. But what an atmosphere! You would see no clear sky from the surface of Venus. The planet is covered constantly by an atmosphere made up mostly of carbon dioxide and sulfuric acid. This dense atmosphere creates a surface hot enough to melt lead and an atmospheric pressure 100 times as great as on Earth. It also prevents all but the largest of bodies from reaching the surface of Venus to create impact craters.

Figure 19–12. Venus is sometimes called the cloud planet because its atmosphere resembles clouds on Earth. The surface of Venus is shown at bottom.

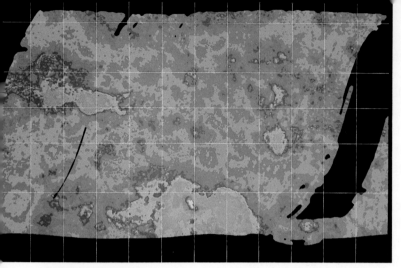

Figure 19–13. Several large Venusian land masses are shown in this relief map created from data collected by the *Pioneer Venus* orbiter.

The surface of Venus is largely covered with plains, as seen in the *Pioneer Venus* relief map at left. The major land masses on Venus are Ishtar Terra and Aphrodite Terra. Ishtar and Aphrodite are two mythical goddesses; in fact, all the features on Venus except one are named for real or mythical women.

Venus probably has an iron-nickel core almost as large as that of Earth. But the planet has no magnetic field, perhaps for reasons similar to those that explain the weak magnetic field of Mercury.

Although scientists are still not sure whether volcanic activity takes place on Venus today, it apparently occurred in the recent past, filling up craters and creating and filling basins and plains. Much of the planet's surface is only about 200 million to 400 million years old, compared with about 3.5 billion years for Mercury.

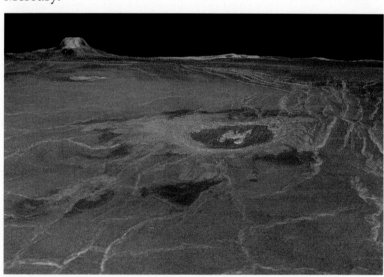

Figure 19–14. *Magellan,* which mapped Venus using radar, could see craters as small as 3 km across. The crater shown in the foreground here is 37 km wide; the other two are larger.

As your tour ship leaves the sulfurous atmosphere of Venus, you long for the relatively clean and breathable atmosphere of Earth. We'll make another brief visit there.

 ASK YOURSELF

What accounts for the basic difference in the surface appearances of Mercury and Venus?

Earth and Moon Revisited

As the tour continues, you fly from Venus to Earth, circling both Earth and its moon. You see that Earth has characteristics both similar to and different from those of the other inner planets. Like the other terrestrial bodies, Earth is a rocky planet whose surface was shaped by meteorite impact and volcanic action early in its history.

However, Earth is also large enough, dense enough, and far enough away from the sun to have an atmosphere, water, and a satellite. A **satellite** is a small body, or moon, orbiting a planet. Earth is the closest planet to the sun that possesses such a companion. Earth's surface temperature is also very moderate. These conditions contributed to the evolution of life on our planet.

As you look down on Earth, you will see some of the ways in which our planet is different from the other inner planets. For example, you will not see hundreds of craters left over from the impact period. As you learned earlier, Earth's surface is constantly changing due to weathering, volcanic action, and tectonic activity. Our planet has the youngest surface in the inner solar system.

Another difference is also present, though you cannot see it. Earth, with its iron-nickel core extending almost halfway to the surface, does generate a magnetic field. Recall that Mercury and Venus have such cores, but little or no magnetic activity. Earth's core is surrounded by a thick mantle of molten rock, which in turn is covered with a solid crust.

Figure 19–15. Active volcanoes are just one process that keeps Earth young and ever changing.

Our moon, about one-fourth Earth's diameter, is nearly 385 000 km from its planet. Density calculations show that the moon may have a small iron-nickel core; however, it has no magnetic field. Data collected from instruments left on the moon by Apollo astronauts indicate the presence of a mantle and a thick crust.

The moon has two types of terrain, as you can see when your tour craft flies over the surface. The older highlands are composed of feldspar rocks. They reflect more light and show up grayish white in the photo below and when you view the moon. The younger *maria,* or "seas," are formed of basalt—hardened lava. These areas reflect less light and show up black or dark in the photo or from Earth. The "seas," of course, are not filled with water; they only resemble seas when viewed from Earth. The moon has never had water, and with such weak gravity (one-sixth that of Earth's), it has never held an atmosphere. How much of the moon's surface is covered with highlands compared with that covered by maria? Try the next activity to find out.

DISCOVER BY *Observing*

On the night of the next full moon, observe the surface of the moon and draw it on a large piece of paper. Be sure to make several observations so that you get as accurate a drawing of the light and dark areas as you can. Then do a rough estimate of the percentage of maria and the percentage of highlands. Compare your drawing with a photograph of the moon's visible side. How close were you? ✐

Figure 19–16. Although the moon has no active weathering, surface features slowly erode over time.

The surface of the moon is marked with craters like those of Mercury. You can see the variety of crater sizes in the photograph. Scientists can determine the age of various craters by examining the condition they are in. For instance, look at the large crater at the top of the photograph. Although there is practically no weathering as such on the moon, erosion does occur over long periods of time. This erosion is due to micrometeoroid impacts, heat expansion and contraction, and gravity. The rims of the large crater have worn away and fallen onto its floor, suggesting that it is much older than the two smaller and sharp-rimmed craters to the south of it.

There are several theories about the origin of the moon. One is that the moon was originally a separate body revolving around the sun that was captured by Earth's gravity. Another theory suggests that at one time Earth spun very fast and flung large parts of itself out into space, where they collected and began to orbit Earth.

Today another theory has taken center stage. Scientists theorize that the young Earth and a large body crashed into each other about 4.6 billion years ago. This impact caused a massive explosion and sent material from both Earth and the other body into space, as shown in the diagrams. Eventually, the colliding bodies formed the new Earth, while the excess material in space clumped together as the moon and was locked into an orbit around Earth. The moon rocks brought back by Apollo astronauts have provided evidence to support this theory.

Figure 19–17. How the moon may have formed

First of all, moon rocks are made up largely of material very similar to that found in Earth's upper mantle. Just such material would have been thrown off Earth in an impact explosion. Second, the moon has no volatiles—materials that evaporate easily, such as water, ammonia, and sodium. All these elements, which are found in other planetary bodies, would have evaporated when they were flung into space after the impact. Scientists continue to evaluate this theory.

Before you head out to Mars, the red planet, your tour ship swings around so you can look down on the far side of the moon. You notice how few maria there are compared with the near side. Scientists hypothesize that the much thicker crust on the far side prevented the massive volcanic flows that formed the maria on the near side. Ten of the craters on the far side are named for *Apollo 1* and *Challenger* astronauts who died in the exploration of space.

 ASK YOURSELF

Which has the older surface, the moon or Earth? How do you know?

Mars and the Asteroids

It is time for your tour ship to visit the planet that has captivated skywatchers on Earth since ancient times. Named for the Roman god of war because of its blood-red color, Mars has been the subject of much speculation, now proven to be in error. For example, there are no artificial canals on Mars, as many people believed for years. And there is no race of aliens living on Mars. For astronomers, however, the real wonders of Mars far outshine these fanciful stories.

The Red Planet   A little more than half the diameter of Earth, Mars is the fourth planet from the sun. Its orbit takes 687 Earth days. Mars is called the red planet because its surface is reddish in color. The planet has a significant but small iron-nickel core and a weak magnetic field.

The surface of Mars includes deserts, dry riverbeds, flood plains, mountains, and the highest volcano in the solar system, Olympus Mons. Another singular surface feature is the Valles Marinaris, shown in the Viking photograph below. The largest and longest canyon in the solar system, Valles Marinaris measures about 4000 km long, the width of the United States. There is a small side canyon in the photo into which Earth's Grand Canyon could fit.

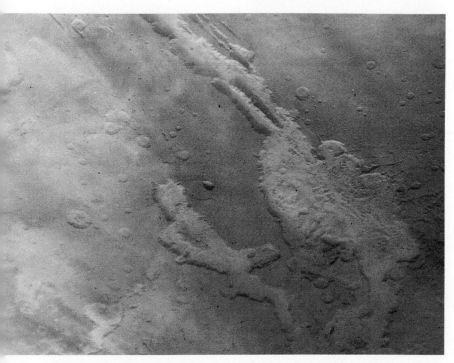

According to scientists, the existence of dry riverbeds and flood plains indicates that water was once present on Mars. Today, Mars is mostly dry, the water having been lost to space. However, there may be water locked in the polar ice caps (along with frozen carbon dioxide) and under the planet's surface. There are also traces of water in the planet's atmosphere, which is almost 96 percent carbon dioxide.

Figure 19–18. Valles Marinaris carves a deep gash across Mars' equatorial region.

Mars has two small satellites, Phobos and Deimos. Both have irregular shapes and may be captured asteroids. **Asteroids** are rocky objects that orbit the sun in a belt between the orbits of Mars and Jupiter. The largest asteroid is Ceres, with a diameter of 1025 km. Large asteroids have a spherical shape; smaller asteroids have irregular shapes, like the two satellites of Mars.

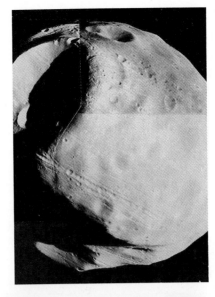

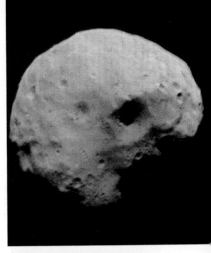

Figure 19–19. Mars' moons Phobos and Deimos. The planet was named for the Roman god of war. He had two attendants—Phobos, the bringer of fear, and Deimos, the bringer of panic.

ASTEROIDS

Space is a

playing field for asteroids.

Dusty and sun-scorched,
orbiting
about the sun,
each has a name:

    Eros,

    Icarus,

    Vesta,

    Flora . . .

Through a great void
these chunks
of rock
and metal
run,

    hurtling,

       colliding

in their endless game.

Myra Cohn Livingston

The Almost Planet   Some scientists think that asteroids represent bodies that failed to combine into a planet. The gravity of the neighboring planet Jupiter was probably too strong to allow the asteroids to form a single body. The asteroids revolve around the sun in individual orbits, often colliding with each other. These collisions break up the asteroids and disturb their orbits.

The only asteroid to have been photographed at close range is Gaspra. Only 13 km long, Gaspra was photographed by the spacecraft *Galileo* in 1991.

Meteorites are asteroid fragments that leave orbit, are captured by Earth's gravity, and fall to Earth. Large ones form craters upon impact, and traces of some of them are visible on Earth. One of the largest ones, Meteor Crater, is in Arizona. It has eroded slowly because the climate of the Southwest is dry and water is a primary source of weathering.

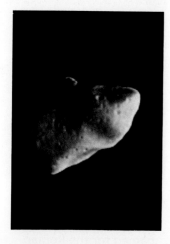

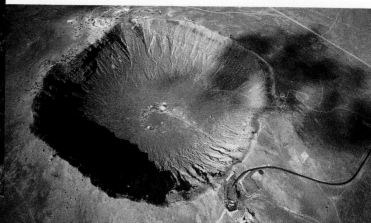

Figure 19–20. The asteroid Gaspra (above). Meteorites (left) form when pieces of asteroids break off, leave orbit, and fall to Earth. Large meteorites form craters such as this one (right) in Arizona.

Perhaps this is a good time to take a break and send postcards to your friends at home. Then it's off to the outer planets!

 ASK YOURSELF

What evidence is there that water once was present on the surface of Mars?

SECTION 2 *REVIEW AND APPLICATION*

Reading Critically

1. How might the moon have formed?

2. Why did the asteroids not form a single planet?

Thinking Critically

3. Why are the surfaces of Mercury and the moon heavily cratered, while those of Venus and Earth are not?

4. Is it possible to say that the age of a planet and the age of its surface are different? Why?

INVESTIGATION

Dating Surface Features

▶ MATERIALS
● paper and pencil

▼ PROCEDURE

1. Examine the photograph of the moon's surface. Notice that there are two basic areas of shading. The lighter areas are high-lands—heavily cratered mountainous areas. They are the older parts of the moon. The darker areas are maria, or seas, that have filled with hardened lava, which causes them to show up darker than highlands. The maria are younger than the highlands and generally have fewer craters.

2. Examine the maria, and find the crater marked "1." Note that you can see its complete outline on top of the mare. This tells you that the crater is younger than the mare on which it sits. If the lava had filled the mare more recently than the impact that formed the crater, the crater would be covered up or filled in. This process of determining relative age is called dating by superposition. That is, the feature that sits on top of another feature is younger. You can use this process with surfaces such as those on the moon, Mercury, and Mars, because little weathering and no tectonic activity has taken place for a long time.

3. Look at the two craters labeled "2" and "3." Make a chart, and indicate which crater appears to be the older of the two.

4. Look at the three craters marked "4," "5," and "6." Enter these three craters on your chart, and indicate which crater appears to be the oldest of the three.

5. Look at the crater marked "7." Indicate on your chart what you think has happened to create it.

▶ ANALYSES AND CONCLUSIONS
Based on what you think has happened to crater "7," is the crater older or younger than the lava flow that created the mare near it? How can you tell? What information does the shape of the crater reveal?

▶ APPLICATION
Could scientists use this process to date features on Earth? Why or why not?

✳ Discover More
Find more photographs taken of the moon, and make a list of what you can tell about the relative age of the features.

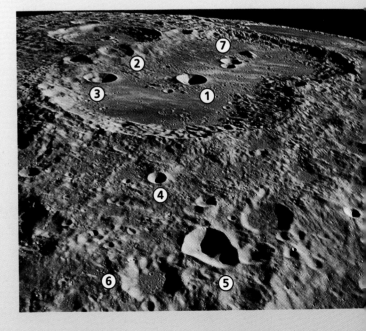

The Outer Planets and Comets

Compare and contrast the structures and atmospheres of the outer planets with those of the inner planets.

List the peculiarities of Jupiter's Galilean moons.

Describe the structure and composition of comets.

Figure 19–21. The layers of Jupiter

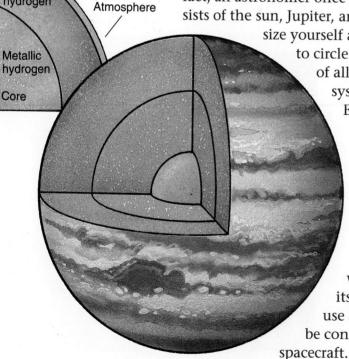

Liquid hydrogen

Atmosphere

Metallic hydrogen

Core

On your way to the outer planets, you recall seeing Jupiter from Earth, shining in the night sky. It is there during much of the year and can be seen through good binoculars or a small telescope. You can also see Jupiter's four largest moons. You have even seen the rings of Saturn through a telescope. Now you will see these gas giants and their colorful satellites and rings up close.

Jupiter and Its Moons

The outer planets include the giant planets Jupiter, Saturn, Uranus, and Neptune, and the smallest planet, Pluto. The giant planets have many features in common. They all consist mainly of light gases and have atmospheres that are largely hydrogen and helium, similar to that of the sun. All four have orbiting ring systems and numerous satellites.

Jupiter is by far the largest planet in the solar system. In fact, an astronomer once commented, "The solar system consists of the sun, Jupiter, and assorted debris." You experience its size yourself as your tour ship seems to take forever to circle the planet. Jupiter contains 71 percent of all the planetary material in the solar system and is 300 times the mass of Earth. The planet is called a *gas giant* because it has no solid crust as Earth and the terrestrial planets have. Beneath the thick atmosphere lies Jupiter's mantle, which extends all the way to its core. The pressure on Jupiter is so great that this material, mostly hydrogen, is liquid. If you could reach the surface of Jupiter without being crushed or poisoned by its atmosphere, you would have to use a boat. But for now you will have to be content with this close-up look from your spacecraft.

Inside Jupiter   The core of this giant planet may have a radius of 10 000 km, larger than Earth's radius. However, the core contains less than one percent of Jupiter's material. The best way to understand Jupiter's immense size is to think of the planet as a failed star. That's right: it could have been a star had it only collected several times more matter while it was forming itself into a planet. Jupiter has the same makeup as the sun after all—mostly hydrogen and helium. So what happened? Very simply, as large as Jupiter is, it is not large enough. To turn on the nuclear furnace that powers a star, there must be a certain kind of matter, which Jupiter has. But there must also be a certain *amount* of this matter, and Jupiter just doesn't have enough.

Jupiter's atmosphere is 17 000 km thick, and although it is not warmed much by the sun, Jupiter itself is very warm. In fact, Jupiter radiates about twice as much heat energy into space as it receives from the sun.

Figure 19–22. Jupiter (right), the largest planet, was named for the chief Roman god. The red spot (left) is an ongoing storm in the planet's atmosphere.

The light-colored bands of Jupiter's atmosphere are hot, rising gases; the dark-colored ones are cooler, sinking gases. These colored bands rotate around the planet at different speeds. A prominent feature of Jupiter's atmosphere is the Great Red Spot south of the equator. The red spot is a swirling mass of gases, with winds probably in excess of 1000 km per hour. As you fly over it, your guide tells you that the storm system is the size of three Earths. Circulation in the red spot is counterclockwise because of a Coriolis force. How do you think the Coriolis forces of Jupiter and Earth compare? The red spot storm has been observed continuously since it was discovered about 300 years ago. Who knows how long it was going on before that?

Hanging Around Jupiter   The four largest of Jupiter's 16 satellites were discovered by the Italian astronomer Galileo. In his honor they are called *Galilean moons*. In order of distance from the planet, they are Io, Europa, Ganymede, and Callisto.

Io, which is slightly larger than Earth's moon, has more volcanic activity than any other body in the solar system. Io's volcanoes spew sulfur and sulfur dioxide from beneath its crust into space. This sulfur material then falls back to the satellite's surface, giving it the brilliant colors of a large pizza. Everyone on the tour will probably want pizza for dinner tonight after seeing Io.

Figure 19–23. Io, often called Jupiter's pizza moon, is the most volcanically active body in the solar system.

Unlike Earth, however, Io's volcanoes are not created by processes within itself. Trapped between Jupiter and the other Galilean moons, Io is continually acted on by the tidal pulls of these bodies. In the next activity, you can find out how this process works.

Discover by Doing

Take a fresh orange, and hold it over a bowl. The orange represents Io. Begin to squeeze the orange over and over, first on one side and then on the other. Each time you squeeze the orange on one side, the opposite side is forced outward (Jupiter and Europa pulling on Io). After a time you will begin to hear the juice (sulfur) moving underneath the orange rind (Io's crust). Soon the juice will start spurting (volcanoes) up through the rind, and eventually the rind will start cracking in places. ✎

As you think about your demonstration, remember that a similar process has been part of Io's life since its beginning. This process keeps the satellite's surface constantly changing.

Europa, the second Galilean moon, is smaller than Earth's moon, and its surface is covered with a thick layer of ice. There are very few impact craters on Europa, suggesting that the surface reseals itself after an impact. Scientists wonder whether there is a melt–freeze–melt process occurring on Europa's surface, perhaps caused by tidal interaction also.

As you can see, Ganymede is a very large moon. In fact, it is the largest satellite in the solar system. Ganymede's surface has some impact craters, but it also has features such as surface grooves that may indicate more recent structural changes. These grooves may be caused by tension in the crust, like the fractures caused by tension in Earth's crust.

Figure 19–24. Jupiter's moons Europa (left), Callisto (center), and Ganymede (right)

Callisto, almost exactly the size of Mercury, is heavily cratered, indicating that its surface has not changed much since its formation. Which of the inner planets has a surface that reminds you of Callisto?

Jupiter has a ring system. It is not as spectacular as that of Saturn, but it is there. The rings consist of micrometer-sized particles of metals, silicates, and ice. The ring system is not visible from Earth; it was discovered and photographed by a Voyager spacecraft. It is best seen on the far side of Jupiter, looking back toward the sun. As your tour ship cruises past Jupiter on its way to Saturn, you should get a good view of the ring system.

 **ASK YOURSELF**

How are Io and Earth similar?

Figure 19–25. Jupiter's ring system as seen by *Voyager 2.*

Saturn, Uranus, Neptune, and Pluto

Saturn, shown in this photograph, is almost 10 times as far from the sun as Earth is. Like Jupiter, Saturn gives off more energy than it receives from the sun. Although only slightly smaller than Jupiter, the planet is much less dense. In fact, Saturn has the lowest density of all the planets. The rocky core of Saturn is small, and the structure of its mantle and atmosphere is similar to that of Jupiter.

Figure 19–26. Saturn was named for the Roman god of agriculture.

Figure 19–27. Saturn's ring system has been observed by scientists since Galileo's time.

Ring–Ring Saturn is the most distant planet you can see from Earth without a telescope. We will see it close up, as the Voyager probes did, and we will also see its moons. Titan, the largest of Saturn's 18 confirmed satellites, is larger than Mercury but smaller than Ganymede. The remaining satellites of Saturn are small, heavily cratered bodies of silicate rock and ice. Scientists had hoped to see Titan's surface from Voyager, but the cloud cover did not lift. You will probably have no better luck.

However, you have good luck with the rings and your tour guide gets you as close as she can to this startling display. Saturn's rings are worthy of a tour in themselves. First observed by Galileo in 1610, the ring system lies on Saturn's equatorial plane and consists of thousands of concentric rings separated by gaps of varying widths. The rings are made of chunks of ice-covered rock ranging in size from less than 1 cm to more than 10 m.

Far Out!　In order to see Uranus, your tour ship has to travel more than twice as far from the sun as it did to reach Saturn. Uranus was named for the Roman god of the sky. The planet is unusual in that its axis is inclined 98° from the plane of its orbit. As we fly over it, you will see that Uranus lies on its side, literally.

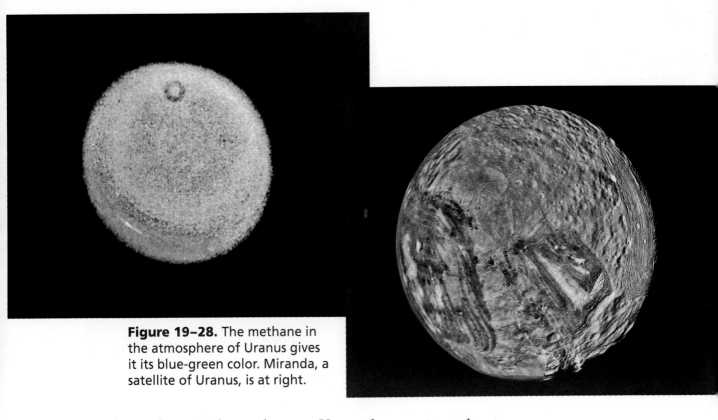

Figure 19–28. The methane in the atmosphere of Uranus gives it its blue-green color. Miranda, a satellite of Uranus, is at right.

A gas planet with a rocky core, Uranus has an atmosphere of mostly hydrogen and helium, with some methane added. The planet's ring system is thin, with particle sizes similar to those in Saturn's rings. Uranus has 15 moons, one of which, Miranda, has a surface that amazes astronomers. It has several different surface features, none of which seems to go with the others. Scientists wonder if Miranda was broken up by impact and then fell back together in this odd configuration.

True Blue   Neptune, our next stop, is almost 30 AU from the sun. With an atmosphere similar to that of Uranus, Neptune really is a blue planet. It is also about the same size as Uranus—about four times the diameter of Earth.

Voyager discovered a great dark spot on Neptune, shown in the photograph. Like the red spot on Jupiter, it is a giant storm circling Neptune in the planet's upper atmosphere. Neptune has four rings and eight moons, six of them discovered by *Voyager 2*. The planet's largest moon, Triton, has active volcanoes, only the third known body in the solar system to feature such geological activity. What are the other two? Triton also has a varied surface; one area is marked in such a way that astronomers have nicknamed Triton the cantaloupe moon.

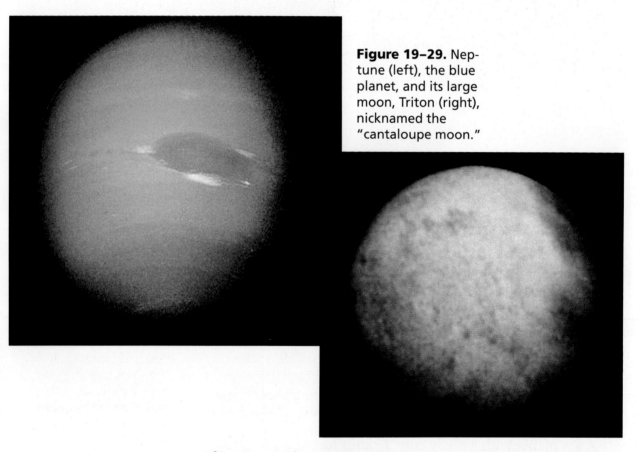

Figure 19–29. Neptune (left), the blue planet, and its large moon, Triton (right), nicknamed the "cantaloupe moon."

Planet or Planets?   Pluto, our last stop, is the outermost planet, averaging 6 billion km from the sun. Pluto is also the only outer planet without a thick atmosphere. Little is known about Pluto except that it has one known satellite, Charon, discovered in 1978. Because Charon is half the size of its planet, which is unusual for a moon, the Pluto-Charon system is known as a double planet.

The Earth-based telescope photograph of Pluto and Charon in Table 19–1 shows them only as blurred dots of light. Scientists can only speculate about what the surface of Pluto looks like, since it is the only planet not to have been visited by an Earth spacecraft. Scientists hope that a spacecraft launched late in this decade will reach Pluto early in the next century. In the meantime, complete the activity below to find out how much we do know about the outer planets and their rings and moons.

Figure 19–30. Pluto can be seen only as a small dot on even the best photographs. In the years to come, perhaps we will discover that Pluto and Charon look like this artist's conception.

DISCOVER BY Researching

Use library resources to find more information about the outer planets, particularly their rings and moons. Write several questions about these bodies that could be used in a class quiz. Ask your questions of your classmates, and then try to answer their questions. Were any of the questions duplicates? Which ones? Which bodies in the outer solar system do you and your classmates find the most interesting? Why? ✎

▼ ASK YOURSELF

Earth, Uranus, and Neptune can all be called blue planets. What causes each one to show so much blue color?

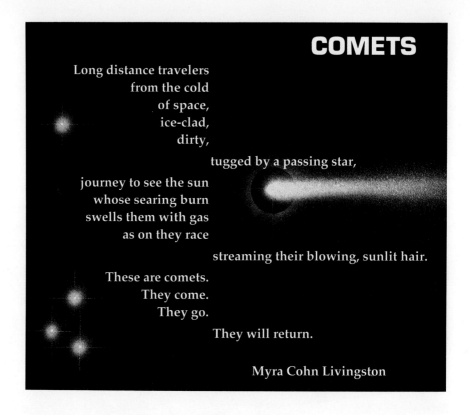

COMETS

Long distance travelers
from the cold
of space,
ice-clad,
dirty,

tugged by a passing star,

journey to see the sun
whose searing burn
swells them with gas
as on they race

streaming their blowing, sunlit hair.

These are comets.
They come.
They go.

They will return.

Myra Cohn Livingston

The Comets

Beyond Pluto are comets—bodies that orbit the sun with extremely long periods. **Comets** consist of silicate rock and metal particles embedded in ices. They are often referred to as dirty snowballs. The ices are frozen gases such as carbon dioxide, methane, ammonia, and water. Comets may be pieces of planetary matter, formed when the solar system was taking shape.

Comets are influenced by the giant planets—mainly Jupiter—and are forced into orbits that take them close to the sun. When a comet approaches the sun, gases escaping from the comet are blown by the solar wind. These gases form a tail, as much as l million km thick and 10 million km long, pointing away from the sun.

The diameter of a typical comet is about 1 km. The body of the comet loses a layer several meters thick each time the comet passes close to the sun. Scientists believe that most comets do not last more than a dozen passages around the sun. Halley's comet, however, has been observed every 76 years since 240 B.C.

How many times has Halley's comet been seen from Earth since then? Since we will not see the comet on this tour, complete the next activity to figure out when you can see it next.

DISCOVER BY *Calculating*

From the information on the period of revolution of Halley's comet, determine how old you will be the next time Halley's comet returns to Earth's neighborhood. Hint: Halley's comet was last seen in 1986. ✏

Your tour of the solar system is over. Well, not quite. You notice we have not visited the sun, the main body of our solar system. We still have not learned to create materials to protect us from the terrible heat and forces of that giant nuclear furnace. You will have to be content with photographs, and those you can see in the next chapter. Meanwhile, we must head back to Earth. Please tell your friends about us if you have enjoyed yourself. We can use the business!

► ASK YOURSELF

Why might astronomers wish to capture some material from a comet?

Figure 19–31. Halley's comet (left) comes near Earth every 76 years. The core of this comet (right) is composed of rocks and ice.

`0:14:57`   `61600 kM`

SECTION 3 *REVIEW AND APPLICATION*

Reading Critically

1. How do the four Galilean moons differ from each other?
2. What are the basic differences between the inner and outer planets?

Thinking Critically

3. How are the ring systems of the outer planets like unformed moons?
4. Why have the outer planets been able to hold on to so much gaseous material, when the inner planets have not?

SKILL Modeling Using Scale Drawings

▶ **MATERIALS**
- string, 30 m • metric ruler • construction paper • scissors • tape • markers

▼ PROCEDURE

1. Table 1 shows the diameter of the sun and each inner planet as well as each planet's mean distance from the sun.

2. Using a scale in which 1 mm = 10 000 km, draw the sun and planets on a piece of construction paper. The size of each planet does not have to be exact. At this scale, you will find that they are all small and that exact dimensions are difficult to represent.

3. Draw a square around the sun and each planet, making sure to label each square.

4. Cut out the squares.

5. Tape the square of the sun to the end of the string.

6. Use the same scale to determine the distance of each planet from the sun, and tape it to the string.

7. Stretch out the string.

▶ **APPLICATION**

Explain how you might use astronomical units instead of kilometers in this activity.

✳ *Using What You Have Learned*

Using the same scale of measurement, how far from Earth is the moon (385 000 km) in centimeters?

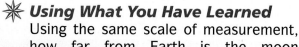

TABLE 1: THE SUN AND INNER PLANETS					
	Sun	**Mercury**	**Venus**	**Earth**	**Mars**
Diameter (km)	1.4×10^6	4900	12 000	12 750	6800
Distance from the sun (km)		5.8×10^7	1.08×10^8	1.5×10^8	2.28×10^8

The Big Idea

The elements of a system must have many similarities to work as a whole. If you look at the solar system as a whole, you see mostly similarities. Each body— sun, planets, moons, asteroids, and comets—was formed from gas and dust. Gravity causes every body to orbit around the sun. Similar processes formed the surfaces and atmospheres of the planets.

Taken separately, however, each body in the solar system is different. The cratered surfaces of Mercury and Callisto are vastly different from the changing surfaces of Earth and Europa. The atmosphere of Jupiter is nothing like that of Mars. Both similarities and differences can be found in every system.

For Your Journal

Look again at the questions you answered at the beginning of this chapter. How have your ideas about the solar system changed? Be sure to write your new ideas in your journal.

Connecting Ideas

Many important forces or actions in the universe have affected our solar system. Give one example of each force or action listed below to show how it has affected one or more bodies of the solar system.

1. **Gravity** _____

2. **Meteor impact** _____

3. **Volcanic activity** _____

4. **Weathering** _____

5. **Atmospheric pressure** _____

> ## Understanding Vocabulary

1. Explain how the terms in each set are related.
   a) lunar eclipse (560), solar eclipse (561)
   b) axis (554), rotation (554), revolution (555), orbit (556)
   c) satellite (569), asteroids (573), meteorites (574), comets (584)

Understanding Concepts

MULTIPLE CHOICE

2. The inner planets are called terrestrial planets because
   a) they underwent massive impact cratering.
   b) they are similar in many ways to Earth.
   c) they are close to the sun.
   d) they orbit around Earth.

3. All of the following affect a planet's surface features except
   a) stars.   c) atmosphere.
   b) water.   d) volcanic activity.

4. The only outer planet that is not a gas giant is
   a) Jupiter.   c) Neptune.
   b) Saturn.   d) Pluto.

5. The tilt of Earth's axis affects all of the following except
   a) the amount of time Earth takes to rotate.
   b) the length of daylight at any place on Earth.
   c) the change of seasons.
   d) the length of night-time at any place on Earth.

6. Jupiter failed to become a star because of its
   a) temperature.
   b) distance from the sun.
   c) mass.
   d) combination of gases.

7. The shape of the orbit of Earth is
   a) a circle.
   b) an ellipse.
   c) an arc.
   d) a sphere.

SHORT ANSWER

8. Why do the surfaces of the moon, Mercury, and Callisto look similar?

9. Which bodies in the solar system have the youngest surfaces? Explain your answer.

Interpreting Graphics

Use the time zone map on page 555 and a U.S. map to answer the following questions.

10. Suppose a salesperson in San Francisco wants to call a client in Boston at 11:00 A.M., Boston time. What time must it be in San Francisco?

11. Suppose you live in Florida. It is 8:00 A.M. If you call your cousin Chris in Alaska, will you be likely to wake him? Explain.

12. Crews of ore carriers that regularly cross which Great Lake from east to west must be concerned about time changes? Explain your answer.

Reviewing Themes

13. *Systems and Structures*
The sizes of the planet Mercury and the asteroid Gaspra are widely different— Mercury is many times the size of Gaspra. However, what structural similarity do they share?

14. *Environmental Interactions*
Explain why the moon does not have sedimentary rocks.

Thinking Critically

15. Explain why a comet's tail always points away from the sun.

16. What happens to the period of a planet's revolution the nearer it is to the sun? Explain the reason for this.

17. Jupiter gives off about twice as much energy as it receives from the sun. Some scientists believe that this energy comes from nuclear reactions within the planet. With this information, discuss the analogy of Jupiter being like a star with its own solar system.

Discovery Through Reading

Carroll, Michael. "The Volcanoes of Other Worlds." *Odyssey* 2 (January/February 1993): 24–29. The volcanoes on Io and Triton are caused by boiling sulfur and frozen nitrogen, respectively. Read these and other fascinating facts about volcanic activity in the solar system.

THE UNIVERSE

Scientists don't know when the universe began. Historians, however, can say when the universe was discovered. It must have been on a night in 1609, when Galileo first used his new telescope to observe the heavens.

To make full use of the telescope as a scientific instrument Galileo did something that no one had done before. He turned it to the sky. . . .

For tens of centuries men had seen the same objects in the sky, and they had come to believe that they had learned all there was to be learned in it. But Galileo saw new depths and a new population in the heavens. Everywhere he turned his telescope he saw stars never seen before, in the most crowded constellations and in the thinly scattered regions. He recognized that the Milky Way is not a long luminous cloud but is made of numberless stars so dim and close together that the naked eye cannot distinguish them.

from *Galileo and the Scientific Revolution*
by Laura Fermi and Gilberto Bernardini

For Your Journal

- What is our nearest star, the sun, really like?
- How is a star born and how will it die?
- How might the position of galaxies explain the origin of the universe?

The Sun—Earth's Star

Human beings and all living things on Earth depend on a "quaking inferno" for life. In the previous chapter you toured the solar system. Now you will make a different journey. This journey must also take place in your mind. Distance and danger make a physical voyage impossible. We will start at the sun.

The Composition and Structure of the Sun

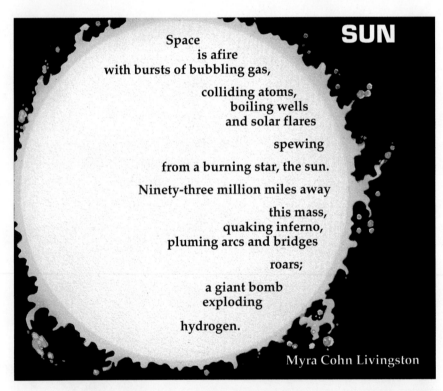

SUN

Space
is afire
with bursts of bubbling gas,

colliding atoms,
boiling wells
and solar flares

spewing

from a burning star, the sun.

Ninety-three million miles away

this mass,
quaking inferno,
pluming arcs and bridges

roars;

a giant bomb
exploding

hydrogen.

Myra Cohn Livingston

You may already be aware of the sun's importance, but consider again the following facts:

1) The planets and other objects are held in their orbits by the sun's immensely strong gravity.
2) Each planet's makeup and atmosphere is directly related to its distance from the sun.
3) Life on Earth is made possible by the constant influx of energy from the sun.

The Sun's Body Astronomers rate the sun as just an average star, and in terms of star characteristics they are right. However, to the inhabitants of Earth, the sun is the most important star. The sun has a diameter of 1 392 000 km, 109 times greater than the diameter of Earth, while its mass is more than 333 000 times that of Earth. In fact, the sun contains about 99.85 percent of all the mass in the solar system.

A dense core of hydrogen and helium makes up the center of the sun. The core has a temperature of about 15 million degrees Celsius. Since no material can exist as a solid at this temperature, this core is made up completely of plasma. Above the core is a thick *radiative layer.* The energy produced in the core warms this layer just as heat from a radiator warms a room. The temperature in this layer averages about 3 million degrees Celsius. Next is the *convective layer,* where energy is transferred by convection in the same way gas bubbles carry energy to the surface of a pot of boiling water.

Figure 20–1. The sun's energy provides the heat and light that sustain all life on Earth (above). The complex structure of the sun produces this energy (below).

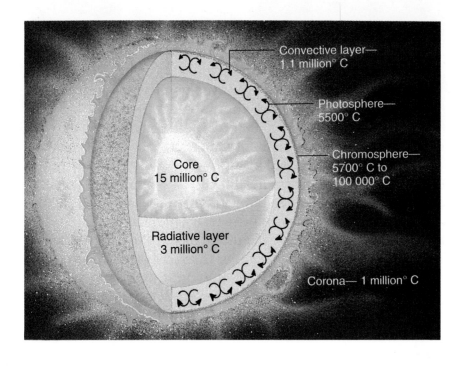

Convective layer— 1.1 million° C

Photosphere— 5500° C

Chromosphere— 5700° C to 100 000° C

Core 15 million° C

Radiative layer 3 million° C

Corona— 1 million° C

The sun's outermost layer is the *photosphere,* or "sphere of light." It forms the visible surface of the sun. The photosphere is similar to the surface of a pan of boiling and bubbling sauce. The temperature of the photosphere is about 5500°C. The surface of the photosphere has many small, bright areas called *granules.* Photographs taken by *Skylab* have revealed that these granules are actually the tops of rising columns of hot gases. Each column is about 1000 km across. The dark areas between granules are sinking areas of cooler gases.

Figure 20–2. Although granules seem to be bright spots, they are really columns of hot gases.

The Sun's Atmosphere   Above the photosphere is the *chromosphere*, a 2500-km thick atmosphere, where the temperature increases to 100 000°C. The chromosphere, or "sphere of color," shows up bright red in a spectroscope, an instrument that records and measures the colors in a spectrum. Above the chromosphere is the *corona*, a layer of thin solar gases that merges into space. The temperature in the corona is about 2 000 000°C. The corona can be seen as visible light only during an eclipse of the sun, when the moon blocks the light of the photosphere. But you just read that the sun has 99.85 percent of the mass of the solar system. How could the moon block out the photosphere? Complete the following activity to see how the tiny moon can accomplish this seemingly Gargantuan task.

Discover by Doing

Look at one of your classmates. Point your thumb as if you were giving a thumbs up sign to someone. Move your thumb back and forth across your line of sight and closer to and farther away from your face. At what point can you block your classmate from your sight with just your thumb? Briefly explain how a small object can block your view of a much larger object. ✏

Although much of the sun's atmosphere is hard to see from Earth, that doesn't mean nothing is happening up there. Intense activity in the sun's atmosphere produces massive *solar prominences*. These erupting clouds of gas may travel vast distances into space before falling back into the sun. The prominence shown here ballooned about 400 000 km into space, the distance from Earth to the moon.

Figure 20–3. The sun's corona is visible to the naked eye during a solar eclipse (left), when the bright photosphere is covered. A solar prominence (right), was photographed in ultraviolet light by a solar telescope.

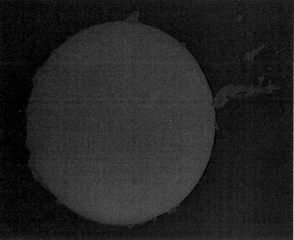

How Hot Is Hot?   You may have noticed that the temperature of the sun drops from the core to the surface and then rises again sharply through the atmosphere and into the corona. Scientists are puzzled by this phenomenon. You may remember that heat cannot naturally flow from a cooler body to a warmer one. Why and how then does the corona get so hot? Scientists are considering several theories for the super-hot temperature of the corona, including one based on activity in the sun's magnetic field.

 ASK YOURSELF

Which part of the sun is visible only during an eclipse?

Motions and Activity of the Sun

The energy that sustains life on Earth at this moment was created about 30 000 years ago in the core of the sun. Nuclear fusion created a form of energy that moved up slowly through the sun to its surface, changing form as it went. Finally, this energy became visible light. About eight minutes ago this energy sped away from the surface of the sun and traveled to Earth. You are reading, perhaps, by light energy created during Earth's Stone Age.

The fusion process that creates this energy in the sun's core goes on several trillion times a second and will continue until the sun dies. Nothing else our solar furnace does is as crucial to us as this process. The following activity will enable you to determine how long the other planets have to wait to receive their energy from the sun's surface.

DISCOVER BY *Calculating*

Using Table 19-1 on page 565, calculate how long it takes for the sun's energy to reach each of the other planets. Multiply the distance in AU by 8 to find the time in minutes.

A Slow Roll   The sun rotates on its axis in the same direction as the planets revolve, that is, counterclockwise as seen from above the sun's north pole. Since the sun is a gaseous body, not all parts rotate at the same speed. The rotational speed decreases toward the sun's north and south poles. At the solar equator, one sun day is equal to about 25 Earth days.

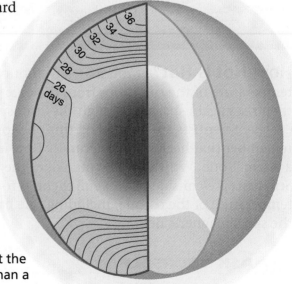

Figure 20–4. A day at the sun's poles is longer than a day at its equator.

"X" Marks the Spot

One characteristic of the sun that has long fascinated astronomers is *sunspots*. Sunspots are dark areas on the sun's surface that are regions of intense activity in its magnetic field. They appear darker than the rest of the sun's surface because they are cooler than their surroundings. Viewing sunspots is always interesting. The following activity will give you some guidelines on viewing and plotting sunspots.

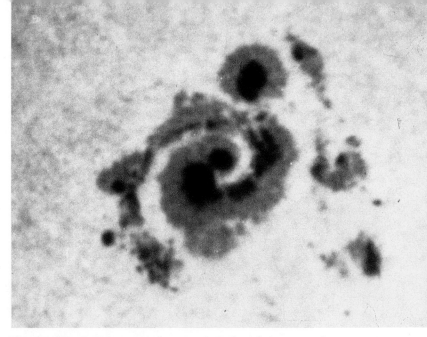

Figure 20–5. Sunspots are associated with increased magnetic activity in the sun.

BY Doing

CAUTION: Do not look directly through the main telescope or the spotter eyepiece. Set up a telescope on a tripod. Without looking through either the main telescope or the spotter eyepiece, point the telescope at the sun and focus the sun's image on a sheet of white construction paper or cardboard. You can set up the same activity using binoculars, only first you must block off either the left or right eyepiece. Use either setup to follow sunspot activity over a period of days or weeks. How long does it take a spot to move across the face of the sun and disappear behind the rim? Record your observations in your journal. ✐

Sunspot formation generally follows an 11-year cycle. During a sunspot minimum, the surface of the sun may be completely free of sunspots. During a sunspot maximum, several major groups may dot the surface of the sun at any given time. The 1980s and early 1990s were a period of intense sunspot activity; the next decade will be quieter. In sunspots, charged particles such as protons, electrons, and ions speed up until they escape from the sun's atmosphere. These particles cause an increase in strength of the solar wind. *Solar flares*, spectacular eruptions that occur on the sun's surface near sunspots, also cause periodic increases in the solar wind.

Figure 20–6. Solar flares interrupt communications on Earth and cause spectacular atmospheric displays called auroras.

The solar wind particles, which reach Earth in about 2.5 days, are responsible for disturbances in Earth's ionosphere. A particularly strong solar wind can disturb Earth's magnetic field, causing interference in radio and television transmissions and in microwave communications. It can also endanger astronauts in space, who are not protected by Earth's atmosphere. When the solar wind reaches Earth's atmosphere, it can produce spectacular displays called *auroras*.

 ASK YOURSELF

Does the heat generated in the sun's core move upward to the surface by conduction or convection? How do you know?

SECTION 1 *REVIEW AND APPLICATION*

Reading Critically

1. How quickly does sunlight reach Earth?

2. How do processes on the sun affect Earth?

Thinking Critically

3. If the photosphere is the only part of the sun that emits visible light, how have astronomers learned about other parts and processes of the sun?

4. Why might scientists study the sun very carefully before scheduling a shuttle flight with astronauts on board?

Other Stars

People long ago believed that the stars were some sort of "fire-folk," sitting in the sky. Our understanding of the stars has changed, but there is still much that is unknown. As our journey of the mind continues deep into space, you cannot help but feel a sense of awe.

Stars and Constellations

Stars are huge masses of glowing gases, primarily hydrogen and helium. They are found everywhere throughout the universe. The **universe** is all the matter and energy that exists: all the stars, planets, dust, gases, and energy in space.

You read in the previous chapter that our solar system was formed from a massive cloud of dust and gas. Other regions of the universe were formed in the same way. Billions of the bodies that formed out of dust and gas became stars. Stars are formed when gravity causes a dust cloud to contract, that is, a clump of matter gradually pulls more and more matter into itself. As the mass increases, the pressure at the center also increases, causing the temperature of the body to rise. When the temperature reaches about 1 million degrees Celsius, nuclear fusion begins. The hydrogen atoms in the gas fuse, creating helium and vast amounts of energy. The mass begins to glow, and the star begins giving off heat and visible light.

Figure 20–7. On a clear night away from city lights, many stars are visible from Earth.

Remember, all life on Earth is warmed and sustained by the giant fusion reactor that is our star, the sun. The fusion process keeps a star shining while it radiates its excess energy out into space. How is the light that a star gives off different from the light given off by other bodies in the universe?

You have learned that our star and solar system were formed about 4.6 billion years ago. Trillions of other stars have been formed since the beginning of the universe. Scientists have recently discovered that new stars are being formed even now. For example, infrared telescopes have detected new stars forming in the Orion nebula. Huge clouds of dust and gas are providing the raw material for new stars there, just as they did for our sun and solar system. Why do you think we cannot yet see these forming stars?

Figure 20–8. New stars are forming in these clouds of cosmic dust.

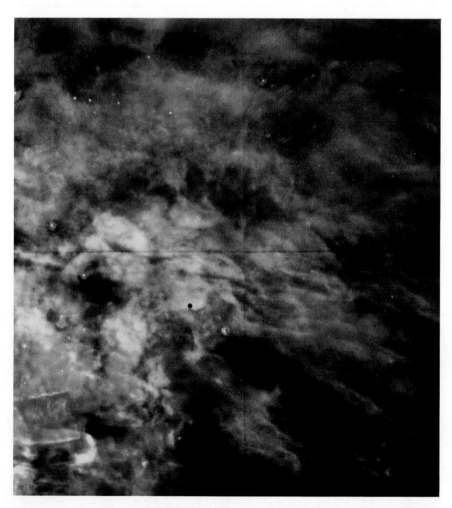

Putting Them All Together
Ancient astronomers looked at the multitude of stars strewn across the sky and saw pictures of animals, humans, and other objects in certain groups

of stars. Rather like the way you might make a dot-to-dot picture, early astronomers filled in shapes around groups of stars. Groups of stars that form such images are called **constellations.** It sometimes takes a great deal of imagination to see the shapes that our ancestors did, but you can always connect the dots in a way that makes sense to you. The ancient Greeks named 48 constellations. Today 88 constellations are recognized, and different areas of the sky are named for the main constellation located in that area.

It is important to remember that although constellations are helpful to observers on Earth, the stars in any given constellation really have no relationship with one another; that is, they are not all close together as they appear to our eyes. For example, find the seven stars of the Big Dipper—part of Ursa Major—shown below. Light from the star that sits at the end of the dipper's handle takes about 180 years to reach Earth, while light from the fourth star in from that one takes only about 30 years. All the others fall somewhere in between.

The diagram shows several other constellations. You can use it to identify some of the major stars in the night sky. Depending upon the time of year, the constellations appear in different parts of the sky at different times of the night. Our ancestors in the Northern Hemisphere made another important discovery in the night sky. The North Star, Polaris, does not appear to move at all. It thus provided a constant directional marker for sailors and travelers in the Northern Hemisphere before there were compasses or other navigational aids.

Figure 20–9. Some groups of stars form identifiable figures in the sky. The capitalized names are constellations, while the others are names of individual stars.

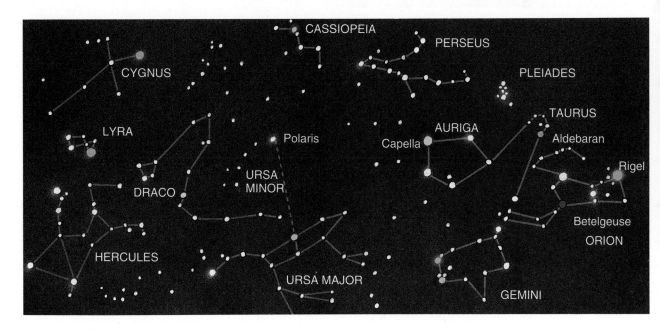

How Bright You Are   Using constellations is one way to locate, identify, and distinguish among stars. Magnitude is a scientific characteristic astronomers use to differentiate among stars. A star's **apparent magnitude** is its brightness as it appears from Earth—not its **absolute magnitude,** or *luminosity.*

The apparent magnitude of a star depends on two things: how much light it emits and how close it is to Earth. A dim star that is nearer Earth may appear more brilliant than a bright star that is farther away. What will happen to the apparent magnitude of stars as you journey out into space?

The ancient Greeks divided stars into six magnitudes. First-magnitude stars appear the brightest. The second, third, fourth, and fifth magnitudes include stars of decreasing brightness. Stars of the sixth magnitude are barely visible to the unaided eye. Every magnitude change means a difference in apparent brightness of 2.512 times. So a first-magnitude star is 100 times as bright as a sixth-magnitude star. Modern telescopes have made it possible to detect stars as faint as those of the 25th magnitude. Negative magnitudes are used for very bright objects. The sun, for example, has an apparent magnitude of –27.

Figure 20–10. The apparent magnitude of stars varies. The bright star shown here looks brighter than the stars around it. This may mean it is closer to Earth, or it may have a greater luminosity than the others.

The constellation Orion, which you can see regularly in the winter and spring, has two very bright stars. Betelgeuse, Orion's right shoulder as he faces you, and Rigel, Orion's left leg, are both 0 magnitude stars. Both stars have similar absolute magnitudes also (–7 and –6) and are similar distances from Earth. Remember that lower positive numbers and all negative numbers indicate very bright stars.

Compare these stars with Sirius, which you can find by following the three stars of Orion's belt to the left. Called the *Dog Star,* Sirius is the brightest star in the sky and is often reported as a UFO. Sirius has an apparent magnitude of –1.45. But it has an absolute magnitude of only +1.4, compared to the –6 and –7 of Betelgeuse and Rigel. Why does Sirius appear to be so much brighter than the other two stars in the group? In the next activity, you can try to locate and identify other bright stars.

Observing

At the same time you are observing Betelgeuse, Rigel, and Sirius, note some other bright stars in the night sky. Try to distinguish three levels of brightness with your naked eye. Make a drawing of where the stars are in the sky, and then find them on a star chart. Which stars are they? Of which constellations are they a part? What are their apparent magnitudes compared with their absolute magnitudes? ✎

The brightest of all stars are supernovas. A **supernova** is a star that has reached the end of its life and has exploded, blowing most of its glowing matter into space. Such a star appears hundreds of billions of times brighter than normal for days, sometimes years. An exploding star is called a supernova because in ancient times it was thought to be a new star; *nova* is Latin for "new." You can see a photograph of supernova 1987A on page 533.

▶ ASK YOURSELF

What two factors prevent us from seeing most of the stars in the universe?

From Red Giants to White Dwarfs

Color is another way to distinguish among stars. The energy output of stars varies tremendously. Some stars emit only one-hundredth as much energy as the sun; other stars emit 100 000 times more energy than the sun. You would not be able to get as close to these stars as you are to the sun. They are just too hot. The more energy a star emits, the hotter it is. The color of a star and its temperature are closely related. The hottest stars, such as Rigel and Sirius, are bluish in color, while the cooler ones, such as Betelgeuse, are reddish.

Figure 20–11. A star's color is based on its temperature. What colors do you see here?

In 1913 Danish astronomer Ejnar Hertzsprung and American astronomer Henry Norris Russell each published a graph showing that the energy emitted by a star is related to its color. They had discovered this characteristic of stars independently, which often happens in science. Their data was used to make a graph known as the *Hertzsprung-Russell diagram*. The diagram is on the next page.

Most of the stars that have been plotted so far form a diagonal curve on the diagram called the *main sequence*. Hot, blue stars are at the top left; cooler, red stars are at the bottom right. Yellow stars are plotted near the middle. Where is the sun on this graph? Is it cooler or hotter than Sirius? than Betelgeuse?

In addition to color differences, the Hertzsprung-Russell diagram shows star masses, that is, how much matter stars have. Star mass increases from the bottom to the top of the main sequence. Star mass affects a star's magnitude. According to the graph, what is the connection?

The largest and brightest stars are called *supergiants*. The smallest stars are called *dwarfs*. Dwarf stars are extremely dense; for example, a white dwarf could be as small as Earth but have as much mass as the sun. Supergiants, on the other hand, are large, but they may be even less dense than Earth's outer atmosphere. You can learn how to use the Hertzsprung-Russell diagram in the next activity.

DISCOVER BY *Doing*

Imagine that you are on a TV quiz show called "Stump the Astronomers." The astronomers are pretty sharp, but you can use the Hertzsprung-Russell diagram to find the answers to the following questions. Write the questions and the answers in your journal.

1. Which named star on the chart has the greatest absolute magnitude? What is this star's temperature, approximately?
2. Red giants have the same color and temperature as red dwarfs. How do these star types differ from one another?
3. What is the range of absolute magnitude for the yellow giants on the diagram?
4. The diagram shows the star Antares to be a red giant. Is this star hot or cool? Explain.
5. What is the absolute magnitude of the sun? How does the sun compare to Vega and Sirius? ✎

The Hertzsprung-Russell Diagram

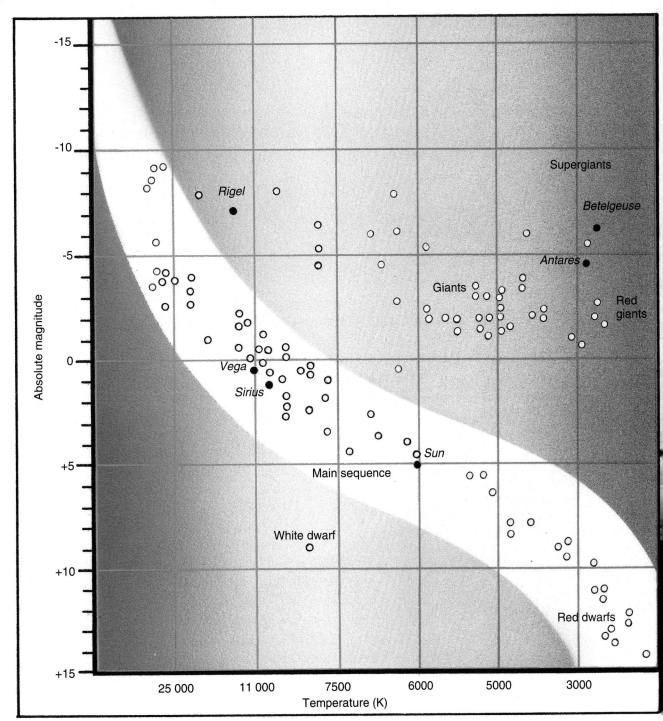

Figure 20–12. The *Hertzsprung-Russell diagram* shows the relationship between a star's color and its energy. What other star characteristic does the diagram show? Temperatures are in kelvin.

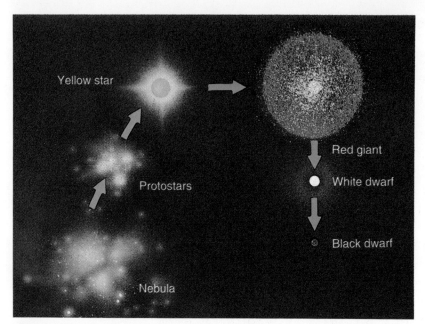

Figure 20–13. Shown here is the life cycle of a main-sequence star.

Scientists think that the sun is about 4.6 billion years old. It probably has enough hydrogen left in its core to continue shining as it does now for another 4.6 billion years. When the sun is about 10 billion years old, all the hydrogen in its core will be used up. The sun's core will contract, its temperature will rise, and its surface will expand outward. After another 100 million years the sun will become a red giant. A *red giant* is a very old star that has expanded and has a relatively low surface temperature. When the sun expands, its surface will cover Venus and Mercury. It will not reach Earth, but its heat will burn all life from our planet.

Eventually the center of the sun will get hotter, its outer layer will blow off, and the sun will become a white dwarf. The nuclear reactions in the center will stop. The sun will become a small star no more than 20 000 km in diameter (about 1.57 times the size of Earth), with a very high density. It will continue shining for billions of years, slowly cooling. When they completely cool off, white dwarfs become cold cinders called *black dwarfs*.

Figure 20–14. A binary star system (top). A nova (bottom) occurs when gas from its binary partner star collects on the hot white dwarf and ignites.

Astronomers estimate that more than half the stars in the universe are part of a binary, or double system, in which two stars are in close orbit around each other. Sometimes a white dwarf is part of such a double-star system. If so, it may undergo a different death. Instead of slowly cooling, it may capture hydrogen from its companion star. When the temperature of the captured hydrogen reaches 10 million degrees Celsius, a nuclear explosion occurs. The white dwarf increases its brightness about 10 000 times and appears as a nova. A nova reaches its maximum brightness in a day or two and then gradually returns to its original intensity.

▼ **ASK YOURSELF**

Will observers out in space be able to see the sun become a nova sometime far in the future? Explain.

Neutron Stars and Black Holes

You've probably guessed that not all stars develop in the same way. A star with a large mass has a different life cycle than a smaller star, such as the sun. When the hydrogen in the core of a large star is depleted, the core can no longer support the weight of the star's outer layers. It may shrink with such violence that the entire star blows up. In some stars, there is a two-way explosion: the core falls inward, and the rest of the star explodes outward. This is called a supernova, and it releases more energy than the sun will produce in its entire lifetime.

During a supernova explosion, 90 percent of the star's mass scatters into space, becoming the matter from which new stars may be born. The matter collects in a *nebula,* the Greek word for "cloud."

The remaining 10 percent, the core of the star, falls inward, becoming a neutron star. A **neutron star** is a very small star, as small as 15 km across. Some neutron stars rotate at high speeds and have strong magnetic fields. They are also very dense. If the Great Pyramid of Cheops in Egypt, which weighs about 3 million metric tons, had the density of a neutron star, it would be the size of a pinhead.

Some neutron stars give off radio waves in a narrow beam, which can be detected by radio telescopes. Because the neutron star is rotating, these waves show up like a blinking light, similar to the beacon of a lighthouse. The first rapidly pulsating neutron star, or **pulsar,** was discovered in 1967. The pulses were so regular that scientists thought for a time that they might be messages from some form of intelligent life.

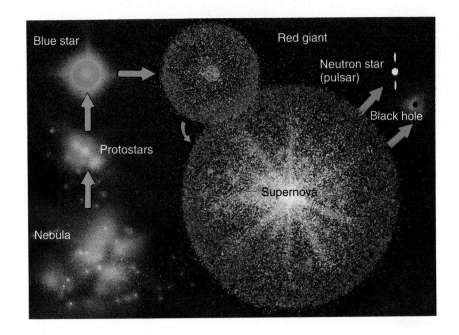

Figure 20–15. Pulsars form at the end of the life of some giant stars.

Figure 20–16. A nebula is the cloud of matter left over from a supernova. This is the Crab nebula, created by a supernova in 1054.

Many hundreds of pulsars have been discovered, but only three emit visible light. One of these is in the Crab nebula. It pulses off and on 60 times each second. Complete the following activity to see how fast that is.

DISCOVER BY *Calculating*

Calculate how many times the Crab nebula pulsar blinks on and off in a minute's time. Would it be difficult for your eyes to register this activity? Why or why not? Use a flashlight that you can turn off and on just by pushing down on a button. How many times can you do this in a five-second period? A three-second period? A one-second period? ✐

If the core mass of a supernova is more than three times greater than the sun's mass, the force of the implosion collapses the core into itself. Here the gravity is so strong that no energy, including light, can escape. Collapsed stars from which energy cannot escape are called **black holes.** Although scientists are fairly certain that black holes exist, there is no hard evidence as yet. Astronomers can "see" what they think are black holes by the actions of stars near a supposed black hole. The immense gravity of a black hole can pull off material from a nearby star, and astronomers can detect this process. Black holes are believed to lie at the center of most galaxies.

▼ ASK YOURSELF

Which group of stars on the Hertzsprung-Russell diagram are most likely to turn into black holes?

SECTION 2 REVIEW AND APPLICATION

Reading Critically

1. How does a star form?
2. What are some of the differences among stars?
3. Why do smaller stars last longer than larger ones?

Thinking Critically

4. We have had telescopes since the early 1600s. Why did it take so long to detect pulsars?
5. What do astronomers mean by saying that "looking at distant stars is looking into the past?"

INVESTIGATION

Demonstrating Magnitude

▶ **MATERIALS**
- small lamps with their shades removed (2)
- a variety of light bulbs of different wattages

▼ **PROCEDURE**

1. Use a very large classroom or a deserted hallway with electric sockets.
2. Have the rest of your group sit in a small area so they all view the demonstration from a similar distance.
3. Put a 60-watt and a 100-watt bulb in each of two lamps, and position the 60-watt lamp closer to the group than the 100-watt lamp. Have them determine which lamp is apparently brighter and which is absolutely brighter.
4. Place both lamps at the same distance from the group. How does this change the apparent brightness of the bulbs? How does it change the absolute brightness?
5. Put 100-watt bulbs into both lamps, and position them in the same manner as in step 3. How do the apparent and absolute brightnesses change?
6. Other groups will use different wattage bulbs to demonstrate the principles of apparent and absolute magnitude. After another group has set up a demonstration, try to determine brightness based on your observations.

▶ **ANALYSES AND CONCLUSIONS**

1. If two stars have the same absolute magnitude, in what situation could your naked eye record this fact?
2. Why does the moon, which is not even a star, look so much brighter than the brightest stars?

▶ **APPLICATION**

Why can't we see even very bright stars during the day?

✳ ***Discover More***

Use the library to find out the names of stars with a magnitude of −1 or brighter.

Galaxies

Define *light-year and* **explain** *its value in astronomy.*

Explain *the importance of the red shift when studying galaxies.*

Reconstruct *the events of the big-bang theory.*

Scientists always ask questions. You've probably had many questions on your journey of the mind. You've had to rethink concepts of size and distance. For the person who studies atomic structure—the world of the atom—a nucleus consisting of protons and neutrons orbited by electrons is a solar system in itself. Our star, solar system, and galaxy are another step up in terms of time and space. And finally, we confront a universe full of galaxies, which are themselves filled with billions of stars and which extend our concept of space and time nearly to infinity.

The Milky Way Galaxy and the Local Group

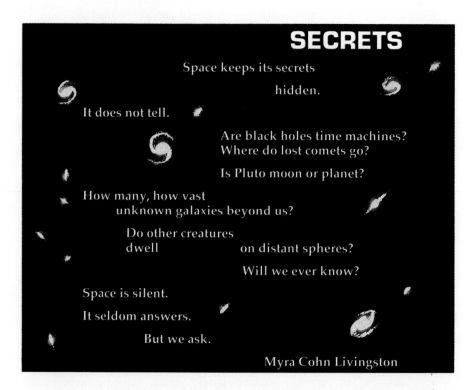

SECRETS

Space keeps its secrets
hidden.

It does not tell.

Are black holes time machines?
Where do lost comets go?

Is Pluto moon or planet?

How many, how vast
unknown galaxies beyond us?

Do other creatures
dwell   on distant spheres?

Will we ever know?

Space is silent.

It seldom answers.

But we ask.

Myra Cohn Livingston

When you took your trip through the solar system, you realized that the distance between planets is vast. If you traveled at the speed of today's spacecraft, you might cruise to the outer planets and back in a lifetime. Yet the distance to Pluto is not even a fraction of the distance between stars, which is measured by the **light-year**—the distance light travels in a year—about 10 trillion km. You would have to suspend time to make a trip to even the nearest star.

In 1838 the German astronomer Friedrich Wilhelm Bessel determined that the star known as 61 Cygni was 11 light-years away from Earth. This means that it takes 11 years for the light from 6l Cygni to reach Earth. There are stars closer to Earth than 61 Cygni. Remember Sirius, the Dog Star? Sirius is only 9 light-years away. Proxima Centauri, the closest star, other than the sun, is about 4.25 light-years distant. Imagine that you could travel 1 million km per hour. How long would it take you to get to Proxima Centauri? The following activity will give you a different sense of time and space.

Figure 20–17. Proxima Centauri, part of the bright-star system shown here, is the closest star, other than the sun, to Earth.

DISCOVER BY *Writing*

The next time it is visible where you live, observe the star Altair in the constellation Aquila the Eagle. Use a book of constellation charts to help you locate it. Altair is 16 light-years away. This means that when you look at Altair, you are seeing light that left the star's surface a couple of years before you were born. Think about what you have been doing as Altair's light has traveled to Earth. What is the fastest you have traveled? When and how? What is the greatest distance you have traveled? When and how? Write some observations in your journal that help describe your awareness of both time and space in your own lifetime. ✎

The stars just mentioned are all part of the Milky Way Galaxy. Our solar system is part of this galaxy also. A **galaxy** is a large grouping of stars. Our galaxy is called the Milky Way Galaxy because of a large band of stars that spreads like a ribbon of milk over our line of sight from Earth. It is harder and harder to observe these days, because of air pollution and light pollution, but if you are out in the country on a clear, moonless night, you should be able to see it. The Milky Way Galaxy we see is made up of billions of stars and nebulas. Yet the Milky Way Galaxy is much larger than that, and it is filled with billions more stars and nebulas than we can see.

What you see as the Milky Way Galaxy is actually only the edge of the galaxy. From the side, our galaxy is shaped like a disk that bulges in the middle. The view from the top of the galaxy, shown in the diagram at the bottom of the page, indicates that our galaxy is a spiral shape as seen from above. Each arm of a spiral galaxy curls around a central point. Galaxies are named for their shapes as seen from Earth. Spiral and irregular galaxies contain much dust and gas, in addition to stars. Elliptical galaxies contain very little dust and gas. Barred galaxies are galaxies whose arms seem to come off a bar rather than a central point, as in a spiral galaxy. Galaxies were classified by the American astronomer Edwin Hubble, for whom the Hubble Space Telescope is named.

Figure 20–18. Galaxies have several shapes (right). The Southern Pinwheel Galaxy (left) in the constellation Hydra, is an example of a barred galaxy.

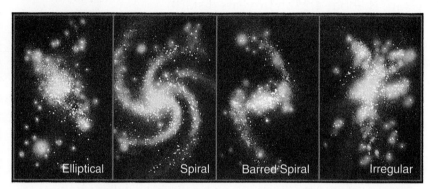

Elliptical   Spiral   Barred Spiral   Irregular

The Milky Way Galaxy is 100 000 light-years across and about 10 000 light-years thick at the center. As you can see in the drawing, the sun is on the inner rim of one arm of the galaxy, about 30 000 light-years from the center and midway between the upper and lower edges of the galaxy. As the Milky Way Galaxy rotates, the sun travels around the center of the galaxy at a speed of 250 km/s, making one complete turn every 250 million years. Complete the next activity in order to "view" the Milky Way Galaxy from different vantage points.

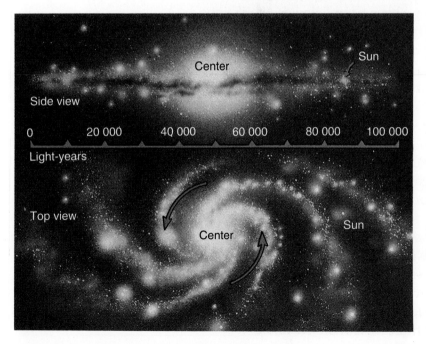

Center   Sun

Side view

0   20 000   40 000   60 000   80 000   100 000

Light-years

Top view

Center   Sun

ACTIVITY

MATERIALS

10–15 objects of the same kind that will sit upright, such as paint jars, soft-drink bottles, glass mugs; sketch pad; pencils

PROCEDURE

1. Gather around a desk with a flat surface.
2. Arrange the objects in a bunch but in no particular order on the desk top. These are "stars" in the Milky Way Galaxy.
3. Kneel down and look at the "stars" from the side, as if you were looking into the center of the galaxy. Make a simple sketch of what you see, and label it "Side View."
4. Stand up and look down over the "stars" as if you were looking down on the spiral arms of the Milky Way Galaxy. Make a second simple sketch of what you see, and label it "Top View."

APPLICATION

1. Which of your sketches shows what appears to be a galaxy with a lot of space between "stars"?
2. Which sketch shows what appears to be a more clustered arrangement of "stars"?
3. Which of your sketches is similar to what you see when you look at the Milky Way Galaxy from Earth?

The Milky Way Galaxy is part of a cluster of galaxies called the *Local Group.* Clusters of galaxies are held in place through their mutual gravitational attraction. There are 31 galaxies in the Local Group, covering a radius of several million light-years. Only three of these galaxies can be seen with the unaided eye.

The Andromeda Galaxy is 2.2 million light-years away and can be seen from the Northern Hemisphere in the constellation Andromeda. The Large and Small Magellanic Clouds can be seen only from the Southern Hemisphere. The Large Magellanic Cloud is 170 000 light-years away and is the site of the spectacular supernova 1987A. The other 27 galaxies in the Local Group are visible only through telescopes or binoculars.

Figure 20–20. The Andromeda Galaxy is one of our nearest galactic neighbors. Like the Milky Way Galaxy, the Andromeda Galaxy is a spiral galaxy.

ASK YOURSELF

Why is the speed of light used as a basis for measuring distances in the universe?

The Great Wall, Quasars, and the Big Bang

In addition to the Local Group, the universe contains billions of other galaxies. In 1985 American astronomers Margaret Geller and John Huchra began mapping the distances and locations of galaxies that are visible from their observatory in Arizona. Their results have shown what is called the Great Wall, a concentration of galaxies about 15 million light-years in thickness, 200 million light-years high, and 500 million light-years wide.

The size of the Great Wall confounds astronomers. They say there is not enough visible or otherwise detectable matter to hold it together through gravitation. This suggests that much of the matter in the universe is still beyond our technological powers of detection. Astronomers call it cold, dark matter, and they hope to discover its nature someday.

Quasars　Another mystery for astronomers is the presence of quasars, first discovered in 1963. *Quasar* stands for "*quasi*-stell*ar* radio source," that is, a source of radio waves that is almost like a star. As it turns out, most quasars do show up in visible light, as in the photograph below. **Quasars** are very small, fast-moving starlike objects. They are believed to be the oldest objects in the universe, perhaps originating near the beginning of time.

Figure 20–21. Quasar 3C-273 "photographed" with a radiotelescope (left); another quasar photographed in false color through a reflecting telescope (below).

　　Quasars emit so much energy that astronomers think they must lie near black holes. Black holes, as you recall, do not emit any energy, but rather swallow matter around them. As this matter heats up it shines brightly. Scientists are working to find the connections among quasars, black holes, and galaxies.

The Big Bang　Scientists have long sought to understand the nature of the world and the universe. One of the questions every astronomer would like to answer is that of the origin of the universe. Quite simply, how did it all begin?

The most accepted theory today that tries to answer this question is called the **big-bang theory.** About 15 billion years ago, the theory states, a massive explosion sent matter and energy traveling out into space. The universe increased in size very quickly for just a fraction of a second, and then slowed down. After billions of years the matter in the universe cooled to a point where it could clump together into stars and other bodies.

Evidence to support the big-bang theory was recently discovered by the Cosmic Background Explorer (COBE). Faint microwave readings from the time of the big bang have been captured by COBE's instruments. The readings show that some of the matter that might have been released in the big bang was just a fraction more dense than the rest. The force of gravity from these denser regions could have drawn them together to form the stars, galaxies, and other bodies in the universe. Complete the following activity to get an idea of how the big bang might have worked.

DISCOVER BY *Doing*

Take an uninflated balloon, and mark it thoroughly with dots using a felt-tipped marker. You can use more than one color if you wish, but make sure the balloon is covered. Then blow up the balloon a little. What happens to the markings? Blow it up some more. Now what happens? You can see that from a very small central point (the uninflated balloon), matter can be sent out in all directions at the same time. ✎

The universe continues to expand, just as it has since the instant of its formation. Our galaxy is rapidly moving, and so are all the galaxies around us. Scientists have discovered that just as sound waves change pitch as they move toward or away from a listener, light changes color as its source moves toward or away from a viewer. Light coming to us from a galaxy that is moving away from us shifts to the red end of the visible spectrum, while light from a galaxy moving toward us shifts to the blue end of the spectrum. All the galaxies far from us have a *red shift*. Therefore, they are all moving away. The astronomer Edwin Hubble further described the movement of galaxies by proving that the farther away they are, the faster they move.

The question is, if the universe is expanding now, will it continue to do so forever? Or might it stop expanding and begin shrinking back to its original size? The answer to this question seems to depend upon how much matter is in the universe, that is, how dense it is. This is why so many astronomers today are working to find out how much more matter is out there, visible or invisible. Perhaps your balloon can expand forever, assuming you had one that would never burst. Or perhaps something will happen to let the air out, and it will shrink back into a small, uninflated "universe." Whatever happens, we will not be here to see it, but the exploration and learning along the way will have been fascinating.

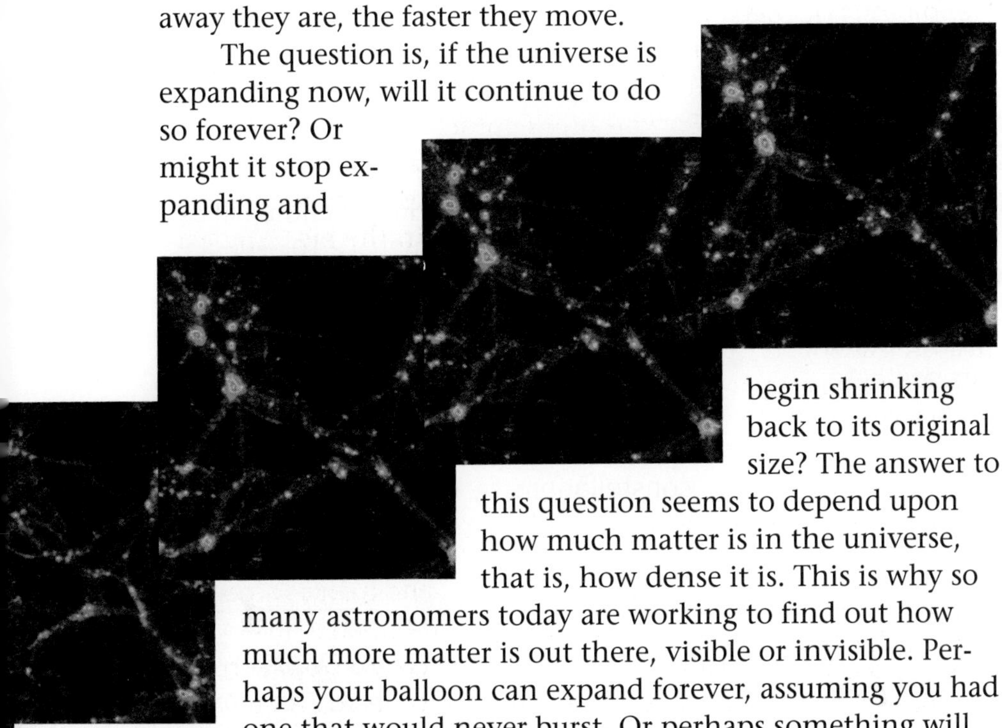

Figure 20–22. A computer simulation of the period following the big bang when, it is theorized, the matter and radiation that eventually formed the universe began to expand

 ASK YOURSELF

Based on what you have learned about magnitude, how can you tell that quasars must be very, very bright?

SECTION 3 *REVIEW AND APPLICATION*

Reading Critically
1. Arrange the following in order from smallest to largest: Great Wall, star, galaxy, universe, Local Group.
2. Why is a quasar not called a star?

Thinking Critically
3. In general, would you use AU or light-years to measure distances in our solar system? Explain.
4. How is the force of gravity connected with the theory of the big bang?

SKILL *Making and Using a Star Chart*

▶ MATERIALS
● compass ● construction paper ● scissors ● tracing paper ● tape ● ruler
● pushpin ● large piece of cardboard

▼ PROCEDURE

1. Measure the radius of the star chart below with a compass, and draw a circle of the same size on construction paper. Cut out the circle with scissors.
2. Trace the star chart on tracing paper, and tape the tracing paper neatly to the circular piece of construction paper.
3. Draw another circle with a radius larger by I cm, and cut it out. Drawing through the center of that circle, use a ruler to divide the circle evenly into 12 slices.

At the rim of the circle, label the slices with numbers *1* through *12* to represent the months in a year.
4. Place the tracing of the star chart on the large circle, and pin both pieces of construction paper, through their centers, to a piece of cardboard.
5. Study the positions of the constellations by comparing their locations relative to the North Star, Polaris.

▶ APPLICATION
Use the star chart to find some of the constellations by positioning your chart in front of you so that the constellations on the celestial sphere match the direction you are facing. Locate the constellations Orion and Hercules.

1. What are the positions of the constellations Hercules and Orion relative to the North Star?
2. Explain why not all constellations in the northern celestial sphere are visible at once.

✳ Using What You Have Learned
Use your star chart to locate as many of the constellations as you can.

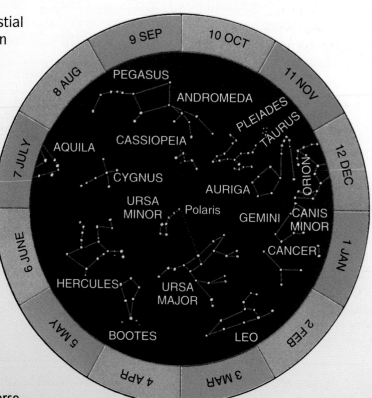

HIGHLIGHTS

The Big Idea

The universe is a collection of systems—from the very small to the very large. From the hydrogen atom whose fusion ignites stars, to galaxies, to the Great Wall, all systems work on some basic principles. Gravity pulls together atoms of matter into small clumps, which in turn form the bodies of the universe. Gravity then keeps these larger bodies attracted to one another, as in the orbital movements of the solar system.

At the same time that some factors remain constant, others are changing. Stars are born, grow old, and die, and as they die they produce the materials that form new stars. Certain forms of life have evolved on the planet Earth and perhaps in other places as well. For the most part, evolution in the universe proceeds slowly. However, in certain cases—a supernova explosion, or an impact between bodies—matter is reorganized instantly and violently to provide the seeds for new growth.

For Your Journal

Look again at the questions you answered at the beginning of this chapter. How have your ideas changed? Be sure to write your new ideas in your journal.

Connecting Ideas

Now that you have learned about Earth's location with respect to a universe containing other planets, stars, and galaxies, take a piece of paper and write down your cosmic address. Start with your own name and street address, and end with the most inclusive location—the universe.

Your Cosmic Address

REVIEW

Understanding Vocabulary

1. Explain how the terms in each set are related.
   a) galaxy (611), constellations (601)
   b) supernova (603), neutron star (607), pulsar (607), black holes (608)
   c) universe (599), big-bang theory (616), light-year (611)

Understanding Concepts

MULTIPLE CHOICE

2. What elements fuse in the nuclear reactions of the sun?
   a) uranium atoms
   b) hydrogen atoms
   c) helium atoms
   d) plutonium atoms

3. The visible surface of the sun is called the
   a) chromosphere.
   b) radiative layer.
   c) sunspot.
   d) photosphere.

4. The most distant known objects in the universe are called
   a) quasars.
   b) galaxies.
   c) supernovas.
   d) spiral galaxies.

5. How old is the universe, according to the big-bang theory?
   a) 4.6 million years
   b) 4.6 billion years
   c) 15 million years
   d) 15 billion years

6. Stars like the sun on the main sequence will eventually end up as
   a) black holes.
   b) black dwarfs.
   c) blue grants.
   d) supernovas.

7. The more massive a main-sequence star is, the
   a) older it is.
   b) more planets it has.
   c) more red it is.
   d) more energy it emits.

8. The leftover core of a supernova may be any of the following except a
   a) black hole.
   b) pulsar.
   c) prominence.
   d) neutron star.

SHORT ANSWER

9. Explain how energy from the sun's core reaches its surface.

10. List the phases of the sun's evolution from its birth to its death.

Interpreting Graphics

Use the Hertzsprung-Russell diagram on page 605 and the constellation diagram on page 601 to answer the questions below.

11. Which star is the hottest among Capella, Betelgeuse, Aldebaran, and Rigel?

12. Of the constellations Perseus, Orion, Cygnus, and Lyra, which does not contain the hottest type of star?

Reviewing Themes

13. *Systems and Structures*
Describe the physical law that keeps planets, stars, and galaxies in place.

14. *Systems and Structures*
How is the distribution of galaxies in the universe similar to that of pores in a sponge?

Thinking Critically

15. What primary characteristic of the sun will keep it from ever becoming a supernova?

16. Why does Venus have an apparent magnitude greater than any star, when it does not even produce its own light?

17. This photograph is an artist's idea of a black hole. In some movies, spacecraft crews travel through a black hole to return to Earth safely. Using your knowledge of black holes, explain why this would not be possible.

Discovery Through Reading

Dickinson, Terence. "The Big Bang." *Odyssey* 1 (December 1992): 4–9. From Steven Hawking to COBE to red shifts, this article presents an overview of what we know about the history of the universe.

CHAPTER 21

THE FINAL FRONTIER

Although the poem may seem a little fanciful to you, it really isn't. It just requires thinking about space from a new perspective. Recently the world celebrated the 500th anniversary of Columbus' arrival in the Americas. Viewed 500 years from now, the space exploration of the late twentieth century may have a similar look. The time of Columbus is often referred to as the Age of Discovery, but that phrase is just as appropriate for twentieth-century space exploration.

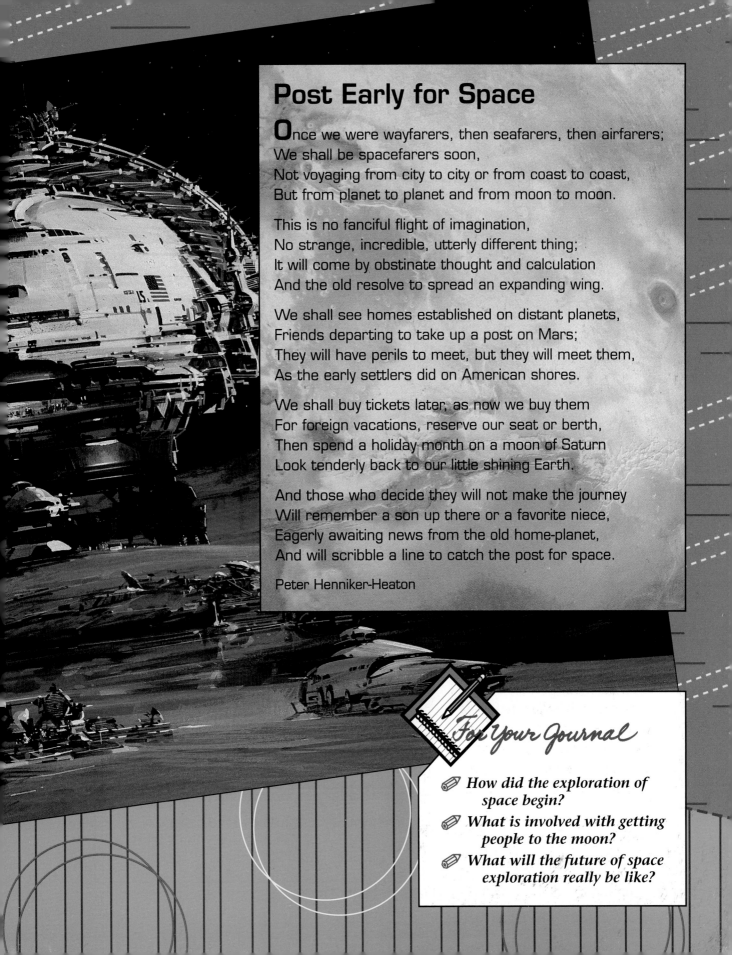

Post Early for Space

Once we were wayfarers, then seafarers, then airfarers;
We shall be spacefarers soon,
Not voyaging from city to city or from coast to coast,
But from planet to planet and from moon to moon.

This is no fanciful flight of imagination,
No strange, incredible, utterly different thing;
It will come by obstinate thought and calculation
And the old resolve to spread an expanding wing.

We shall see homes established on distant planets,
Friends departing to take up a post on Mars;
They will have perils to meet, but they will meet them,
As the early settlers did on American shores.

We shall buy tickets later, as now we buy them
For foreign vacations, reserve our seat or berth,
Then spend a holiday month on a moon of Saturn
Look tenderly back to our little shining Earth.

And those who decide they will not make the journey
Will remember a son up there or a favorite niece,
Eagerly awaiting news from the old home-planet,
And will scribble a line to catch the post for space.

Peter Henniker-Heaton

For Your Journal

- How did the exploration of space begin?
- What is involved with getting people to the moon?
- What will the future of space exploration really be like?

The Road to Space

State *the principle on which rocket propulsion is based.*

Evaluate *the contributions made by Tsiolkovsky and Goddard to rocketry.*

Name *the four stages of "firsts" in the early space program.*

I t is important for scientists, and everyone else for that matter, to ask questions. Sir Isaac Newton must have asked many questions about why and how things move. The answer to some of his questions resulted in Newton's laws of motion.

Rocket Propulsion

Rocket travel is based on Newton's third law of motion, which states that for each action there is an equal and opposite reaction. Try this activity to find out what that really means.

ACTIVITY

How can you demonstrate Newton's third law of motion?

MATERIALS
ring stands (2); string, 10 m; plastic straw; balloon; tape

PROCEDURE
1. Place the ring stands on desks on opposite sides of the room.
2. Tie one end of the string to the top of one of the ring stands.
3. Thread the other end of the string through the plastic straw. Stretch the string tight and tie that end to the other ring stand.
4. Inflate the balloon but do not tie the neck. Hold it closed and tape the balloon to the straw.
5. On the count of "5, 4, 3, 2, 1, launch!" release the balloon and observe its movements.

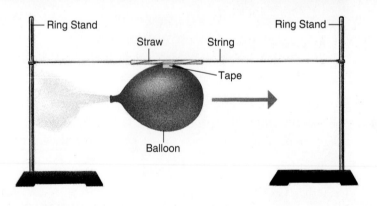

APPLICATION
1. How does this activity illustrate Newton's third law? What provides the force to move the balloon?
2. How do you think Newton's third law of motion applies to rockets?

The activity demonstrated the basic principle of rocket propulsion. The air rushed backward out the neck of the balloon, and the balloon shot forward. These two actions are equal and opposite.

The results, however, often appear unequal because objects have different masses. Therefore, we must consider **momentum**, the product of an object's mass and velocity. An object with a small mass would need a high velocity to equal the momentum of a large object with a low velocity.

For example, when a cannon is fired, the cannonball goes forward while the cannon itself is propelled backward—action-reaction. But because the cannon has so much more mass, it does not travel as fast or as far. Likewise, when a rocket is launched, hot gases rapidly exit the back of the more massive rocket causing it to move forward.

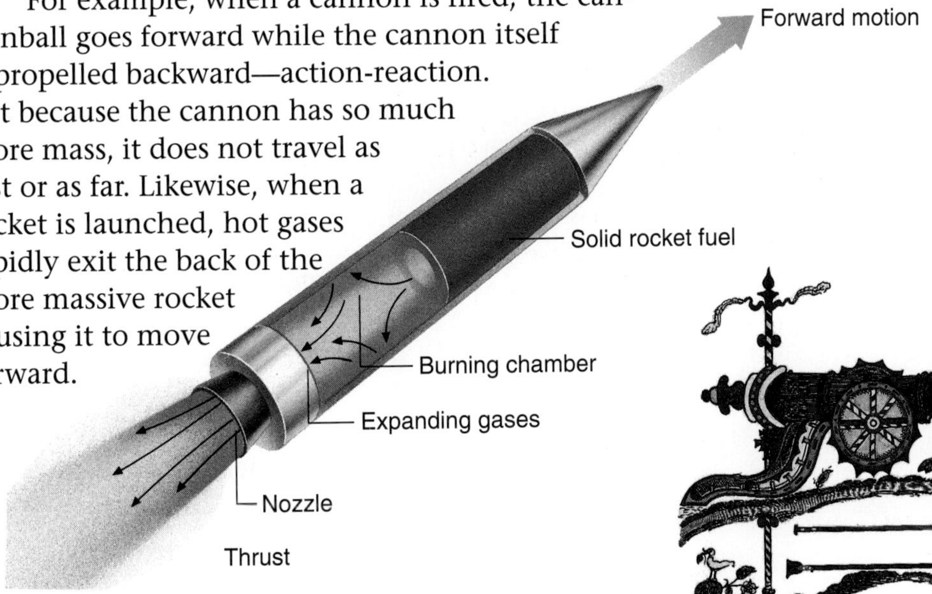

Forward motion

Solid rocket fuel

Burning chamber

Expanding gases

Nozzle

Thrust

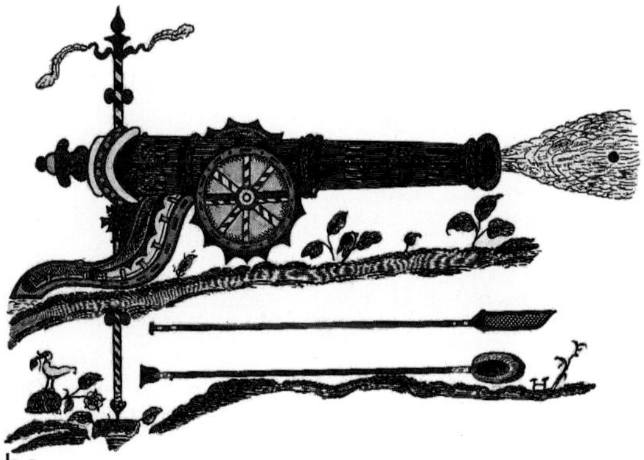

Figure 21–1. Exploding gunpowder causes the cannon's backward motion, while exhaust gases cause the rocket's forward motion.

Have you ever watched sprinters starting a race? Racing involves Newton's third law and involves momentum, just as rocket launching does. You can examine this relationship in the next activity.

DISCOVER BY *Doing*

Crouch down in the "ready" position for a race. Does your body have momentum in this position? Why or why not? Now leap up and run. What are your legs doing? What does their action cause the rest of your body to do? If your starting position represented a loaded cannon, which action would represent the explosion? the moving cannonball? ✎

▶ ASK YOURSELF

Compare a sprinter beginning a race to the launching of a rocket.

Rocket Pioneers

Figure 21–2.
Although Tsiolkovsky's model rocket looks like something out of a Jules Verne novel, it was not science fiction.

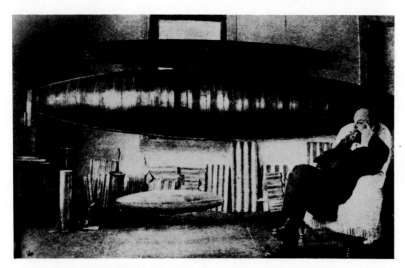

One of the first people to study the use of rockets for space travel was Konstantin Tsiolkovsky (1857–1935), a self-taught Russian scientist and inventor. Like all scientists, Tsiolkovsky had many questions. Perhaps he asked himself, "How much thrust is necessary to cause a rocket to escape Earth's gravity? What kind of fuel will provide that much thrust?"

Using Newton's third law of motion, Tsiolkovsky worked out the basic principles of rocket propulsion. He proposed a combination of liquid hydrogen and liquid oxygen as a rocket fuel, years before this mixture came to be used commonly. He developed the concept of multistage rockets and predicted that jet planes would replace those driven by propellors.

In the 1920s Tsiolkovsky designed a spaceship, support systems for astronauts in the spaceship, and pressurized suits to enable them to leave their ships and construct space stations. Though he did not make or test rockets himself, Tsiolkovsky laid the groundwork for the development of space flight.

At about the same time, American physicist Robert Goddard (1882–1945) was actually experimenting with rockets. He helped develop liquid fuels, and in 1926 launched the world's first gasoline-fueled rocket, shown here. Goddard wanted to know how much fuel would be needed to provide the necessary thrust without blowing up the rocket. He tested many rockets, some of which did in fact blow up. He also worked on the development of systems related to rocket stability and steering.

In 1944, Germany used a powerful, self-steering rocket called the *V-2* to bomb London. After World War II, scientists began looking at peaceful applications of rocket science. What followed was an exciting and intense period of discovery of what lay beyond Earth's atmosphere.

Figure 21–3. Robert Goddard launched this gasoline-fueled rocket in 1926.

 ASK YOURSELF

What fuel did Tsiolkovsky propose that rockets use?

Out of This World

When Columbus sailed west more than 500 years ago, he was looking for a shortcut to the Far East. He thought he knew what was out there, but he knew nothing for sure. In the 1950s, space exploration was a similar question mark to scientists. They thought they had some idea of what was out there, but they were not sure. Like Columbus, scientists worried about travel into the unknown.

On October 4, 1957, the former Soviet Union captured the attention of the world by launching an *artificial satellite* into orbit around Earth. The satellite, *Sputnik 1,* carried instruments to measure the density and temperature of the upper atmosphere. The United States launched its own satellite, *Explorer 1,* on January 31, 1958. The satellite was 2.05 m long and carried instruments to measure temperature and cosmic rays.

Figure 21–4. *Sputnik 1* (left) and *Explorer 1* (above)

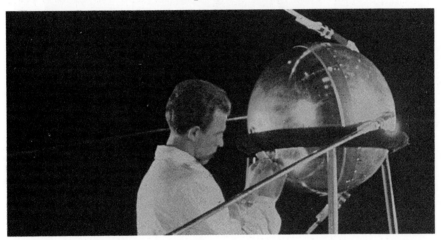

Space exploration became an exciting and frenzied dash toward new discoveries. On April 12, 1961, Yuri Gagarin, a lieutenant in the Soviet Air Force, became the first person to travel into space. Gagarin orbited the earth once, reaching an altitude of 327 km. Ten months later Alan Shepard and Virgil (Gus) Grissom made two suborbital flights for the United States. Then on February 20, 1962, John Glenn was launched into space. Glenn made three Earth orbits. All three of these flights were part of the Mercury program.

From then on, the pace was fast and furious. In June 1963, Valentina Tereshkova, a Soviet cosmonaut, became the first woman in space. Then, in March 1965, Soviet Lieutenant Colonel Alexei Leonov took the first spacewalk. Leonov, dressed in a protective spacesuit, left his pressurized cabin and floated for 10 minutes in space. "Take it easy, take it easy, I told myself. Do not move too quickly," Leonov recalled of his adventure.

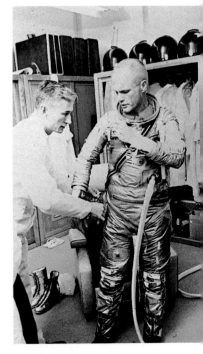

Figure 21–5. John Glenn preparing for flight

Eleven weeks later Gemini program astronaut Edward White walked in space for the United States. "This is fun," he kept repeating, until Gus Grissom at Mission Control finally ordered him back into the spacecraft. But soon everyone was reminded that the space race was not a game. On January 27, 1967, White, Grissom, and Roger Chaffee were killed by a fire that swept through the *Apollo 1* spacecraft as it sat on the launch pad. They were training for a test flight in preparation to the first trip to the moon. Their tragic deaths had a sobering effect on space exploration, and the trip to the moon was delayed for 20 months.

Figure 21–6.
Astronaut Edward
White: "This is fun!"

 ASK YOURSELF

Why did the pace of space exploration increase so rapidly after October 4, 1957?

SECTION 1 *REVIEW AND APPLICATION*

Reading Critically

1. What was the major difference between the rocketry efforts of Tsiolkovsky and Goddard?

2. What was the purpose for most of the early rocket-science research?

Thinking Critically

3. Give two examples of Newton's third law of motion in everyday life. Use examples that are different from those in the book.

4. How did each new plateau of "firsts" in the space race provide important information about what was possible for spacecraft and crews?

▶ MATERIALS
- paper and pencil

▼ PROCEDURE

Mercury

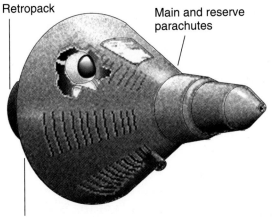

Retropack

Main and reserve parachutes

Heat shield

Gemini

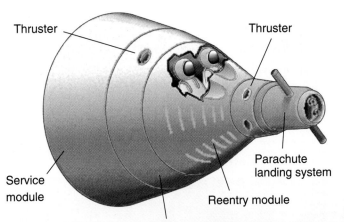

Thruster

Thruster

Service module

Parachute landing system

Reentry module

Retrorocket section

1. Look at the diagrams of the Mercury and Gemini spacecraft. Mercury carried John Glenn and his comrades into space, while Gemini was the training program for the moon landings. Notice that both diagrams are about the same size. How can you tell that one spacecraft is actually larger than the other?

2. Finding answers or explanations through observation is called making inferences. You infer when you provide an explanation to fit your observations. When you look at the number of astronauts and observe their size in comparison to their craft, you can make the inference that Gemini is clearly the larger craft.

3. Based on your observations, make inferences to answer the following questions:

   a. Can either of the spacecraft leave orbit and return to Earth and land on its own? What part of each diagram helps you infer the answer to this question?

   b. Is either spacecraft able to change its orbit? If so, which one? Which of the diagrams help you infer the answer to this question?

▶ APPLICATION
Based on your inferences so far, which spacecraft would you prefer to pilot as an astronaut? Why?

✳ Using What You Have Learned
Think up a question based on the information in the diagrams. Be sure the answer can be inferred from the information in the diagrams. Ask a classmate your question.

Living and Working in Space

I believe this nation should commit itself to achieving the goal, before this decade is out, of landing a man on the moon and returning him safely to Earth. No single space project in this period will be more exciting, or more impressive to mankind, or more important for the long-range exploration of space; and none will be so difficult or expensive to accomplish.

John F. Kennedy

On to the Moon

In a stirring speech, President John Kennedy committed the United States to placing an astronaut on the moon before the end of the 1960s. In preparation, many spacecraft were launched, and Americans learned to live and work in space.

The scientists at the National Aeronautics and space Administration (NASA) worked under intense pressure and tight deadlines. The design of both spacecraft and spacesuits had to be improved. Because of the earth's strong gravity, both spacecraft and crews were subject to the tremendous force of acceleration during liftoff. Astronauts had to be protected from this force, from the temperature extremes of space, and from cosmic particles and radiation. Astronauts underwent extensive training sessions to prepare them for the hostile conditions they would encounter in space.

Figure 21–7. A Gemini astronaut trains in an underwater setting that simulates the microgravity of space.

Following an October Earth-orbit test, *Apollo 8,* led by Frank Borman, made the first moon flight in December 1968. The spacecraft did not land, but orbited the moon 10 times and sent back televised views of the lunar surface. Each time *Apollo 8* passed behind the moon, Mission Control held its breath as radio contact with the craft was lost. Would *Apollo 8* come back? Try the next activity to see what problems had to be overcome before humans could go to the moon.

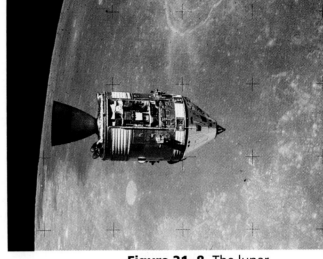

Figure 21–8. The lunar command module

DISCOVER BY *Researching*

In the early days of space exploration, there were many unanswered questions. Would you have gotten into a spacecraft with so many questions unanswered? Make a list of the problems that would have to be solved before humans could explore the hostile environment of space. Research the solutions that NASA found to those problems using library resources.

The task of landing on the moon was much trickier than orbiting it. Three separate craft were needed. The command module carried the astronauts to the moon and back. The service module supplied oxygen, power, and fuel to the command module. The lunar excursion module (LEM) could place two astronauts on the moon's surface. The LEM included an engine to allow it to lift off from the moon and return to the command module.

The crew of *Apollo 11* lifted off on July 16, 1969, atop a Saturn 5 rocket, the most powerful rocket built to date. Three days later they reached the moon. Neil Armstrong and Edwin (Buzz) Aldrin walked on the moon's surface on July 20, while Michael Collins remained in orbit around it.

Figure 21–9. "That's one small step for a man, one giant leap for mankind."

631

Figure 21–10. Plaque left by *Apollo 11* crew

The last lunar mission, *Apollo 17*, was completed in December 1972. No one has been to the moon since then. Because there is almost no erosion on the moon, the footprints of all who have been there remain. Scientists have discussed the possibility of making the moon the first permanent space colony.

Apollo astronauts have had varied reactions to their voyages. Michael Collins of *Apollo 11* says of the moon, "It didn't seem like a very friendly or welcoming place." James Irwin of *Apollo 15* remembers, "When I got on the moon, I felt at home." While Eugene Cernan, who flew on *Apollo 10,* takes a more analytical approach. He says, "We spent most of the way home discussing what color the moon was." How do you think you would feel about being on the moon? In the next activity you can express those feelings.

DISCOVER BY *Writing*

Imagine that you are an astronaut who has traveled into space or landed on the moon. What aspect of the experience would interest you the most? What do you think your strongest feelings would be? Write a short journal entry to describe your point of view.

▶ ASK YOURSELF

Would the force of acceleration on the *Apollo 11* astronauts have been greater or less as they lifted off from the moon than it was when they left Earth? Why?

Figure 21–11. Soviet scientists have occupied various space stations since 1971.

Space Stations

"'Not a day without a discovery' was our motto during the mission. If we were unable to make a discovery in our experiments, then we would discover what was for lunch."

Vladimir Kovalyonok

While the United States went to the moon, the former Soviet Union concentrated on space stations. The first space station, *Salyut 1,* was launched in April 1971. Unlike previous spacecraft, *Salyut 1* was built to remain indefinitely in Earth orbit.

On July 15, 1975, an American Apollo spacecraft joined with a Soviet Soyuz spacecraft 225 km above Earth. The three American astronauts and two Soviet cosmonauts visited each other through a passage connecting the two spacecraft. They jointly conducted several experiments.

Spaceports   In September 1977 a new space station, *Salyut 6,* was launched by the former Soviet Union. The station had two docking ports. On December 11, 1977, *Soyuz 26* docked at the rear port, and on January 11, 1978, *Soyuz 27* docked at the front port. For the first time three spacecraft were linked in space, with a total of six people on board.

On January 20, 1978, the first unpiloted ferry craft was launched by the Soviets to resupply *Salyut 6.* Resupplying by crewless flights meant that the crews on *Salyut 6* could stay in space indefinitely. *Salyut 6* was in operation until the summer of 1986, when it was replaced by *Mir,* which means "peace", a permanently crewed station. In 1987 and 1988 two Soviet cosmonauts lived aboard *Mir* for a year and a day.

The Soviet Union has had a crew on board the space station ever since. With the collapse of the Soviet Union, however, the future of their space program is uncertain. One crewman of *Mir* extended his stay in orbit for 13 months because the Russian government had no money to retrieve him.

The only American space station to date was *Skylab,* which operated from May 1973 to February 1974. Astronauts aboard *Skylab* performed useful experiments in weightlessness and broke new ground in solar astronomy. *Skylab*'s orbit eventually deteriorated, causing it to fall back to Earth in 1979.

Figure 21–12. The Apollo-Soyuz spacecraft was linked by a common passage.

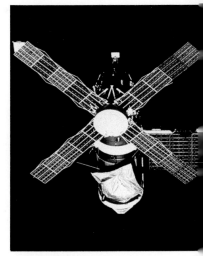

Figure 21–13. With space stations such as *Mir* (left) and *Skylab* (above), astronauts have relatively comfortable workplaces.

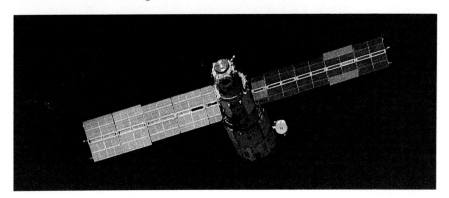

Figure 21–14.
Astronauts exercise at the Johnson Space Center in preparation for space station duty.

Space Life   Life in a space station is very different from that of a short trip in a cramped spacecraft. Crews on a space station must develop a pattern of living. They have more choices of food, for example, including the option of re-heating frozen dishes. They also have more room to live and play in. Fuel cells that produce electricity for the space station also produce pure water for drinking and bathing and for cooling the station.

Space-station living revealed a major health problem, however, that was never encountered on short missions. On Earth the human body's constant resistance to gravity helps keep muscles and organs in shape. Short-term exposure to weightlessness causes no negative effects to astronauts. But scientists discovered that when humans are exposed to long-term weightlessness, muscle tone and abilities of crews to perform their duties significantly deteriorates. Crews also experience reduced heart and blood vessel functioning. Space-station astronauts are put on strict exercise plans to help counteract the effects of living with no apparent gravity.

Space-station inhabitants also find themselves with time on their hands. Boredom creeps in when routines become established. Operating a space station does not consume every minute of the day. Recreation is encouraged. Chess is a Russian favorite, while Americans play cards.

Figure 21–15.
Vladimir Kovalenok and Aleksandr Ivanchenkov play chess on *Salyut 6.*

The future of space stations is uncertain. In 1988 the United States, Canada, Japan, and nine members of the European Space Agency agreed to construct an international space station. The Congress of the United States continues to debate funding for America's space station. But scientists fear that there will not be nearly enough support for the development of such a station. In the meantime, American astronauts will share use of Russia's space station, *Mir.*

 ASK YOURSELF

What important contributions have the space programs of the United States and the former Soviet Union made to our knowledge of the universe?

The Space Shuttle

"Houston, I think we've got a satellite."

Daniel Brandenstein

Figure 21–16. A successful satellite rescue

Rescuing and repairing stranded satellites is all in a day's work for a space-shuttle crew. The space shuttle was a logical development in the U.S. space program. A reusable craft could be deployed to ferry astronauts to and from space stations and to carry industrial, scientific, and military cargos into space. The final design consisted of an orbiter, two solid-fuel boosters, and one disposable liquid-fuel tank. The orbiter is about the size of a small airliner. It is 37 m long and has a 23 m wingspan.

The shuttle performs as three space vehicles in one. It blasts off from Earth as a rocket, orbits the planet as a satellite, and lands as an airplane. After launch the boosters and fuel tank detach and fall into the ocean. The boosters are recovered by a ship for reuse.

Columbia, the first shuttle, roared into space on April 12, 1981, and returned five days later. On board with John Young, who had flown two Gemini missions and two Apollo missions and landed on the moon, was first-time astronaut Robert Crippen. Although the flight was uneventful, Young said, "If you're not nervous, you don't understand what's happening."

Figure 21–17. A space-shuttle orbiter being readied at Kennedy Space Center in Florida.

Other shuttles have been named *Challenger, Atlantis, Discovery,* and *Endeavour.* Shuttle crews have deployed astronomical devices, such as the Hubble Space Telescope and *Spacelab,* an orbiting laboratory built by the European Space Agency. They have launched, retrieved, repaired, and reorbited satellites and conducted numerous experiments.

Unfortunately, the shuttle program has not been without tragedy. On January 28, 1986, *Challenger* blew up 75 seconds into its flight, killing all seven astronauts. It was later determined that the explosion was caused by a booster problem that space officials had detected but had left uncorrected. No space shuttles flew for more than two years while the entire shuttle program was reorganized.

Figure 21–18. The *Challenger* astronauts: Christa McAuliffe, Gregory Jarvis, Judith Resnik, Michael Smith, Ronald McNair, Dick Scobee, and Ellison Onizuka.

 ASK YOURSELF

How is the space shuttle different from the other American space vehicles?

SECTION 2 *REVIEW AND APPLICATION*

Reading Critically

1. What two major environmental problems face the designers of spacecraft?
2. What kinds of activities can be done with space shuttles that cannot be completed with other spacecraft?

Thinking Critically

3. How might cooperation between the former Soviet Union and the United States have benefited space exploration?
4. List several examples of space technologies that have Earth applications.

Space Science

The early exploration of space is a dramatic story. But the secrets of the universe have barely begun to be revealed. How has space exploration really affected our daily lives? You have always lived in a world in which spacecraft have orbited Earth or probed the solar system. In fact, many of your everyday activities depend on functions performed by satellites and other spacecraft.

Objectives

Explain *how satellites improve communications systems on the earth.*

Evaluate *how* Landsat *photos document human activities and their effects on the earth.*

Summarize *the major scientific advances produced by space exploration.*

Communications Satellites

Modern communications systems allow for almost instantaneous transmission of information across countries and continents, and to crews and satellites in space. These systems, like all communications technology, are extensions of our sight, hearing, and speech. Satellite systems are more efficient than ground systems. They avoid ground-based interference caused by barriers such as mountains and buildings, and even the curvature of the earth.

A *communications satellite* is launched into a very high orbit around Earth. Communications satellites relay radio, television, telephone, and other signals from one ground system to another, from ground to satellite, and from satellite to satellite.

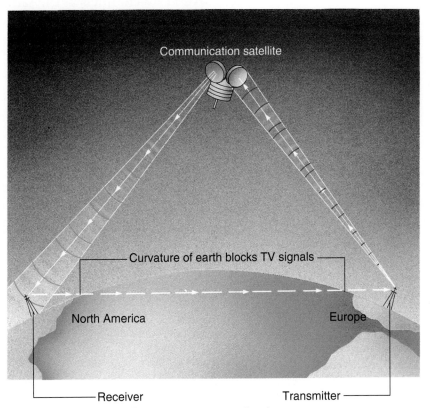

Figure 21–19. Satellites carry communications beyond the curve of the earth.

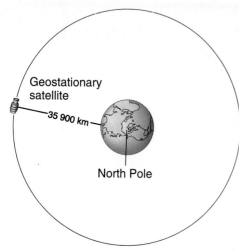

Geostationary satellite

35 900 km

North Pole

Figure 21–20. Clarke proposed this orbit 15 years before a satellite was even sent up.

The United States launched the first communications satellite, *Echo 1,* in 1960. *Echo 1* was simply an aluminum-coated, helium-filled balloon. Voice and television communications sent from one land station bounced off *Echo 1's* metal skin and were received by another land station.

In 1945 English scientist and writer Arthur C. Clarke proposed that three satellites, equally spaced in orbits 35 900 km above the equator and moving with the earth, could be used to form a worldwide communications network. Today there are many satellites in these orbits, called *Clarke orbits.* A satellite in a Clarke orbit, as shown in the diagram, is **geostationary,** because from the ground it does not appear to move from its position. In other words, it orbits Earth at the same speed as Earth rotates.

Two years after *Echo 1,* the United States put a more sophisticated satellite, named *Telstar 1,* into orbit. This satellite could receive and transmit signals, not just reflect them. Today's satellites receive, strengthen, and transmit signals. The Intelsat satellite in the photograph can relay 12 000 telephone conversations and several TV channels at the same time.

Figure 21–21. Today's satellites have names such as Intelsat.

Intelsat is also the name of a communications network formed by 140 nations in 1964. Today this network controls about 18 satellites and 600 Earth stations.

Our communications system has expanded enormously because of satellites. You couldn't view so many cable channels without them. Today it is possible for individual homeowners to purchase small satellite dishes that fit on their roofs and tune into hundreds of satellite transmissions.

 ASK YOURSELF

Explain why a satellite receiving dish always points in the same direction and does not have to track a satellite.

Weather Satellites

Television viewing options have improved tremendously due to communications satellites. But is variety and convenience reason enough to spend billions of dollars launching satellites? How else do satellites directly affect people? You would probably agree that the weather can have a direct impact on people. People can be killed in hurricanes or blizzards if no warning is given. To alert people to severe weather as well as to gather other information about Earth's weather systems, satellites were developed that could photograph storms and provide information about other weather phenomena. These satellites then relay the information and photos to Earth.

Nine early-weather satellites, called Tiros, were launched by the United States between 1960 and 1965. These satellites observed Earth using visible light and infrared radiation. Their sensors could distinguish between different clouds, oceans, and land.

In the late 1960s, Tiros satellites were replaced by the Nimbus group, which were launched into *polar orbits*. Polar orbits are those that pass over the North and South Poles. There they continuously circle the earth, moving from pole to pole. As Earth rotates beneath them, the satellites photograph a different strip of the planet during each orbit. The Nimbus group photographs a complete picture of Earth every 24 hours.

Figure 21–22. The Nimbus satellites are in polar orbits.

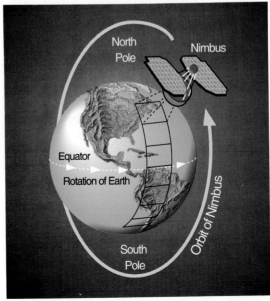

Today, GOES circle in geostationary orbits around the equator. GOES stands for *Geostationary Operation Environmental Satellites*. GOES can record the movements of clouds, the buildup of ice and snow on the land, and the heat coming from the earth, the seas, and the clouds. The pictures you see on TV weather reports come from GOES. Weather satellites help meteorologists make accurate forecasts of weather conditions. In the next activity you can see how accurate these forecasts are.

Figure 21–23. Satellite photos help weather forecasters track and predict blizzards, (right) hurricanes, (left) and other storms.

 DISCOVER BY *Observing*

Keep a two-week record of the daily forecasts made by television weather reports and your newspapers. Observe the weather directly to see how accurate these forecasts are. What might cause a prediction based on satellite data to turn out to be incorrect? ✐

Figure 21–24. Holes in the ozone layer are shown by the pink colors.

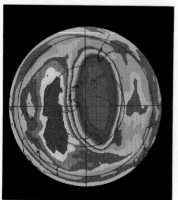

Data from GOES provided the first warnings scientists received of damage to the ozone layer. Nimbus satellites have tracked the growing danger for more than a decade. This photograph shows the ozone hole over Antarctica in 1992. Weather and other satellites continue to warn us of the fragile nature of Earth's support systems.

▼ **ASK YOURSELF**

If weather satellites stay in the same orbit continuously, how can they provide complete photographic data for the entire Earth?

Observing the Earth

SATELLITES

Monitors of steel,
these space detectives seek
clues to the beginning of our galaxy.
Informers of the energy of stars, of gamma rays;
weighted with sensors,
They listen, watch, and speak
of radiation, solar flares,
atmospheric density.
Stalking magnetic fields, they serve out their days.

Myra Cohn Livingston

Many scientists have become attached to these lines. The ideas confirm what many have suspected—that the more we explore space, the more we learn about Earth. The value of observing Earth from space became clear when the first piloted space flights began. Photographs taken by early astronauts showed features of Earth that were not clear from the ground. You can see this effect for yourself. Imagine you are standing at ground level looking at a corn field, and all the corn in front of you is dead. You assume the entire corn crop has failed. Yet, if you climb to a higher point and look down on the field, you notice that only a small strip of corn is dead. You see that this strip is next to a road along which toxic spraying has taken place recently. Won't the information gathered from this new perspective change your viewpoint?

The first satellite specifically designed to photograph Earth was *ERTS 1*, launched into a polar orbit on July 23, 1972. *ERTS*, which stands for *Earth Resources Technology Satellite*, was later named Landsat. Five Landsats have been launched since 1972.

The first application of Landsat was in cartography—the process of map making. Landsat photographs can record objects as small as 30 m wide. In addition, Landsat pictures of crop areas have made it possible to forecast production and to establish a worldwide food watch. Landsat photos are also crucial in monitoring dust storms, forest and oil fires, clear-cutting of forests, and air pollution. Landsat photos have given us a clearer picture of those human activities that are damaging Earth.

Figure 21–25. It's all a matter of perspective with Landsat.

Data from Landsat has also been used to locate mineral resources. The structure of the earth's upper crust can be studied as a means of identifying subsurface mineral deposits. The color of bare rock and the pattern of vegetation where bedrock is covered often reveal the structure and mineral composition of the rocks.

The changing snow cover in middle and high latitudes is also monitored by Landsat as a way of forecasting spring floods. The ice cover in Antarctica, Greenland, and many mountainous areas is monitored as a way of predicting possible sea level changes.

Figure 21–26. Landsat photos of New York (left), Chicago (center), and Los Angeles (right)

Although Landsat photos can be color-enhanced in different ways to show a variety of topographies and resources, one element remains constant. Human activity shows up in straight lines, squares, and rectangles, while natural forms tend to be curved and irregular. Given this distinction, point out the centers of human population in each map shown here. The next activity can help you see how useful Landsat photographs are.

DISCOVER BY *Calculating*

Find a variety of Landsat photos in a book collection or in science magazines. Choose two or three, and find the corresponding maps in an atlas. Using the scale on the atlas maps, calculate the amount of area covered by each Landsat photograph. Then identify features in the Landsat photos that are identified on the map. ✎

A satellite called *Seasat 1* was launched in 1978 for the specific purpose of studying the oceans. *Seasat 1* measures wave heights, the speed and direction of currents, and the surface temperature of the oceans. The satellite also records the distribution of sea ice.

Recently a joint French/U.S. satellite, TOPEX/Poseidon, has been put into orbit. The satellite will observe the oceans for three to five years, monitoring the oceans' role in global climatic changes.

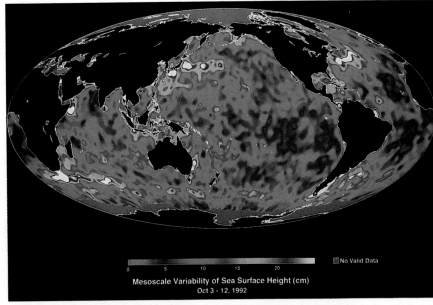

Mesoscale Variability of Sea Surface Height (cm)
Oct 3 - 12, 1992

Figure 21–27. Scientists created this map of ocean water levels from data gathered by the TOPEX/*Poseidon* satellite.

 ASK YOURSELF

How could scientists monitor the effects of a continuing drought using Landsat maps?

Exploring the Solar System

"In all the history of mankind, there will be only one generation that will be first to explore the solar system."

Carl Sagan

Dr. Sagan is right. The first generation to explore space was your parents' generation, but the exploration is continuing into your generation, too. Not only are new exploration vehicles on their way to Jupiter, Mars, and the sun, but the data collected by *Voyager, Magellan,* and others will help us learn new things about our neighbors in space for years.

Both the United States and the former Soviet Union have sent space probes, such as *Viking* and *Venera,* into the solar system. **Space probes** are spacecraft that are launched into the solar system rather than into Earth orbits, as satellites are. *Viking 1* sent this photo from Mars in the 1970s.

Figure 21–28. Viking photos showed us Mars' features, such as *Olympus Mons,* the largest volcano in the solar system.

The space probes launched during the past 30 years have provided scientists with a wealth of information about the solar system. This information has been used by scientists to study the origin and evolution of Earth and its sister planets. It has caused many people to push for the establishment of a human colony on the moon or Mars as soon as possible. Many factors, not only scientific but political and economic, will affect the realization of this dream. In the next activity, you can find out more about space probes.

DISCOVER BY Researching

Research the missions of the Viking probes to Mars. Find out what kinds of experiments were used in the search for life on Mars. Write your findings in your journal. ✎

Figure 21–29.
Photographs taken by *Voyager* probes revealed more than a thousand rings around Saturn.

The scientific information from space probes is valuable and extensive, although it is just the tip of the iceberg. Some of the discoveries made by space probes are summarized in Table 21-1.

ASK YOURSELF

What space probes found active volcanoes in the solar system?

SECTION 3 *REVIEW AND APPLICATION*

Reading Critically
1. Describe geostationary and polar orbits, and explain the advantages of each.
2. What kinds of data are planetary probes able to gather?

Thinking Critically
3. Name at least three ways in which satellites affect your daily life.
4. Give two examples of ways in which the technology discussed in this chapter extends the senses of human beings.

Table 21-1 **Space Probes**

Mariner   The United States launched a series of Mariner probes in the early 1960s. *Mariner 4* first photographed Mars in 1965, revealing craters that had not been visible from Earth. *Mariner 9* first orbited Mars in late 1971, photographing a dust storm on the surface and the two Martian moons. *Mariner 10* flew past Mercury in 1974 and 1975, revealing the planet's heavily cratered surface. *Mariner 10* was also the first probe to visit two planets—it measured Venus on its way to Mercury.

Venera   A series of Venera probes was launched by the former Soviet Union. *Venera 9* and *Venera 10* landed on Venus in 1975, taking one photograph of the surface before being destroyed by Venus's punishing atmosphere. *Venera 13* and *Venera 14* landed on Venus in 1982, where they took four photographs, including the first color picture of the surface. They made other measurements of the planet as well.

Viking   *Viking 1* and *Viking 2,* launched by the United States, orbited Mars in 1976, taking more than 50 000 photographs. Landers were sent to the surface to make direct observations. They photographed the surface, recorded quakes and weather conditions, analyzed Martian soil, and conducted experiments to search for life on the planet.

Voyager   Two Voyager spacecraft were launched by the United States in 1977 to explore the outer planets. In 1979 and 1980 *Voyager 1* and *Voyager 2* passed close to Jupiter, discovering many new satellites and a ring system. Data showed that one of Jupiter's large moons, Io, has active volcanoes. In 1980 and 1981, the Voyagers crossed the path of Saturn and found that the planet's three rings, as seen from Earth, are really a series of more than 1000 small ringlets. Voyager data also showed that Titan, Saturn's largest moon, has an atmosphere of mostly nitrogen. *Voyager 2* continued to Uranus and Neptune, reaching them in 1986 and 1989 respectively. It found that Uranus has 10 more satellites than were previously known, and one of its moons, Miranda, has surface features of astonishing variety. At Neptune, *Voyager 2* found a huge storm system and a volcanic moon, Triton. Data from the Voyagers will keep scientists busy for decades.

Magellan   *Magellan* was launched from the space shuttle in 1989 and began mapping the surface of Venus in August 1990. *Magellan* used radar imaging because of Venus's thick, cloudy atmosphere. Results of mapping to date show a varied surface of volcanoes, faults, valleys, and possible evidence of active volcanoes.

Galileo   Launched in 1989, *Galileo* is to reach Jupiter in 1995; it will study the planet and its large moons.

INVESTIGATION

Interpreting Landsat Maps

▶ **MATERIALS**
- Landsat map ● paper ● pencil ● atlas

▼ **PROCEDURE**

1. Examine the Landsat map. Notice that the exact map location has not been specified.
2. Recall that Landsat maps look much like relief maps, with different colorations. Vegetation is usually various shades of red. Earth features are brown, white, black, and shades in between. Water is usually blue or black. Clouds and snow are white. Natural areas are irregular or curved, while areas of human activity are more regular, with squares and straight lines.
3. Use the map and the letters to answer the following questions:
   a. What natural feature is labeled A?
   b. What natural feature is labeled B?
   c. Why is the area to the south of A colored red?
   d. What vegetation-covered landform is located at D?
   e. There are two mountainous areas in the image. What kind of landform lies between them? Which letter identifies this landform?
   f. There is much evidence of human activity in this image. Identify these areas.
   g. Which of them are more likely to be cities?
   h. Given the character of the land on the map, what do you think takes place in the other areas of human activity?
   i. Why do you think the deep red color is concentrated in only a few areas?
4. Using a U.S. atlas, find physical and political maps of California. Identify San Francisco Bay, the San Joaquin Valley, and the Sierra Madre.

▶ **ANALYSES AND CONCLUSIONS**

How does the Landsat map differ from the maps in the atlas?

▶ **APPLICATION**

1. This image was taken recently. If it had been taken 200 years ago, how might it have differed?

2. If environmentalists suspected large-scale deforestation, and wanted evidence to support their suspicions, how might they use a Landsat image?

✳ *Discover More*

How does information about the same area collected in different forms help expand our knowledge of the area?

The Big Idea

Space exploration requires human beings to change the way they live and think in order to travel to new places and gather knowledge. We use the physical laws of the universe to learn more about our world. This process involves formulating theories and making and experimenting with models. It also involves taking the risk that the models will fail and that people will sometimes die in the attempt to reach new worlds. Once the physical problems have been overcome, we must learn to change the way we think about our world based on our new discoveries.

How have your ideas about space exploration changed? How did space exploration begin? What will the future be like? Think about these questions, and write your revised answers in your journal.

Connecting Ideas

Copy the concept map into your journal. Complete the map using what you have learned about space exploration.

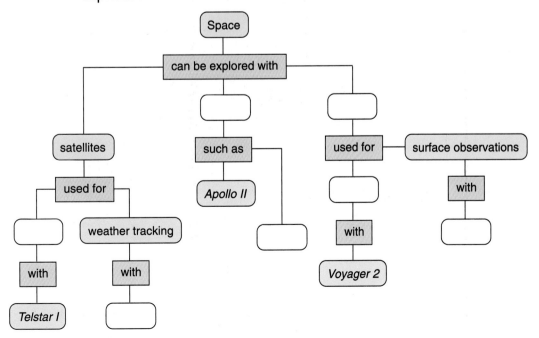

Understanding Vocabulary

1. Explain how the terms in each set are related.
   a) Newton's third law (624), momentum (625)
   b) geostationary orbit (638), polar orbit (639)
   c) artificial satellite (627), space probe (643)

Understanding Concepts

MULTIPLE CHOICE

2. Which of the following people physically pioneered space?
   a) Robert C. Clarke
   b) Konstantin Tsiolkovsky
   c) Yuri Gagarin
   d) Robert Goddard

3. Each of these space probes was designed primarily as a flyby photographic and sensory probe except
   a) *Mariner.*   c) *Galileo.*
   b) *Voyager.*   d) *Viking.*

4. During a long-term assignment as a space-station crew member, the most severe challenge you would face physically is
   a) cleanliness.
   b) malnutrition.
   c) decreased muscle tone.
   d) temperature regulation.

5. Probably the most significant practical advancement represented by a space shuttle is that a space shuttle is
   a) very complex.
   b) capable of large payloads.
   c) reusable.
   d) capable of carrying a large crew.

6. Of the following space achievements attained by the former Soviet Union, which first affected the United States?
   a) the first space walk
   b) the first human being to orbit Earth
   c) the first space station in orbit
   d) the first satellite to be placed in orbit

SHORT ANSWER

7. Summarize the missions of *Mariner, Venera, Viking, Voyager, Magellan,* and *Galileo* space probes.

8. Explain the purpose of the *Apollo 11* mission.

Interpreting Graphics

9. Use the terms *rocket fuel, nozzle,* and *expanding gases* to label each part of this rocket.

10. Use the illustration to explain rocket propulsion.

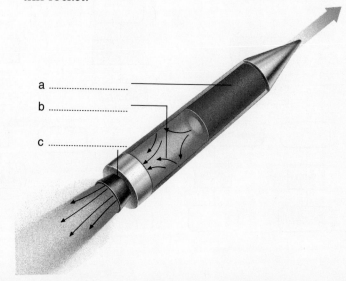

a ....

b ....

c ....

Reviewing Themes

11. *Technology*
Crewed and uncrewed space exploration
has been developed at the same time.
Debate the need for developing spacecraft
that can support human life when robots
can travel, analyze, experiment, and
retrieve materials as effectively as human
beings and at no risk to human life.

12. *Systems and Structures*
Compare the systems, structures, and
purposes of Apollo spacecraft with space
shuttle orbiters.

Thinking Critically

13. Discuss why there has been no further
crewed exploration of the moon since the
early 1970s.

14. Justify the merits of space exploration in
light of the monetary costs of the space
program.

15. Predict some possible results of a successful
cooperative international space station
construction effort.

16. Speculate as to what our present state of
communications and technology would be
if there had not been a race to the moon in
the 1960s.

Discovery Through Reading

Ambruster, Ann, and Elizabeth A. Taylor. *Astronaut Training.* Franklin Watts, 1990. Are you
thinking of becoming an astronaut? Read
this book to find out more about the selection and training of astronauts.

SCIENCE IN SPACE

Since their beginnings, space exploration programs have produced a bonanza of technological benefits. The race to put astronauts on the moon yielded commercial benefits in computers and other areas. The National Aeronautics and Space Administration (NASA) has hailed the space shuttle and proposed space laboratory as vehicles that will brighten our future through their space research.

Space shuttle missions have produced some promising discoveries.

Promising Discoveries

Space shuttle missions have produced some promising discoveries in the science of materials. For example, crystals and polymers can be grown more effectively in the microgravity, or near zero gravity, of space. Several experiments have produced delicate protein crystals that would collapse from their own weight if grown in Earth's gravity. Space-grown crystals

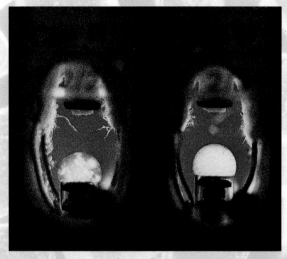

Space-grown crystals may provide research advantages (above); the International Microgravity Lab in the shuttle (right)

may have significance in understanding plants' conversion of light to energy or the function of enzymes in human biology.

Other space research is designed to generate commercial profit. In one experiment in materials science, a polymer film to be used to filter industrial products was grown. A polymer membrane grown in microgravity has more uniform "pores," or filtering holes, than one grown under the warping effects of gravity. The membrane's pores were smaller and consistent in size, thus providing a more efficient filter.

Unpredictable Results

Overall, the results of space science research have been unpredictable. Some scientists have called for more basic research into microgravity and its effects on materials and organisms. Without a better basic understanding of space science, they say, research on uses and applications will be hit or miss. Other scientists believe that governments will continue to support space research only if its products can be applied in medicine or industry.

The bright new promise of space science rests on an age-old question: Should science be concerned with the pursuit of knowledge for itself, or with the application of knowledge to improving human life? ◆

The International Microgravity Lab

SPACE CADETS

Mission control at Space Camp

by Russell Ginns from *3-2-1 Contact*

"This is the flight director," says the voice over the radio. "Prepare to separate solid rocket boosters."

You press two flashing yellow buttons. Then you hear the sound of two small explosions as the rockets separate from your ship. The whole cabin shakes.

The pilot taps you on the shoulder. She points out the window as the Earth seems to shrink and fall away.

"T plus eight minutes and 10 seconds," a voice crackles.

"The shuttle is now moving at a speed of 3,700 feet per second," says another voice.

Are you really flying in the space shuttle, 100 miles above the Earth? No—but you're doing the closest thing to it, as you complete your final mission at U.S. Space Camp in Huntsville, AL.

Each year, thousands of kids from all over the U.S. travel there to learn about outer space and the U.S. space program. And the neatest part of all: Campers get hands-on astronaut training.

Space Camp is for kids in grades four through six. Space Academy is a similar program for kids in grades seven through twelve. Both are five

and a half days of lessons, tours and practice missions that give kids a chance to experience what it's really like to be an astronaut.

BUSY SCHEDULE

From early morning until late at night, kids at Space Camp are very busy. "Sometimes they bring their skateboards or radios to camp," trainer Bridget Damberg told me. "But I've never seen anyone who had time to use them."

Starting at six in the morning, campers go to one training session after another, with just a little time in between to eat

The Space Habitat

Like walking on the moon

meals and phone home. They attend lectures about the history of space suits and learn all about the different parts of a space shuttle. They build and launch their own model rockets and tour the Space and Rocket Center, the largest space museum in the U.S. Each evening, campers watch movies about the history of space exploration.

Much of the equipment at Space Camp is similar to what real astronauts train on.

The "Five Degrees of Freedom" simulator was actually used by the early astronauts to practice moving about in zero-gravity. Campers sit strapped into a chair that can swivel in any direction: up, down, left, or right. The whole chair floats on a cushion of air. So the slightest push in any direction can start it sliding across the floor.

Holding a wrench in one hand, the trainee gets pushed towards a wall that has a bolt screwed into it. The object is

to unscrew the bolt without touching the wall. "Until you get the hang of it, you start spinning while the bolt stays still," camper Danny Shaw, 11, told me.

Another simulator is called the "Microgravity Training Chair." It's a seat that hangs from the ceiling by a system of pulleys and springs. By adjusting it for each person, the wearers weigh one-sixth of their normal weight. "It's just like walking on the moon," says Danny.

But walking on the moon isn't easy. If you are strapped into the Microgravity Chair and try walking normally, you won't go forward. Instead, you'll bounce up and down in the same place. "The easiest way to walk on the moon is to pretend you're

jogging in slow motion, swinging your arms and taking giant steps," trainer Paul Crawford tells the campers.

"That works," Danny agrees, "but it looks kind of silly."

Kids at Space Academy also go for a spin in the "Multi-Axis Training

653

Space Camp training center

Simulator"—and they really go for a spin! The Multi-Axis is a system of large metal rings, one inside the other. The astronaut sits in the center. When the instructor starts up the motor, the astronaut spins randomly in three different directions—all at once.

"It looks like it could make you want to throw up!" Camper Robin Lundesh, 14, told me. "But it really isn't so bad." Because the astronaut's stomach is always in the very center of the spin, there is never enough force to make anyone feel sick. However, you do get a good idea of what it's like to be spinning out of control in space. "You get very confused about which way is up," says Robin.

THE FINAL MISSION

The main event of a week at Space Camp or Space Academy is the team mission. During a two-hour simulation, the team acts out an entire space shuttle flight. This includes takeoff, launching a satellite, docking with a space station, landing, and any emergency actions that have to be done along the way.

Each camper gets a different assignment. Some are sent to mission control, where they help direct the shuttle during takeoff and landing and keep track of life support systems. Other campers are assigned to the cockpit where their main job is to fly the shuttle. Still others are payload specialists, who release satellites into orbit and make repairs outside the ship while in outer space.

"At first, I was mad because I didn't get to be the pilot," says Sean Allen, 13. He was chosen to be the mission's Weather and Tracking Officer. "But my job turned out to be a big challenge. I had to keep track of everything that was happening during the launch. And at the end I decided where the shuttle had to land."

During the mission, each person follows a script that tells him or her what to do and say at certain times. Meanwhile, the trainers sit at a computer away from everyone and keep track of how the mission is running. They also create problems that the campers have to solve.

"There are 30 things that we can throw at the team," says trainer Tammy Motes. "We usually give them at least five or six."

For example, the trainers could decide to signal that the shuttle's cabin pressure is

654

dropping. It's up to the pilot or the life supports systems officer to notice that the warning light is flashing. They must let the flight director know that something is wrong. Then, the flight director has to look through a book to find out what should be done, and radio back with instructions.

"While all this is going on, it's still up to the crew members to complete their other jobs," says Tammy Motes. "If they spend too much time fixing things and don't launch the satellite when they are supposed to, the whole mission is a flop."

BACK HOME
Once the campers have completed their final mission, they head to the graduation ceremony, where they receive a Space Camp Diploma and a pair of wings to pin on their shirts. After that, they head home.

Many of them will come back next year, moving up from Space Camp to Space Academy. Others may come back someday to take part in the more difficult Space Academy level II. And maybe some of them, just maybe, will go on to become the astronauts of tomorrow. ◆

Three directions all at once

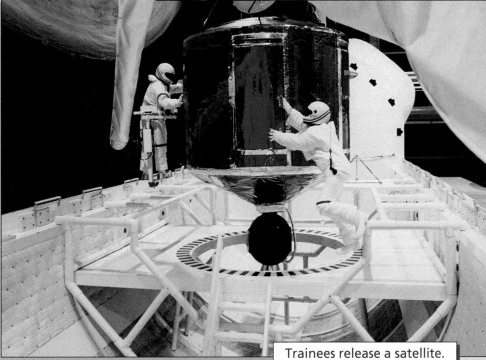

Trainees release a satellite.

Caroline Lucretia Herschel (1750—1848)

Caroline Herschel did not conform to the expectations of her day. Instead of continuing her training to become a concert vocalist, she became an astronomer.

Born in Hanover, Germany, Caroline Lucretia Herschel's interest in science and astronomy developed through lessons given by her brother William Herschel. As a result, she eventually abandoned her music career.

Herschel's career in astronomy started in 1782, when William gave her a small refracting telescope to observe comets in the sky. Four years later Herschel became the first woman recognized for the discovery of a comet. She ultimately discovered a total of eight comets. Her accomplishments included being the first woman appointed as assistant to the court astronomer, and receiving the Gold Medal of the Royal Astronomy Society in 1828. At the age of 96, Herschel also received the Gold Medal for Science from the king of Prussia.

Despite deteriorating eyesight in her old age, Herschel continued her study of astronomy. Her career and discoveries encouraged other women to pursue the study of science. ◆

Mae Jemison (1956—)

A young girl watched the early Gemini and Apollo missions on TV with great interest. The little girl from the south side of Chicago wanted to be an astronaut. Her dream was encouraged by a woman on the popular TV series, *Star Trek*.

Mae C. Jemison was born in Decatur, Alabama. Her family moved to Chicago, where she grew up. Jemison excelled there in science and math at Morgan Park High School. She won a scholarship to Stanford University, where she majored in chemical engineering and African and African-American studies.

After receiving her bachelor's degrees, Jemison went to medical school at Cornell University. In 1981, she earned her medical degree. She spent the next two years as a Peace Corps medical officer in Sierra Leone and Liberia, in Africa.

After serving in the Peace Corps, Jemison began a medical practice in Los Angeles and applied to become an astronaut. NASA selected her in 1987 and on September 12, 1992, Jemison's dream came true when she became the first African-American woman in space. Jemison has since left NASA to pursue interests in teaching, mentoring, and helping disadvantaged students to participate in science and technology. ◆

GUION BLUFORD, Astronaut

Guion Bluford knows the importance of research and experimentation in space. He is a space shuttle astronaut for NASA in Houston, Texas.

How did you first become interested in becoming an astronaut?

I love to fly in airplanes, so I became a pilot in the Air Force. In fact, I enjoy flying so much that I decided to become an aerospace engineer. I became interested in the space program when the space shuttle was being developed. I had the opportunity to apply for the program, and I did so in 1977. I was accepted and began training for my first mission.

What is your role as a member of a space-shuttle crew?

I am a mission specialist astronaut. Mission specialists are responsible for running

the experiments during a mission. We learn about the equipment needed to run the experiments, how to use it, and how to fix it in case it breaks down. We work with the pilot and the commander of the mission.

What types of experiments do you perform in space?

Most experiments fall into one of two categories: materials-processing experiments and life-science experiments. Materials-processing experiments involve making new materials in a near-zero gravity environment. These materials include new metals that could not be made under the full force of gravity. Some life-science experiments involve growing plants in near-zero gravity. We study how near-zero gravity affects the growth of plants.

What other plans does NASA have for experimenting in space?

In addition to the space shuttle program, NASA is planning to operate a space station. We hope to have the space station completed by the end of the decade. The space station will be a laboratory where we hope to run many experiments and learn about our planet and the universe. Eight astronauts will live and work in the space station for a period of three months. They will then be replaced by another team of eight astronauts. In this way, the space station will always be in use.

What is the most challenging part of your job?

Mission specialists run many experiments during the course of a mission. Sometimes we run as many as 76 different experiments. Learning all the information for the experiments is the most challenging part of the job. ◆

DISCOVER MORE

For more information about astronauts and other NASA jobs, write to the
NASA Public Affairs,
AP 4
Johnson Space Center
Houston, TX 77058

RADIO TELESCOPES

The optical telescopes of today are capable of revealing cosmic objects that ordinarily would not be visible to the unaided eye. These large refracting and reflecting telescopes give astronomers a detailed view of objects never seen before. Despite their advanced capabilities, optical telescopes are "blind" to many structures that exist in our universe. There are events in the cosmos that they cannot "see" or detect.

Radio image of a spiral galaxy

Stanford's radio telescope in California

Beyond Visible

Refracting and reflecting telescopes rely on visible light for their observations. Visible light is only a small fraction of the entire range of electromagnetic radiation that floods the vacuum of outer space. Besides visible light, there are gamma rays, X-rays, ultraviolet and infrared rays, radar microwaves, shortwaves, and radio waves given off by cosmic bodies. To capture these other important forms of energy, the radio telescope was developed.

Main Components

There are three main components that make up a radio telescope. First, there is a large reflecting surface that collects and focuses on incoming radiation. Second, there is an electronic receiver that detects and amplifies cosmic signals. Radio telescopes must have powerful amplifiers, since the radio waves that reach Earth are extremely weak. Third, there is a data display device that is used to view and record information.

The operation of the radio telescope is similar to that of the reflecting telescope. Electromagnetic energy strikes a reflecting concave surface and is focused on the receiver. Like visible light in an optical telescope, the electromagnetic energy is reflected by a mirror and focused through the eyepiece.

Precision Counts

The reflecting surface of a radio telescope must be precisely measured to take into account the frequency and wavelength of the energy it is designed to capture. Radio telescopes have a curved shape that must focus all radiation to a single point or else the image will be blurred.

Today, astronomers are constructing large radio telescopes that can detect large fields of radiation. Construction of these giant structures is difficult to engineer accurately. So astronomers have worked out a method of combining the efforts of a series of individual radio telescopes to form what is called a *very large array*, or VLA. One of these arrays in New Mexico includes a series of 27 radio dishes that can achieve a resolution equal to that of a single dish 40 km wide.

Worthwhile Effort

A large complex of computers is required for a VLA to integrate the information from the several radio telescopes. Astrophysicists believe that this kind of effort is worthwhile, since the most distant objects of the cosmos can be studied using an array. ◆

REFERENCE SECTION

SAFETY GUIDELINES

Participating in laboratory investigations should be an enjoyable learning experience. You can ensure both learning and enjoyment from the experience by making the laboratory a safe place in which to work. Carelessness, lack of attention, and showing off are the major causes of laboratory accidents. It is, therefore, important that you follow safety guidelines at all times. If an accident should occur, you should know exactly where to locate emergency equipment. Practicing good safety procedures means being responsible for your classmates' safety as well as your own.

You will be expected to practice the following safety guidelines whenever you are in the laboratory.

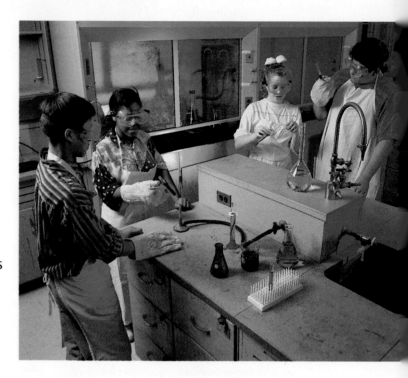

1. **Preparation** Study your laboratory assignment in advance. Before beginning your investigation, ask your teacher to explain any procedures you do not understand.
2. **Neatness** Keep work areas clean. Tie back long, loose hair and button or roll up long sleeves when working with chemicals or near an open flame.
3. **Eye Safety** Wear goggles when handling liquid chemicals, using an open flame, or performing any activity that could harm the eyes. If a solution is splashed into the eyes, wash the eyes with plenty of water and notify your teacher at once. Never use reflected sunlight to illuminate a microscope. This practice is dangerous to the eyes.

4. **Chemicals and Other Dangerous Substances** Some chemicals can be dangerous if they are handled carelessly. If any solution is spilled on a work surface, wash the solution off at once with plenty of water.
   • Never taste chemicals or place them near your eyes. Never eat in the laboratory. Counters and glassware may contain substances that can contaminate food. Handle toxic substances in a well-ventilated area or under a ventilation hood.
   • Never pour water into a strong acid or base. The mixture produces heat. Sometimes the heat causes splattering. To keep the mixture cool, pour the acid or base slowly into the water.
   • When noting the odor of chemical substances, wave the fumes

toward your nose with your hand rather than putting your nose close to the source of the odor.

• Do not use flammable substances near a flame.

5. **Safety Equipment** Know the location of all safety equipment, including fire extinguishers, fire blankets, first-aid kits, eyewash fountains, and emergency showers. Report all accidents and emergencies to your teacher immediately.

6. **Heat** Whenever possible, use an electric hot plate instead of an open flame. If you must use an open flame, shield the flame with a wire screen that has a ceramic center. When heating chemicals in a test tube, do not point the test tube toward anyone.

7. **Electricity** Be cautious around electrical wiring. Do not let cords hang loose over a table edge in a way that permits equipment to fall if the cord is tugged. Do not use equipment with frayed cords.

8. **Knives** Use knives, razor blades, and other sharp instruments with extreme care. Do not

use double-edged razor blades in the laboratory.

9. **Glassware** Examine all glassware before heating. Glass containers for heating should be made of borosilicate glass or some other heat-resistant material. Never use cracked or chipped glassware.

• Never force glass tubing into rubber stoppers.

• Broken glassware should be swept up immediately, never picked up with the fingers. Broken glassware should be discarded in a special container, never into a sink.

10. **Unauthorized Experiments** Do not perform any experiment that has not been assigned or approved by your teacher. Never work alone in the laboratory.

11. **Cleanup** Wash your hands immediately after any laboratory activity. Before leaving the laboratory, clean up all work areas. Put away all equipment and supplies. Make sure water, gas, burners, and electric hot plates are turned off.

Remember at all times that a laboratory is a safe place only if you regard laboratory work as serious work.

The instructions for your laboratory investigations will include cautionary statements when necessary. In addition, you will find that the following safety symbols appear whenever a procedure requires extra caution:

 Wear safety goggles

 Biohazard/disease-causing organisms

 Electrical hazard

 Wear laboratory apron

 Flame/heat

 Rubber gloves

 Sharp/pointed object

 Dangerous chemical/poison

 Radioactive material

LABORATORY PROCEDURES

READING A METRIC RULER

1. Examine your metric ruler. The numbers on it represent lengths in centimeters. The usual metric ruler is about 30 cm long. There are 10 marked spaces within each centimeter, which represent tenths of centimeters (0.1 cm).

2. To measure the width of a piece of paper, place the ruler on the paper. The zero end of the ruler must line up exactly with one edge of the paper. Look at the other edge of the paper to see which of the marks on the ruler is closest to that edge. In Figure A, for example, the edge of the paper is nearest to the second line beyond the 7. Therefore, the width of the paper is 7.2 cm.

3. The edge of the paper might fall exactly on one of the centimeter marks. In Figure B, the edge is just on the 5-cm mark. The width of this paper is 5.0 cm. You must write in the .0 to indicate that the measurement is accurate to the nearest tenth of a centimeter; that is, it is more than 4.9 cm and less than 5.1 cm.

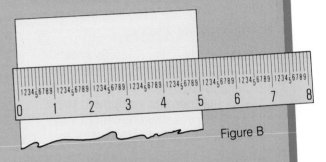

Figure B

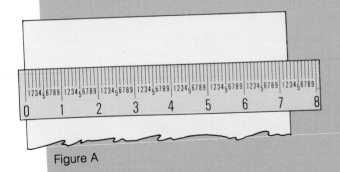

Figure A

4. Sometimes you may want to make a reading with more accuracy. It is possible to estimate readings to the nearest hundredth of a centimeter, but you must be very careful. Look at Figure A again. You can guess the number of tenths in the distance between the marks. The edge of the paper is about 3 tenths of the space between 7.2 and 7.3. The best estimate, then, is that the width of the paper is 7.23 cm.

5. In Figure C, the edge of the paper falls exactly on the 8.6 mark. If you are taking careful readings, accurate to the nearest hundredth of a centimeter, you must record the width as 8.60 cm.

6. Note the general rule: You can estimate scale readings to the nearest tenth of a scale division. If the scale is marked in tenths, you can estimate the hundredths place but never more than that.

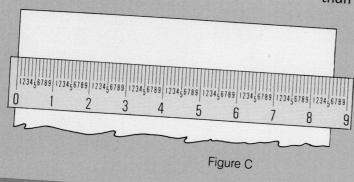

Figure C

CONVERTING SI UNITS

In SI, it is easy to convert from unit to unit. To convert from a larger unit to a smaller unit, move the decimal to the right. To convert from a smaller unit to a larger unit, move the decimal to the left. Figure D shows you how to move the decimals to convert in SI.

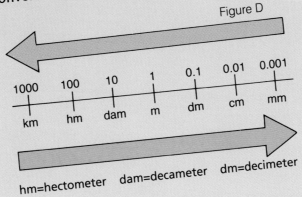

Figure D

hm=hectometer   dam=decameter   dm=decimeter

LABORATORY PROCEDURES

SI Conversion Table

SI Units		Converting SI to Customary		Converting Customary to SI	
Length				1 mile	= 1.609 km
kilometer (km)	= 1000 m	1 km	= 0.62 mile	1 yard	= 0.914 m
meter (m)	= 100 cm	1 m	= 1.09 yards	1 foot	= 0.305 m
			= 3.28 feet	1 foot	= 30.5 cm
		1 cm	= 0.394 inch	1 inch	= 2.54 cm
centimeter (cm)	= 0.01 m	1 mm	= 0.039 inch		
millimeter (mm)	= 0.001 m				
micrometer (μm)	= 0.000 001 m				
nanometer (nm)	= 0.000 000 001 m				
Area		1 km^2	= 0.3861 square mile	1 square mile	= 2.590 km^2
		1 ha	= 2.471 acres	1 acre	= 0.4047 ha
square kilometer (km^2)	= 100 hectares	1 m^2	= 1.1960 square yards	1 square yard	= 0.8361 m^2
hectare (ha)	= 10 000 m^2			1 square foot	= 0.0929 m^2
square meter (m^2)	= 10 000 cm^2			1 square inch	= 6.4516 cm^2
		1 cm^2	= 0.155 square inch		
square centimeter (cm^2)	= 100 mm^2				
Mass				1 pound	= 0.4536 kg
		1 kg	= 2.205 pounds	1 ounce	= 28.35 g
kilogram (kg)	= 1000 g	1 g	= 0.0353 ounce		
gram (g)	= 1000 mg				
milligram (mg)	= 0.001 g				
microgram (μg)	= 0.000 001 g				
Volume of Solids		1 m^3	= 1.3080 cubic yards	1 cubic yard	= 0.7646 m^3
1 cubic meter (m^3)	= 1 000 000 cm^3		= 35.315 cubic feet	1 cubic foot	= 0.0283 m^3
		1 cm^3	= 0.0610 cubic inch	1 cubic inch	= 16.387 cm^3
1 cubic centimeter (cm^3)	= 1000 mm^3				
Volume of Liquids				1 gallon	= 3.785 L
		1 kL	= 264.17 gallons	1 quart	= 0.94 L
kiloliter (kL)	= 1000 L	1 L	= 1.06 quarts	1 pint	= 0.47 L
liter (L)	= 1000 mL	1 mL	= 0.034 fluid ounce	1 fluid ounce	= 29.57 mL
milliliter (mL)	= 0.001 L				
microliter (μL)	= 0.000 001 L				

READING A GRADUATE

1. Examine the graduate and note how the scale is marked. The units are milliliters (mL). A milliliter is a thousandth of a liter and is equal to a cubic centimeter. Note carefully how many milliliters are represented by each scale division on the graduate.

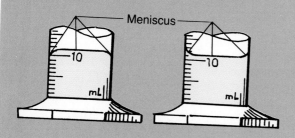

2. Pour some liquid into the cylinder and set the cylinder on a level surface. Notice that the upper surface of the liquid is flat in the center and curved at the edges. This curve is called the *meniscus* and may be either upward or downward. In reading the volume, you must ignore the curvature and read the scale at the flat part of the surface.

3. Bring your eye to the level of the surface and read the scale at the level of the flat surface of the liquid.

USING A LABORATORY BALANCE

1. Make sure the balance is on a level surface. Use the leveling screws at the bottom of the balance to make any necessary adjustments.

2. Place all the countermasses at zero. The pointer should be at zero. If it is not, adjust the balancing knob until the pointer rests at zero.

3. Place the object you wish to mass on the pan. **CAUTION: Do not place hot objects or chemicals directly on the balance pan, because they can damage its surface.**

4. Move the largest countermass along the beam to the right until it is at the last notch that does not tip the balance. Follow the same procedure with the next largest countermass. Then move the smallest countermass until the pointer rests at zero.

5. Determine the readings on all beams and add them together to determine the mass of the object.

6. When massing crystals or powders, use a piece of filter paper. First, mass the paper; then add the crystals or powders and remass. The actual mass is the total minus the mass of the paper. When massing liquids, first mass the empty container, then mass the liquid and container. Finally, subtract the mass of the container from the mass of the liquid and the container to get the mass of the liquid.

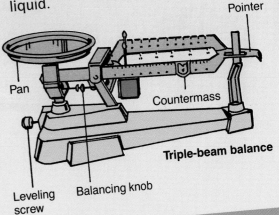

Triple-beam balance

USING A BUNSEN BURNER

1. Before lighting the burner, observe the locations of fire extinguishers, fire blankets, and sand buckets. Wear safety goggles and an apron. Tie back long hair and roll up long sleeves.

2. Close the air ports of the burner and turn the gas full on by using the valve at the laboratory outlet.

3. Hold the striker in such a position that the spark will be just above the rim of the burner. Strike a spark.

4. Open the air ports until you can see a blue cone inside the flame. If you hear a roaring sound, the ports are open too wide.

5. **CAUTION: If the burner is not operating properly, the flame may burn inside the base of the barrel. Carbon monoxide, an odorless gas, is released from this type of flame. Should this situation occur, immediately turn off the gas at the laboratory gas valve. Do not touch the barrel of the burner.** After the barrel has cooled, partially close the air ports before relighting the burner.

6. Adjust the gas-flow valve and the air ports on the burner until you get a flame of the desired size with a blue cone. The hottest part of the flame is just above the tip of the blue cone.

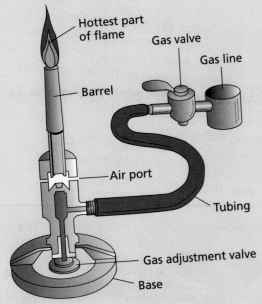

Hottest part of flame

Gas valve

Gas line

Barrel

Air port

Tubing

Gas adjustment valve

Base

FILTERING TECHNIQUES

1. To separate a precipitate from a solution, pass the mixture through filter paper. To do this, first obtain a glass funnel and a piece of filter paper.
2. Fold the filter paper in fourths as shown. Then open up one fourth of the folded paper. Put the paper, pointed end down, into the funnel.

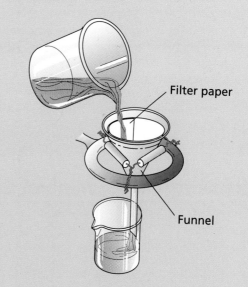

Filter paper

Funnel

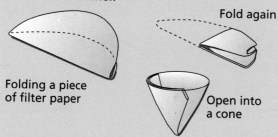

Fold again

Folding a piece of filter paper

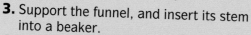

Open into a cone

3. Support the funnel, and insert its stem into a beaker.
4. Stir the mixture to be filtered and pour it quickly into the filter paper within the funnel. Wait until all the liquid has flowed through.

5. You may wish to wash the solid that is left in the filter paper. If this is the case, pour some distilled water into the filter paper. Your teacher will tell you how much water to use.

USING REAGENTS

1. For safety reasons, it is important to learn how to pour a reagent from a bottle into a flask or a beaker. Begin with the reagent bottle on the table.
2. While holding the bottle steady with your left hand, grasp the stopper of the bottle between the first and second fingers of your right hand. Remove the stopper from the bottle. **DO NOT** put the stopper down on the table top.
3. While still holding the stopper, lift the bottle with your right hand and pour the reagent into your container.
4. Replace the bottle on the table top and replace the stopper.

If you are left handed, reverse the instructions.

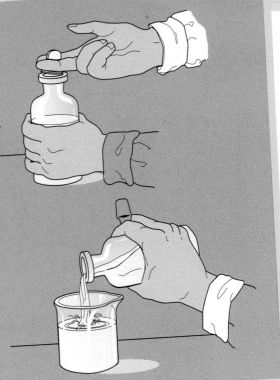

MINERAL CLASSIFICATION KEY

DIRECTIONS: Examine your mineral sample. First, determine whether the sample has a metallic luster, which places it in Category I, or a nonmetallic luster, which places it in Category II. If the luster is metallic, determine the mineral's streak. If the streak is black (Choice A), determine the mineral's hardness. If the streak is yellow or red, move to the Key of Mineral Groups.

If the luster is nonmetallic, determine whether the mineral is light colored (Choice A) or dark colored (Choice B). With each choice, determine the mineral's hardness, then move to the proper group in the Key of Mineral Groups.

Key for Determining Mineral Groups

I. Metallic luster

    A. Streak black, greenish-black, or dark gray

        1. Hardness 3 or less ....Group 1

        2. Hardness between 3 and 6..Group 2

        3. Hardness 6 or over ....Group 3

    B. Streak yellow to brown..Group 4

    C. Streak red..Group 5

II. Nonmetallic luster

    A. Mineral white or light colored

        1. Hardness 3 or less ....Group 6

        2. Hardness between 3 and 6..Group 7

        3. Hardness 6 or over ....Group 8

    B. Mineral dark colored, black, or green

        1. Hardness 3 or less ....Group 9

        2. Hardness between 3 and 6..Group 10

        3. Hardness 6 or over ....Group 11

Key of Mineral Groups

DEFINITIONS: The following terms are used in the Key of Mineral Groups.

    Cleavage—

        basal—one direction

        cubic—three directions, 90° angles

        diamond—two directions, less than 90° angles

        rhombohedral—three directions, less than 90° angles

        square—two directions, 90° angles

        octahedral—four directions

        dodecahedron—six directions

    Concretionary—looks like chunks of different-colored minerals glued together; luster is earthy or dull, color of samples

    Saline—tastes like salt

    Malleable—will bend without breaking

Group 1
Galena—cubic cleavage, very high specific gravity
Graphite—basal cleavage, greasy feeling

Group 2
Chalcopyrite—dark brassy color

Group 3
Magnetite—magnetic, black color
Pyrite—light yellow color

Group 4
Limonite—no cleavage, rarely has metallic luster

Group 5
Copper—malleable, reddish color
Hematite—brittle, black to reddish in color, can also have appearances of glitter

Group 6
Bauxite—no cleavage, earthy, concretionary, tan, white color
Calcite—rhombohedral cleavage, hardness 3
Gypsum—nonelastic, hardness 2, basal cleavage not often evident
Halite—cubic cleavage, saline taste
Kaolinite—no cleavage, white color, earthy, smells like clay when moist
Muscovite—perfect basal cleavage
Serpentine—variety of asbestos, separates into silky fibers, although this may not always be evident
Talc—may show basal cleavage, nonelastic, greasy feeling

Group 7
Amphibole—hornblende, variety of tremolite, diamond-shaped cleavage, white to gray color
Apatite—no cleavage, brown or green color
Barite—three cleavage planes, two at right angles, high specific gravity
Dolomite—rhombohedral cleavage
Fluorite—octahedral cleavage, hardness 4
Malachite—no cleavage, green color
Sphalerite—dodecahedral cleavage, resinous luster

Group 8
Corundum—no cleavage, may show parting, hardness 9
Orthoclase—feldspar, two cleavages at right angles, hardness 5, tan or pink color

Plagioclase—feldspar, two cleavages at 86° and 94°, striations, blue-gray to white color
Quartz—no cleavage, hardness 7, conchoidal fracture
Tourmaline—no cleavage, pink, blue, or green color

Group 9
Bauxite—no cleavage, earthy, concretionary
Biotite—perfect basal cleavage, elastic, black color
Chlorite—perfect cleavage, nonelastic, green color
Graphite—basal cleavage, greasy feeling
Hematite—red streak, high specific gravity
Limonite—no cleavage, brown streak
Muscovite—perfect basal cleavage, elastic
Talc—basal cleavage, nonelastic, greasy feeling

Group 10
Amphibole—hornblende, diamond-shaped cleavage, two planes not equal to 90°
Apatite—no cleavage, brown or green color, hardness 5
Azurite—no cleavage, blue color
Fluorite—octahedral cleavage, hardness 4
Hematite—red streak, high specific gravity
Limonite—no cleavage, brown streak
Malachite—no cleavage, green color, green streak
Pyroxene—resembles hornblende, square cleavage, two planes at 92°
Serpentine—no cleavage, green, waxy luster
Sphalerite—dodecahedral cleavage, resinous luster

Group 11
Corundum—no cleavage, hardness 9
Garnet—no cleavage, red color
Olivine—olive-green color, never occurs with quartz
Orthoclase—feldspar, two cleavages at right angles, hardness 6, white, tan, or pink color
Plagioclase—feldspar, two cleavages at 86° and 94°, striations, white to blue-gray color, hardness 6
Quartz—no cleavage, hardness 7
Tourmaline—no cleavage, usually black color

ROCK CLASSIFICATION KEY

DIRECTIONS: Examine your rock sample. First, determine if it belongs in Category I by asking yourself the question, "Does the rock have layers or clastic particles?" If you decide the answer is yes, then determine whether choice A or B applies to your rock sample.

If you decide that the rock does not belong in Category I, then go to Category II. Continue through the Categories until you find one that fits your rock sample. Then read the detailed choices to determine the name of your rock sample.

I. Does the sample have layers, or clastic particles glued together? If it does, go to Choice A. If it does not, go to Category II.

   A. Does the rock react to HCl? If it does, go to Example 1. If it does not, go to Choice B.
      1. Limestone—not shiny; whole surface reacts
      2. Calcareous sandstone—sand grains present
      3. Limestone conglomerate—pebble-sized particles; HCl reacts with matrix
      4. Marble—shiny

   B. If the rock does not react to HCl and it is not foliated, go to Example 1. If the rock is foliated, go to Category II.
      1. Conglomerate—rounded pebbles
      2. Breccia—sharp-edged pebbles
      3. Sandstone—sand-sized grains
      4. Shale—clay-sized grains; frequently shows thin layers
      5. Bituminous coal—black color which rubs off on your hands

II. Is the rock foliated or banded? If the sample is foliated or banded, go to Choice A. If the sample is not foliated or banded, go to Category III.

   A. Banded
      1. Gneiss—shows layers of different-colored minerals

   B. Foliated
      1. Schist—distinct mica-type layering with wavy surfaces
      2. Phyllite—shows some "mica shine" but looks less foliated than schist

III. Does the rock contain holes? If it does, go to Choice A. If it does not, go to Category IV.

   A. Scoria—dark color, dense and heavy

   B. Pumice—light color, tan, brown, or red, lightweight

IV. Does the rock have luster like glass? If it does, it is obsidian or anthracite. If it does not, go to Category V.

   A. Obisdian—hard, shows conchoidal fracture, black to gray color

   B. Anthracite coal—very lightweight, black color, glassy luster

V. Can mineral crystals be identified? If yes, the rock is granite. If not, go to Choice A.

   A. Rhyolite—light color, dull luster

   B. Basalt—dark, dull luster

   C. Obsidian—shiny luster like glass

Granite

Basalt

Pumice

Rhyolite

Obsidian

Scoria

Limestone

Conglomerate

Breccia

Sandstone

Shale

Siltstone

Coquina

Chert

Rock salt

Marble

Quartzite

Slate

PHYSICAL MAP OF THE WORLD

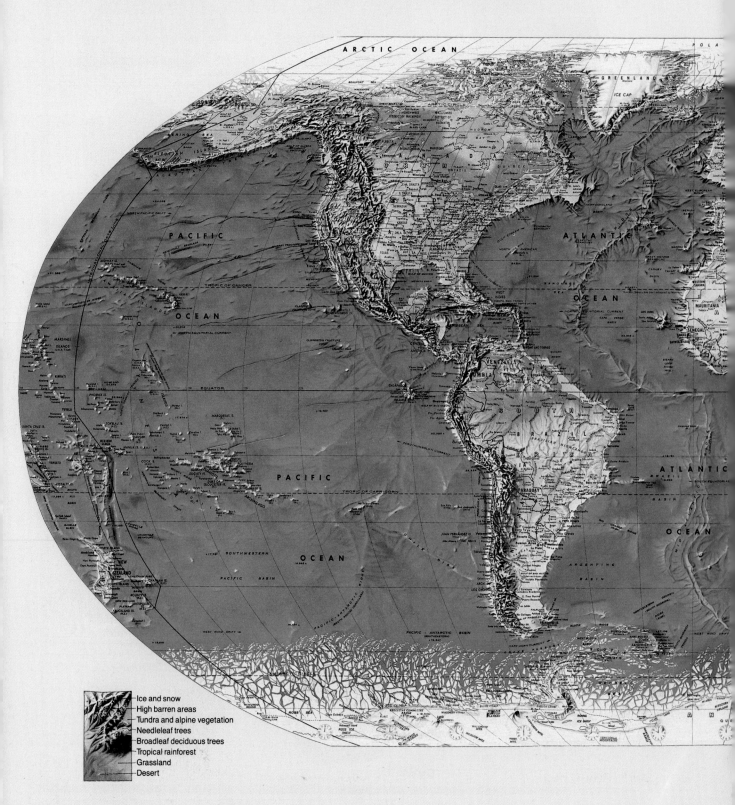

Ice and snow
High barren areas
Tundra and alpine vegetation
Needleleaf trees
Broadleaf deciduous trees
Tropical rainforest
Grassland
Desert